Land Use Competition

Human-Environment Interactions

VOLUME 6

More information about this series at http://www.springer.com/series/8599

Jörg Niewöhner · Antje Bruns
Patrick Hostert · Tobias Krueger
Jonas Ø. Nielsen · Helmut Haberl
Christian Lauk · Juliana Lutz
Daniel Müller
Editors

Land Use Competition

Ecological, Economic and Social Perspectives

Springer

Editors
Jörg Niewöhner
Institute of European Ethnology
Humboldt-Universität zu Berlin
Berlin
Germany

and

Integrative Research Institute THESys
Humboldt-Universität zu Berlin
Berlin
Germany

Antje Bruns
Governance and Sustainability Lab
Trier University
Trier
Germany

Patrick Hostert
Department of Geography
Humboldt-Universität zu Berlin
Berlin
Germany

and

IRI THESys
Humboldt-Universität zu Berlin
Berlin
Germany

Tobias Krueger
Department of Geography
Humboldt-Universität zu Berlin
Berlin
Germany

and

IRI THESys
Humboldt-Universität zu Berlin
Berlin
Germany

Jonas Ø. Nielsen
Department of Geography
Humboldt-Universität zu Berlin
Berlin
Germany

and

IRI THESys
Humboldt-Universität zu Berlin
Berlin
Germany

Helmut Haberl
Institute of Social Ecology
Alpen-Adria University
Klagenfurt, Vienna
Austria

Christian Lauk
Institute of Social Ecology
Alpen-Adria University
Klagenfurt, Vienna
Austria

Juliana Lutz
Institute of Social Ecology
Alpen-Adria University
Klagenfurt, Vienna
Austria

Daniel Müller
Halle (Saale)
Leibniz Institute for Agricultural Development
in Transition Economies
Berlin
Germany

and

IRI THESys
Humboldt-Universität zu Berlin
Berlin
Germany

Human-Environment Interactions
ISBN 978-3-319-33626-8 ISBN 978-3-319-33628-2 (eBook)
DOI 10.1007/978-3-319-33628-2

Library of Congress Control Number: 2016938662

Cover Illustration: Steep terrace farming at the Yangtze River, China. Photo taken by Zhanli "Jerry" Sun

Printed on acid-free paper

This Springer imprint is published by Springer Nature
The registered company is Springer International Publishing AG Switzerland

Preface

The basis for this volume has been a two-week workshop at Humboldt-Universität zu Berlin in September 2014: the KOSMOS Summer University FutureLand, supported by Humboldt-Universität through funds from the German government's Excellence Initiative. Graduate students and postdocs from physical and human geography, agricultural economics, anthropology and sociology brought their thick knowledge and experience of particular cases, regions and phenomena to Humboldt's Anthropology Department Integrative Research Institute THESys (www.iri-thesys.org) to enter into conceptual debate with a series of renowned international keynote speakers. Conceptual sessions, keynote lectures and debates around empirical material made for busy and productive two weeks. I thank everyone who took part in this endeavour through input, organisation, advice or any other form of support. We hope that everyone learned at least as much as we did from the experience.

The graduate students and postdocs took the results of the debate back to their home departments and produced, over the course of a year and often in co-authorship with colleagues from their departments and their study regions, the series of chapters that you have in front of you. Reading through the volume as a whole, you will notice how far lead authors shifted from their original disciplinary speciality into a broader understanding and contextualisation of their respective cases. We hope that processes and texts like this will help to shape a new generation of scientists who insist on and develop their focused expertise while staying mindful and respectful of other perspectives and developing boundary concepts between different thought styles.

We also thank Wiebke Hampel for producing the typeset manuscript under severe time constraints and everyone involved at the Springer Publishing Company, particularly Fritz Schmuhl and Human-Environment Interactions series editor Emilio F. Moran, for their patient support.

Berlin, January 2016 Jörg Niewöhner

Contents

1 **Land Use Competition: Ecological, Economic and Social Perspectives** 1
Jörg Niewöhner, Antje Bruns, Helmut Haberl, Patrick Hostert, Tobias Krueger, Christian Lauk, Juliana Lutz, Daniel Müller and Jonas Ø. Nielsen

Part I Going Beyond Distal Drivers in Land Use Competition

2 **Conceptualizing Distal Drivers in Land Use Competition** 21
Jörg Niewöhner, Jonas Ø. Nielsen, Ignacio Gasparri, Yaqing Gou, Mads Hauge, Neha Joshi, Anke Schaffartzik, Frank Sejersen, Karen C. Seto and Chris Shughrue

3 **At a Distance from the Territory: Distal Drivers in the (Re) territorialization of Oil Palm Plantations in Indonesia** 41
Anke Schaffartzik, Alina Brad, Melanie Pichler and Christina Plank

4 **The Transformation of Land-Use Competition in the Argentinean Dry Chaco Between 1975 and 2015** 59
Nestor Ignacio Gasparri

5 **Mind the GAP: Vietnamese Rice Farmers and Distal Markets** 75
Mads Martinus Hauge

6 **The Role of Maps in Capturing Distal Drivers of Deforestation and Degradation: A Case Study in Central Mozambique** 91
Yaqing Gou

7 **Nuts About Gold: Competition for Land in Madre de Dios, Peru** 111
Neha Joshi

Part II Land Use Competition and Ecosystem Services

8 Competition for Land-Based Ecosystem Services: Trade-Offs and Synergies . . . 127
Daniel Müller, Helmut Haberl, Lara Esther Bartels, Matthias Baumann, Marvin Beckert, Christian Levers, Florian Schierhorn, Jana Zscheischler, Petr Havlik, Patrick Hostert, Ole Mertz and Pete Smith

9 Contested Land in Loliondo: The Eastern Border of the Serengeti National Park Between Conservation, Hunting Tourism, and Pastoralism . . . 149
Lara Esther Bartels

10 How the Collapse of the Beef Sector in Post-Soviet Russia Displaced Competition for Ecosystem Services to the Brazilian Amazon . . . 165
Florian Schierhorn, Alex Kramer Gittelson and Daniel Müller

11 Of Trees and Sheep: Trade-Offs and Synergies in Farmland Afforestation in the Scottish Uplands . . . 183
Marvin Beckert, Pete Smith and Stephen Chapman

12 Land Use Competition Related to Woody Biomass Production on Arable Land in Germany . . . 193
Jana Zscheischler, Nadin Gaasch, David Butler Manning and Thomas Weith

13 Land-Use Competition in the South American Chaco . . . 215
Matthias Baumann, María Piquer-Rodríguez, Verena Fehlenberg, Gregorio Gavier Pizarro and Tobias Kuemmerle

Part III Understanding the Interactions of Land and Food Systems

14 The Future Is Made. Imagining Feasible Food and Farming Futures in an Unpredictable World . . . 233
Christian Lauk and Juliana Lutz

15 Exploring a 'Healthy Foodshed': Land Use Associated with the UK Fruit and Vegetables Supply . . . 247
Henri De Ruiter, Jennie I. Macdiarmid, Robin B. Matthews and Pete Smith

16 Strengthening City Region Food Systems: Synergies Between Multifunctional Peri-Urban Agriculture and Short Food Supply Chains: A Local Case Study in Berlin, Germany . . . 263
Beatrice Walthall

17 Agribusiness and Family Farming in Brazil: Competing Modes of Agricultural Production 279
Andreas Mayer

18 Local Food Systems and Their Climate Impacts: A Life Cycle Perspective 295
Michaela C. Theurl

Part IV Waterscapes: Competing for Land and Water

19 A Water Perspective on Land Competition 313
Antje Bruns, Tobias Krueger, Bruce Lankford, Fanny Frick, Catherine Grasham and Christina Spitzbart-Glasl

20 Travelling Through the Densu Delta: Location, Place and Space in the Waterscape 333
Fanny Frick

21 Competing Narratives of Water Resources Management in Ethiopia 347
Catherine Fallon Grasham

22 Competition in Transition: An Exploration of Water and Land Use in the Wien River Valley Through the Eyes of Mid-Nineteenth-Century Engineers 363
Christina Spitzbart-Glasl and Gudrun Pollack

Chapter 1
Land Use Competition: Ecological, Economic and Social Perspectives

Jörg Niewöhner, Antje Bruns, Helmut Haberl, Patrick Hostert, Tobias Krueger, Christian Lauk, Juliana Lutz, Daniel Müller and Jonas Ø. Nielsen

Abstract This chapter introduces competition as a heuristic concept to analyse how specific land use practices establish themselves against possible alternatives. We briefly outline the global importance of land use practices as the material and symbolic basis for people's livelihoods, particularly the provision of food security

J. Niewöhner (✉)
Institute of European Ethnology and Integrative Research Institute THESys, Humboldt-Universität zu Berlin, Berlin, Germany
e-mail: joerg.niewoehner@staff.hu-berlin.de

A. Bruns
Governance and Sustainability Lab, Trier and IRI THESys, Trier University, Trier, Germany
e-mail: brunsa@uni-trier.de

H. Haberl · C. Lauk · J. Lutz
Institute of Social Ecology, Alpen-Adria University, Vienna, Austria
e-mail: helmut.haberl@aau.at

C. Lauk
e-mail: christian.lauk@aau.at

J. Lutz
e-mail: julia.lutz@aau.at

D. Müller
Leibniz Institute for Agricultural Development in Transition Economies, Halle (Saale) & IRI THESys, Berlin, Germany
e-mail: mueller@iamo.de

P. Hostert · T. Krueger · J.Ø. Nielsen
Department of Geography & IRI THESys, Humboldt-Universität zu Berlin, Berlin, Germany
e-mail: patrick.hostert@geo.hu-berlin.de

T. Krueger
e-mail: tobias.krueger@hu-berlin.de

J.Ø. Nielsen
e-mail: jonas.ostergaard.nielsen@hu-berlin.de

J. Niewöhner et al. (eds.), *Land Use Competition*,
Human-Environment Interactions, DOI 10.1007/978-3-319-33628-2_1

and well-being. We chart the development over time from research on land cover towards research on drivers of land use practices as part of an integrated land systems science. The increasingly spatially, temporally and functionally distributed nature of these drivers poses multiple challenges to research on land use practices. We propose the notion of 'competition' to respond to some of these challenges and to better understand how alternative land use practices are negotiated. We conceive of competition as a relational concept. Competition asks about agents in relation to each other, about the mode or the logic in which these relations are produced and about the material environments, practices and societal institutions through which they are mediated. While this has centrally to do with markets and prices, we deliberately open the concept to embrace more than economic perspectives. As such competition complements a broadening of analytical attention from the 'who', 'what' and 'when' to include prominently the 'how' and 'why' of particular land use practices and the question to whom this matters and ought to matter. We suggest that competition is an analytically productive concept, because it does not commit the analyst to a particular epistemological stance. It addresses reflexivity and feed-back, emergence and downward causation, history and response rates—concepts that all carry very different conceptual and analytical connotations in different disciplines. We propose to make these differences productive by putting them alongside each other through the notion of competition. Last not least, the heuristic lens of competition affords the combination of empirical and normative aspects, thus addressing land use practices in material, social and ethical terms.

Keywords Relational perspective · Land cover · Global change · Scaling · Interdisciplinarity

1.1 The Global Relevance of Land Use Practices

Land is essential for sustaining human existence and development on Earth. People's livelihoods are largely land-based or are affected by land-based activities. People live on land and land provides them with food, energy and the material and symbolic basis for social–ecological development and welfare. Yet this utility derived from land and land use is highly unevenly distributed across the Earth's surface—an uneven distribution that leads to ethically and ecologically untenable effects: human starvation and disease, irreversible damage to ecosystems and biodiversity, the permanent lack of energy to sustain a dignified everyday life.

In principle, the Earth's surface provides enough land to sustain current and future generations. However, land is a limited resource and considered a planetary boundary. While most studies suggest that this boundary has not been reached (Rockström et al. 2009; Steffen et al. 2015), the pressure to use land efficiently and effectively is mounting. Yet what is considered efficient and effective use of land, what suitable, feasible or just, and for whom and on the basis of what kind of evidence, experience or belief system, is anything but trivial. Alternative land uses therefore constantly compete with each other. The dynamics of these processes are increasing in speed, interconnectedness and complexity and are shifting for a number of reasons.

The demand for land is increasing as the world's population grows. The United Nations Department of Economic and Social Affairs has revised its median global demographic projection for 2100 to 11.2 billion people. Africa will likely exceed 4 billion people, thus almost catching up with Asia, and accounting for more than half of the population growth between 2015 and 2050 (United Nations 2015). Food consumption increasingly shifts to more livestock-based diets with higher resource demands (Kastner et al. 2012) New non-food uses have entered the arena, e.g. bioenergy, carbon storage, biodiversity conservation. And land has rapidly become a significant asset class for major investors such as pension and sovereign wealth funds. As a result, the number of actors competing for land and particular land uses has increased and land use competition has become part of global ecological, trade, finance, information and people flows. These flows are rapidly increasing in speed and number of participating agents and sites. Land use competition has thus become one of *the* central arenas within which the effects of global change on human–environment systems are negotiated.

Actual land use practices and their drivers are key to better understanding the dynamics in these arenas. Land use practices are highly local and for many people on this planet they are the very site of existential struggles to make a living (Martinez-Alier 2002). Yet they are at the same time highly embedded in a complex global network of driving forces, reaching from climate dynamics through financial flows to transnational trade networks or diasporic relations. It is in the everyday practices of competing over how land and land-based resources may—or should (not)—be used that the complex dynamics of human–environment relations crystallize. These changing practices in turn drive global environmental changes, e.g. in climate, biodiversity, and other realms, that again feedback on people and their livelihoods. Understanding these dynamics is a key challenge for science and governance alike.

1.2 Land: Matter, Markets and Meaning

Land is a biophysical entity. It has an Euclidian extent and biophysical properties. It can be categorized into types of land cover and mapped with different resolutions. Yet land is also used, owned and traded. It is territory and it is a good that can be commodified in various ways. It is a material resource and the basis for all kinds of productive activities and housing. It is often allocated to people, regulated and administered, but it is also often an open access resource or used in a variety of customary ways by individuals or groups without formal property rights. Last not least, land is a source of meaning. It is home, it is part of landscapes and it is symbolically loaded. It is an element of belief systems, religious or otherwise, and it is an anchor for memories and heritage as well as for hopes and aspirations. Competing over the use of land thus takes many different forms.

1.2.1 From Land Cover to Global Change: The Loss of Innocence

The biophysical characteristics of the Earth's surface are classified into land cover classes. Increasingly fine-grained analyses of high-resolution remote sensing data deliver information about types of cover as well as the spatial and temporal dynamics of change (e.g. Hostert et al. 2015). These studies raise questions of land use that cannot be answered from data on land cover alone. Instead, land system science expands the research agenda integrating several natural sciences from physical geography to ecology, supported by a number of international research platforms (Gutman 2004; Verburg et al. 2013; Verburg et al. in press).

Land use dynamics, however, are shaped simultaneously by biophysical, ecological, economic and sociocultural drivers. Data and knowledge is needed on the actors involved in land use, their reasons for using land in particular ways and the rationales for decision-making. Large-scale changes in land use patterns over time are now linked to global trade and financial flows to demonstrate the increasingly complex interactions across the globe between changing demands, modes of production and distribution as well as land and resource use (Garrett et al. 2013; Lambin et al. 2001; Meyfroidt et al. 2013). Questions of governance of land use competition, i.e. of the mechanisms of achieving an efficient, legitimate and just distribution of access to and resources from land across people and time, have gained substantial attention (Verburg et al. in press). Increasingly, the role of environmental social sciences and humanities is considered to better contextualize biophysical and economic development within the relevant social and moral orders.

Land system science in this broad sense has become an integral part of global change research (Turner et al. 2007). This raises two major new challenges. Firstly, global change is fundamentally a societal issue. At its heart lies the question how people live together on this planet, how they organize production and consumption and the related resource flows (Fischer-Kowalski and Haberl 2007), how that affects land use (Krausmann et al. 2003; Kastner et al. 2012) and how all of this is changing. This question is fundamentally about the manifold entanglement of nature and culture and about the dynamics of human–environment relations (Palsson et al. 2013). The very framing of relevant research problems in this context already requires a multitude of disciplinary perspectives rooted in very different epistemological and ontological assumptions (O'Brien 2010). Conducting research in this area then requires a portfolio of methods operating on different spatial and temporal scales (Young et al. 2006). Yet knowledge from such vastly different methods and thought styles does not simply add up to an integrated whole. Instead, much research will be required on the intersections of these different approaches. The challenge is as much empirical as it is conceptual.

Secondly, research on the dynamics of human–environment relations under conditions of global change cannot be entirely disentangled from normative questions about how researchers think people *ought* to live together. This holds true for natural and social sciences alike and it has at least three consequences: (1) In order to understand each other, the empirical sciences need to work closely with the normative sciences, namely philosophy and law, particularly environmental ethics and political philosophy as well as international and environmental law. (2) Such collaborative research must not be a sequential endeavour, where a reality described by the natural sciences in a first step is evaluated in ethical terms thereafter. Rather a more symmetrical approach must entail an opening up and analysis of the normative and social theoretical assumptions inherent in empirical work. (3) Research on global change is seeking to intervene explicitly in the governance of the very change it is trying to understand and explain. This seems appropriate to many given that global change is a paradigmatic example of a "wicked problem" (Rittel and Webber 1973). The notion of transformation has been suggested to cover this duality of research to understand transformation and research to affect transformation (WBGU 2011). Such transformative research is explicitly working towards particular futures. Choosing one path towards a particular future over another can be informed by different kinds of empirical data and predictive models. Yet it is never determined by data. Value-laden choices abound in transformative research. They need to be explicated and assessed in their consequences with the help of the normative disciplines as well as through an opening up of the scientific community and its knowledge practices to civil society and policy-making. The co-production of knowledge between science and society to legitimate transformative research and ideally arrive at a shared ownership for a common world is the continuation of a long-standing critique of the ivory tower model of scientific expertise (Krueger et al. 2016).

1.3 The Case for Land Use Competition: Going Beyond Drivers

Research on the dynamics of land systems has begun to reframe its object of research in the context of global change. Patterns of land use practices are increasingly framed as the outcome of a complex web of driving forces that operate across many spatial and temporal scales. Within the land system science community, the concept of telecoupling has been put forward as a conceptual framework to deal with this increasingly spatially distributed and interactive nature of driving forces (Eakin et al. 2014; Liu et al. 2014). It has already been put to useful effect in focusing attention on connections that emerge between hitherto seemingly independent human–environment systems. Hence, it makes the analysis sensitive to connections that were not expected within the dominant framework of governance and it draws out new spatial configurations of sending and receiving systems particularly in the context of rapidly changing urban–rural relations (Seto et al. 2012).

Yet the challenge of reconceptualizing land use change in the context of global change goes further (see also Friis et al. 2015):

- Metabolic aspects of land use practices and lifestyles need to be articulated, localized and quantified.
- Actor networks and their complex interactions and power relations need to be understood without losing sight of the material and ecological components of the dynamics.
- Driving forces can only be disentangled when the analysis explicitly considers the institutional and infrastructural contexts through which these forces are mediated.
- Better understanding of vertical and horizontal shifts in market integration is necessary to shed light on the effects of increasingly globalized value chains particularly in agricultural trade.
- The role of power, knowledge and agency is crucial in understanding shifts in control over land use decisions.
- A highly visible discussion about social–ecological notions of equity needs to be developed.

What is required is a process that brings into dialogue different epistemological frameworks and the expertise from economic geography, institutional analysis, ecological economics, political ecology, social ecology, environmental anthropology and ethics and integrated land change science (Chap. 2 and Sect. 1.1).

We propose 'land use competition' (Haberl 2015; Lambin and Meyfroidt 2011; Smith et al. 2010) as one concept that holds the potential to further this agenda. We begin by defining competition in simple terms: Competition occurs when two or more agents strive for a goal that only one can attain or that not all can attain to the desired degree. Competition unfolds in constellations where an increase in one agent's ability or desire to attain that goal brings about a decrease in other agents' ability to do so. Thus, competition refers to the mode of solving antagonisms between agents or processes in the production of social or material order.

Competition takes the focus beyond drivers of land use because it is inherently and explicitly a relational concept. Competition asks about agents in relation to each other, about the mode or the logic in which these relations are produced and about the societal institutions through which they are mediated. As such competition complements, a broadening of analytical attention from the 'who', 'what' and 'when' to includes prominently the 'how' and 'why' of particular land use practices and the question to whom this matters and ought to matter (Lambin and Geist 2007). Importantly, competition is principally indifferent to the type of agents that compete. They can be human actors, but need not be restricted to humans (Law and Hassard 1999). Discourses, policies, ideologies and knowledge about land use change may also be said to compete in the context of land use (Latour 2005). The nature of the competing agents will have an impact on the nature of competition, on the practices, processes and infrastructures of competition as well as on its outcomes. The notion of competition is thus perfectly suited to relate material and social elements within human–environment systems. One dominant mode in which alternative land use practices are negotiated is the market. Therefore, our analytical concept deliberately has strong ties with the discipline of economics. While the understanding of competition in economics varies according to disciplinary school of thought, competition is commonly seen to be dependent on a number of prerequisites, first and foremost the existence of a market, but also an antagonistic relationship between at least two actors, the complementarity of incentivizing and coordinating mechanisms of (capitalist) markets and, last not least, the existence of goals.

Yet competition is a concept widely used in different disciplines (Chap. 8). These ties into different methodologies and different theoretical traditions are its strength. They allow a broad range of topics and foci to be pursued while retaining a joined conceptual base. The following aspects of land use competition are of particular relevance to land use competition in the context of land use and land use change:

1. In anthropology and sociology, competition has not so much been the object of explicit theorizing (Wetzel 2013 for an overview) as a constant analytical dimension of and a lens for understanding processes of social ordering. In this sense, it is often paired and contrasted with conflict, where competition solves antagonistic processes and interests within existing conventions, social and moral orders, whereas conflict operates 'out of order'. Three main directions of investigation can be discerned: (a) particularly sociology is concerned at the macro-level of empirical investigation whether competition as a mode of social interaction stabilizes or destabilizes social order. Investigations usually rely on large samples and comparative analyses between different forms of competition within or across societies (Hall and Soskice 2001). (b) Closely related but pursued at the micro-level of everyday practices, anthropology is interested in the 'how' of competing. The theoretical vanishing point is the question how social–ecological order is being produced and reproduced (or not). The approach is largely ethnographic and qualitative, investigating social conduct as well as practices and infrastructures of competition, e.g. markets, traders and trading technologies such as algorithms as they operate *in everyday action* (Ouma et al. 2013; Knorr-Cetina and Bruegger 2002; Callon 2012). It is particularly in this reading that land use competition can act as a boundary concept facilitating dialogue between econometric analyses and the sociocultural histories and ethnographic analyses of markets and their consequences in everyday action (Clark et al. 2011). And it can help to analytically grasp the manifold interactions between increasingly fast transnational and global market flows and 'local' social–ecological systems. (c) A third direction is shared by sociology and anthropology and is concerned with the subject positions and forms of collectivity afforded by competitive social processes. What sort of subjects do market economies produce, how is economic capital and agency distributed

across populations and what forms of collectivity and individuality does this afford?

2. Political science is interested in competition within structures of governance. This stretches from political theory to empirical analyses of concrete processes of governing. Of particular interest in the context of this volume is the question how political competition facilitates market integration as well as the coordinating mechanisms of markets (e.g. Kosec et al. 2015). The interaction between economic and political competition is an important field. In many emerging market economies, e.g. in post-socialist Europe or in sub-Saharan Africa, corruption, family ties, old political and regulatory networks or tribal structures play a considerable role. It will not be sufficient to consider this panoply of real-world economic action as 'informal' or 'imperfect' vis-à-vis the ideal of market dynamics at the heart of neoclassical theory. It might be the more prudent strategy to unpack the notion of informal markets also in the context of post-colonial theories about different forms of modernity and multiple cosmopolitanisms (Breckenridge et al. 2000).
3. The affinity of social theory and ecology has been noted ever since sociologists started to use biological and organic metaphors for descriptions of social groups if not long before (Park 1915). Most recently, many discussions about ecological and historical materialism (Hughes 2000) and the rise of political ecology have taken this link far beyond metaphor and into the heart of studies of land use change. Other disciplinary junctures such as cultural ecology (Sahlins and America 1964), historical ecology (Balée 2013) or environmental history (Grove 1995) further attest to the fruitful intersection of material and social thought in ecology. In general, ecology discusses competition as a phenomenon of interactions between organisms within one species (intraspecific competition) as well as between different species (interspecies competition). Competition usually emerges about access to resources (depletable) within the framework conditions set by environmental factors (non-depletable). Intraspecific competition is density-dependent and limits population growth, whereas interspecific competition is the main process structuring biological communities, a hugely important factor affecting the course of evolution (Cain 2008, and Chap. 8).
4. Moreover, competition is scale-bound in that competition occurs at the level of direct interaction between individual organisms. Increasingly sophisticated models of competitive processes, for example in co-evolutionary approaches to niche construction, allow for different strategies in competing and model the different forms of hierarchical sociality that emerge from such strategic competitive interactions (Odling-Smee et al. 2003). At the other end of the scalar spectrum, macro-ecological investigations of species distribution, patterning or invasion tend to pay less attention to actual competitive processes either because they are solely focused at levels above concrete entity interaction or because model or computational restraints force them to abstract from micro-interaction. Hence, competition is sensitive to scale and the analytical understanding and choice of scale heavily impacts on the analysis of competition. In sum,

competition in ecology is a relational concept that describes a prominent mode of species interaction as a prerequisite for discriminate relations between organisms.

5. Competition is not in itself committed to a single epistemological stance. Studies of land use competition draw on very different epistemologies from forms of positivism to constructivism. While explicit debates about epistemology—or ontology for that matter—are rare, people stand their epistemological ground and debates often erupt when that ground appears to come under threat. Conflicts between, for example, climate scientists and political ecologists over the quality of climate models can rapidly turn into unnerving entrenchment. At the same time, however, these debates are absolutely vital for global change research. Finding ways of turning them into constructive debates and learning from them is an important task. We propose competition as a helpful concept here because it does not pre-empt epistemological conflict. Instead, it may act as a boundary concept linking different thought styles (Star and Griesemer 1989). Using competition as an analytical lens within a case study, for example, allows to focus on issues of power asymmetries inherent in structures of governance or on price levels within market dynamics. Competition does not solve potential conflicts between the two perspectives or suggest priorities. Yet it offers conceptual common ground to continue discussion.
6. Competition enables the combination of empirical and normative analyses. Most analysts of the dynamics of land use practices would concede that any particular land use is historically and socially contingent, i.e. could have developed into a different state over time and could be organized differently under different circumstances. Few would defend land use as some kind of natural given, particularly under conditions of global change where the idea of 'nature' as an untouched state seems to be an anachronism. A prominent perspective in anthropology and the social sciences analyses the processes of negotiation between these alternative social–ecological orders (Mol 2002; Boltanski and Thévenot 1999): How do agents reason and legitimate 'their' order of things vis-à-vis alternatives? Do they even 'see' alternatives and, if so, how do they relate to them and why? And how do they pursue a particular path? 'Seeing' here is not only a matter of individual perspective, but also an infrastructural phenomenon (Scott 1998). The type of vision is heavily dependent on the instruments used to see: remote sensing enables a vision different from econometrics or surveys. People, markets and governing institutions are heavily reliant on such instruments to (re-)present the world to them and suggest certain pathways. How do these negotiations between alternative pathways operate in everyday life and what sources of justification and legitimacy do agents employ in competition? Competition as a relational concept is ideally suited to analyse these differences in vision and how they relate to each other. It enables a close alignment of empirical analysis—how do agents justify their actions—and normative assessment: What are the ethical premises of particular justifications, are they coherent and how do they fare within larger normative frameworks relevant to land use competition.

1.3.1 Scales and Scaling

The above discussion of different notions of competition has already demonstrated how understandings of competition may vary with analytical stance. Land use competition may be understood as the direct competition of two agents on a single piece of land. Yet land use competition may also be heavily mediated by economic and political infrastructures, contexts, mechanisms and instruments. These discussions always run the risk of conceptual confusion between scale and level of analysis as well as between scale, aggregation and resolution. In research on land use competition, this potential confusion is ever-present. Widespread terminology such as 'global drivers' or 'local case studies' is pragmatic, yet does not necessarily stand up to closer scrutiny. Furthermore, epistemological differences among researchers compound conceptual confusion. We do not pretend that we can solve a debate about scale here that has been pursued for many decades (Marston et al. 2005; Manson 2008; Cumming et al. 2006). Yet we can try to clarify our own usage of terms.

Within a Euclidian universe, entities differ in spatial extent. We may therefore refer to some entities as local, meaning small and in one single space, and to others as global, meaning large and covering a significant amount of space relative to the planet as a whole. Spatial extent is, of course, relative to some agreed upon yardstick, yet even constructivists today accept that this material dimension of the world may assert itself in relevant ways.

We only know the spatial extent of an object, because we measure it in some way. Hence, the object is in practice always a mixture of something out there and us looking at it with a particular method. When we use concepts such as 'agro–forest frontier' in land system science, we mean something that we know through a particular set of methods. This is the constructivist stance that informs much of this volume. It is not social constructivism in the sense that such frontiers only exist, because scientists talk and write about it. Neither is it positivist in the sense that such frontiers cannot be known independently of the observing apparatus used to study it, i.e. somehow objectively represented. It is constructivist in that the phenomenon agro–forest frontier is part of the material world as known through particular methods. In science studies, the phrase 'reality kicks back' has been coined to alert social and natural scientists alike to the fact that the material world contributes to its being known through scientific method (Barad 1999).

Importantly, looping effects exist between the world and how it is known and represented (Hacking 2006). Take remote sensing operating with different temporal and spatial resolutions as a case in point. If we, for example, use 1-km resolution daily data, we will analyse very different agricultural phenomena in highly variable landscapes of sub-Sahel Africa, compared to using fine-scale (e.g. 10 m) data—that is only available 3 times a year. The same applies to optical versus radar sensors that 'see different worlds'. The resulting maps are potentially used to inform farmers or decision-makers at World Bank—with very specific and probably different conclusions. Hence, 'seeing like a geographical information system'

(Scott 1998; Law 2009) alerts us to the fact that the choice of scale in scientific knowledge production is reflected in how we represent and intervene into the world.

This includes scalar terminology. Working at the 'regional' scale, for example, constructs the phenomenon of the region—and, already confusing, this could be a world region as in many integrated assessment models or a far more local area of some degree of geographical, political or cultural coherence. Using the term sets boundaries in particular ways, because it implies the use of particular methods and hence particular types of data. So does research at 'national' scale. A nation state is a political unit and very different from saying $\times$ km^2. It is obvious that the choice of scale needs to reflect the object of research and the research question. What we want to highlight here is that the choice of scale contributes in significant ways to the framing of the phenomenon under investigation and thus to the problematization (Rabinow 2004) of an issue. Climate change is one phenomenon when investigated through an Earth system model, and it is quite another when investigated ethnographically in a village in northern Burkina Faso (Nielsen et al. 2012). This is not to say that the two are not related in many ways. We just want to emphasize that a looping effect exists between choice of scale, problematization of land use competition, actual land use change and interventions into land use. This means being reflexive about and taking responsibility for methodological choices and their social–ecological implications.

Last but not least, many land use-related phenomena are complicated with respect to their spatial extent. They run across different scales. Take the farmer somewhere in sub-Saharan Africa arguing in the local market with a potential customer that his wheat is now more expensive, because it said on the news that drought and fires in Russia have increased global wheat prices. It is obvious that this social situation cannot be adequately analysed in either local or transnational terms. It is both. It is what some sociologists today call a "synthetic situation", i.e., a situation that is constructed from different spatially distributed elements that are brought into an interactive social situation with the help of different kinds of technology (Knorr-Cetina 2009). Synthetic situations rarely occur on a single scale. They are neither local nor global. They are distributed across different scales and research methods need to take that into account by "following" situations through different scales and conducting multi-sited investigations (Marcus 1995).

The land science community has responded to this multi-scalar challenge by investigating teleconnections and telecoupling, i.e. the spatial, temporal and functional distribution of drivers of land use practices (Friis et al. 2015 and Sect. 1.2.1). Importantly, telecoupling does not only alert us to events in different places and on different scales. It alerts us to the processes that connect these events. It complements the space- and place-based understandings of scale by adding a process-based or relational perspective. Such processual understandings of processes of 'scaling' help to capture how competition unfolds across different scales. Importantly, we note that physical processes tend to scale differently from social processes. For many intents and purposes, a tree, a coppice and a large forest exhibit many common properties. An individual person, a family, a social or ethnic group and a society, on the other hand, are organized in categorically different ways. Institutions and norms, dynamics of social interaction, systems of exchange and reciprocity all change radically along discriminate forms of sociality. Society is not simply 'more people' than a family or a village. Fundamentally different logics of social practice and order develop. Further, questions of representation arise and social processes are reflexive in (self-)conscious ways that differ from the feedback dynamics in physical systems. The pueblo system in the Chaco in Argentina, for example, operates as a local and highly distributed social system within a shared ecology (see Chaps. 4 and 13). When approached by an agri-business operating at a regional scale and wanting to buy several plots of land, farmers' concerns lack collective representation at such a regional level. Upscaling from the local farm in this context is an intricate political process that requires social organization, a political culture and suitable power structures to be successful.

Anthropology and the social and cultural sciences have thus insisted ever since Durkheim that sociality is an emergent phenomenon *sui generis*, the properties and dynamics of which cannot be captured at the level of the individual (Durkheim and Lukes 1982). At the same time, it has become very clear in social–ecological research that social and physical dynamics cannot be strictly separated in analyses of transformation processes. The biophysical structures of society are entangled in many ways with their social processes and moral orders (Fischer-Kowalski and Haberl 2007). Nature and culture do not neatly split along received disciplinary traditions (Ingold 2004) and social and ecological variability as well as their interactions may be driven by very different underlying processes and dynamics. Great care thus needs to be taken in social–ecological research when, for analytical purposes, assuming properties of systems and dynamics such as scales, hierarchical structures or agent-based rules (Young et al. 2006). Reflexivity and feed-back, emergence and downward causation, history and response rates all carry very different conceptual and analytical connotations in different disciplines that need to be taken into account in social–ecological research on land use competition.

1.3.2 Interdisciplinarity: Alternatives to Integrative Theories

We set out land use competition as a boundary object (Star and Griesemer 1989) that enables links between many different theoretical and methodological traditions. We are not suggesting that land use competition ought to form the basis of some kind of integrative land system theory. We believe that any attempt at integration is futile and not productive. Research on land use competition thrives on epistemological, methodological and theoretical differences. Our goal is to find ways of exploring tensions rather than to level them.

We take our cue here from analyses of interdisciplinarity in the sociology of science and scientific knowledge. Barry and Born (2013) observe three modes of interdisciplinarity:

- the integrative–synthesis mode where two or more disciplines converge into a new perspective;

- the subordination–service mode where one discipline supports research within the framework of another;
- the agonistic–antagonistic mode where two or more disciplines argue over the ontological status of a common object of research.

Whereas the first two modes are very much self-explanatory, the third deserves explication. Take 'land' as a perfect example. We have discussed above that land is a biophysical entity, commodity and source of meaning, all at the same time and depending on the observer. Hence, while geographers, economists and anthropologists share 'land' as the object of their research, they assign a different ontological status to this object. The agonistic–antagonistic mode now suggests that the different disciplines argue with each other trying to change the respective other's ontological stance with respect to a shared research object, thereby engendering reflexivity about disciplinary philosophies and methods on the part of the researchers that may lead to novel research questions, objects and practices and, at its most powerful, a reshaping of the originating disciplines themselves.

We suggest that research on land use competition pursues all three modes of interdisciplinarity depending on the research problem at hand. In some cases, particularly across the great divide of generalizing and interpretive epistemologies, however, it might well be most fruitful to begin in agonistic mode and keep the tensions transparent and productive rather than striving too soon for integrative frameworks that lack analytical sharpness. In other cases, analyses of land use competition have already begun to move forward in an integrative mode, for example, by bringing closer together geographical, economic and governance perspectives in the formation of land system science (Ledford 2015; Turner et al. 2007, Verburg et al. in press).

1.4 Land Use Competition in This Volume: Generative Differences

This volume comprises four sections each pursuing a different aspect of land use competition:

- Conceptualizing distal drivers in land use competition,
- Competition for land-based ecosystem services: trade-offs and synergies,
- The future is made: imagining feasible food and farming futures in an unpredictable world and
- A water perspective on land use competition.

This is, of course, not a comprehensive coverage of the topic of land use competition. Neither is it just an eclectic list of subthemes. The four sections address four areas of concern that currently appear particularly pertinent to researchers in the field. The concern with the notion of the 'distal driver' in section one is borne out of the observation that increasingly land use change is driven by factors that are spatially, temporally or functionally remote from the actual territory in question. The section aims to alert research to these different facets of 'distalness' and to the fact that attending to these facets means attending to different epistemologies. Section two differentiates the trade-offs between different ecosystem services. Rather than reproducing current polarized controversies around food vs. fuel or feed, the chapter systematizes forms of competition and discusses possible synergistic effects between different land uses with regard to their contribution to ecosystem services. Section three and four explore the manifold and deep intersections between land science, food production and water governance. Section three discusses how research on land use competition is implicated in constructing futures. Using research on agro-food systems, the section demonstrates the importance of understanding place-based dimensions of food production vis-à-vis the more established regional and planetary dimensions. Section four complements this expansion of perspective by addressing land–water intersections. Building on the concept of waterscapes, the authors suggest that the study of power as a material *and* social phenomenon ought to be central to investigations of land–water governance.

The authors' disciplinary backgrounds span from ecology and geography via agricultural economics to sociology and anthropology. We have not aimed at integrating this diversity into a collection of research papers that promote a single, coherent theory of land use competition. Rather, the different sections use their respective case study material to make inroads into the spaces in between different thought styles and matters of concern.

The attentive reader will recognize the epistemological and political tensions between the lines and between the sections. They reflect very real differences in perspective between individual researchers and between disciplinary approaches. The editorial team has not been concerned with leveling these differences. Rather we have tried to edit the chapters such that the reader may appreciate the differences as generative. It is in this diversity that we feel the real value of research lies on land use competition under conditions of global change.

References

Balée, W. L. (2013). *Advances in historical ecology*. New York City: Columbia University Press.

Barad, K. (1999). Agential Realism. Feminist interventions in understanding scientific practices. In M. Biagioli (Ed.), *The science studies reader*. New York: Routledge.

Barry, A., & Born, G. (2013). *Interdisciplinarity: Reconfigurations of the social and natural sciences*, New York: Routledge.

Boltanski, L., & Thévenot, L. (1999). The sociology of critical capacity. *European Journal of Social Theory, 2*, 359–377.

Breckenridge, C. A., Chakrabarty, D., Bhabha, H., & Pollock, S. (2000). Cosmospolitanisms. *Public Culture,* 12.

Cain, M. L. (2008). *Ecology*. San Francisco, CA: W.H. Freeman.

Callon, M. (2012). Framing finance: the boundaries of markets and modern capitalism. *Contemporary Sociology-a Journal of Reviews, 41*, 665–666.

Clark, W. C., Tomich, T. P., Van Noordwijk, M., Guston, D., Catacutan, D., Dickson, N. M., & Mcnie, E. (2011). Boundary work for sustainable development: Natural resource management at the Consultative Group on International Agricultural Research (CGIAR). In *Proceedings of the National Academy of Sciences*.

Cumming, G. S., & Cumming, D. H. M., & Redman, C. L. (2006). Scale mismatches in social-ecological systems: Causes, consequences, and solutions. *Ecology and Society,* 11.

Durkheim, E., & Lukes, S. (1982). *Rules of sociological method,* New York: Free Press.

Eakin, H., Defries, R., Kerr, S., Lambin, E. F., Liu, J., Marcotullio, P. J., et al. (2014). Significance of telecoupling for exploration of land use change. In S. Kc & A. Reenberg (Eds.), *Rethinking global land use in an urban era*. Cambridge: MIT Press.

Fischer-Kowalski, M., & Haberl, H. (Eds.). (2007). *Socioecological transitions and global change. Trajectories of social metabolism and land use*. Cheltenham: Edward Elgar.

Friis, C., Nielsen, J. Ø., Otero, I., Haberl, H., Niewöhner, J., & Hostert, P. (2015). From teleconnection to telecoupling: taking stock of an emerging framework in land system science. *Journal of Land Use Science*, 1–23.

Garrett, R. D., Lambin, E. F., & Naylor, R. L. (2013). Land institutions and supply chain configurations as determinants of soybean planted area and yields in Brazil. *Land Use Policy, 31*, 385–396.

Grove, R. H. (1995). *Green imperialism: colonial expansion, tropical Island Edens and the origins of environmentalism,* pp. 1600–1860.

Gutman, G. (2004). *Land change science: observing, monitoring and understanding trajectories of change on the earth's surface*. Springer Science & Business Media.

Haberl, H. (2015). Competition for land: a sociometabolic perspective. *Ecological Economics, 119*, 424–431.

Hacking, I. (2006). *Kinds of people: moving targets*. London: British Academy Lecture.

Hall, P. A., & Soskice, D. (2001). *Varieties of capitalism: the institutional foundations of comparative advantage: the institutional foundations of comparative advantage*. Oxford: Oxford University Press.

Hostert, P., Griffiths, P., Linden, S. V. D., & Pflugmacher, D. (2015). Time series analyses in a new era of optical satellite data. In: Kuenzer, C. (Ed.), *Remote sensing time series*. Dordrecht: Springer.

Hughes, J. (2000). *Ecology and historical materialism*. Cambridge: Cambridge University Press.

Ingold, T. (2004). Beyond biology and culture. The meaning of evolution in a relational world. *Social Anthropology, 12*, 209–221.

Kastner, T., Rivas, M. J. I., Koch, W., & Nonhebel, S. (2012). Global changes in diets and the consequences for land requirements for food. *Proceedings of the National Academy of Sciences, 109*, 6868–6872.

Knorr-Cetina, K. (2009). The synthetic situation: interactionism for a global world. *Symbolic Interaction, 32*, 61–87.

Knorr-Cetina, K., & Bruegger, U. (2002). Global microstructures: the virtual societies of financial markets. *American Journal of Sociology, 107*, 905–950.

Kosec, K., Haider, H. S., Spielman, D. J., & Zaidi, F. (2015). The effects of political competition on rural land: evidence from Pakistan. *IFPRI Discussion Paper,* 1441, http://ebrary.ifpri.org/cdm/ref/collection/p15738coll2/id/129182.

Krausmann, F., Haberl, H., Schulz, N. B., Erb, K.-H., Darge, E., & Gaube, V. (2003). Land-use change and socio-economic metabolism in Austria—Part I: driving forces of land-use change: 1950–1995. *Land Use Policy, 20*, 1–20.

Krueger, T., Maynard, C., Carr, G., Bruns, A., Mueller, E.N. & Lane, S. (2016). A transdisciplinary account of water research. *Wiley Interdisciplinary Reviews: Water, 3*, 369–389. doi:10.1002/wat2.1132.

Lambin, E., & Geist, H. (2007). Causes of land-use and land-cover change. *Retrieved from* http://www.eoearth.org/view/article/150964.

Lambin, E. F., & Meyfroidt, P. (2011). Global land use change, economic globalization, and the looming land scarcity. *Proceedings of the National Academy of Sciences, 108*, 3465–3472.

Lambin, E. F., Turner, B. L., Geist, H. J., Agbola, S. B., Angelsen, A., Bruce, J. W., et al. (2001). The causes of land-use and land-cover change: moving beyond the myths. *Global Environmental Change, 11*, 261–269.

Latour, B. (2005). *Reassembling the social: An introduction to actor-network-theory*. Oxford: Oxford University Press.

Law, J. (2009). Seeing like a survey. *Cultural Sociology, 3*, 239–256.

Law, J., & Hassard, J. (1999). *Actor network theory and after,* Oxford [England]; Malden, MA, Blackwell/Sociological Review.

Ledford, H. (2015). How to solve the world's biggest problems. *Nature, 525*, 308.

Liu, J., Hall, V., Moran, E., Nagendra, H., Swaffield, S.R., & Ii, B. L. T. (2014). Applications of the telecoupling framework to land-change science. In: S. Kc, & A. Reenberg (Eds.), *Rethinking global land use in an urban era.* Cambridge: MIT Press.

Manson, S. M. (2008). Does scale exist? An epistemological scale continuum for complex human-environment systems. *Geoforum, 39*, 776–788.

Marcus, G. E. (1995). Ethnography in/of the World System: The emergence of multi-sited ethnography. *Annual Review of Anthropology, 24*, 95–117.

Marston, S. A., Jones, J. P., & Woodward, K. (2005). Human geography without scale. *Transactions of the Institute of British Geographers, 30*, 416–432.

Martinez-Alier, J. (2002). *The environmentalism of the poor: A study of ecological conflicts and valuation*. Edward Elgar Publishing.

Meyfroidt, P., Lambin, E. F., Erb, K. H., & Hertel, T. W. (2013). Globalization of land use: distant drivers of land change and geographic displacement of land use. *Current Opinion in Environmental Sustainability, 5*, 438–444.

Mol, A. (2002). *The body multiple: Ontology in medical practice*. Durham: Duke University Press.

Nielsen, J. Ø., D'haen, S., & Reenberg, A. (2012). Adaptation to climate change as a development project: a case study from Northern Burkina Faso. *Climate and Development, 4*, 16–25.

O'brien, K. (2010). Responding to environmental change: A new age for human geography? *Progress in Human Geography.*

Odling-Smee, F. J., Laland, K. N., & Feldman, M. W. (2003). *Niche construction: The neglected process in evolution*. Princeton: Princeton University Press.

Ouma, S., Boeckler, M., & Lindner, P. (2013). Extending the margins of marketization: Frontier regions and the making of agro-export markets in northern Ghana. *Geoforum, 48*, 225–235.

Palsson, G., Szerszynski, B., Sörlin, S., Marks, J., Avril, B., Crumley, C., et al. (2013). Reconceptualizing the 'Anthropos' in the Anthropocene: Integrating the social sciences and humanities in global environmental change research. *Environmental Science and Policy, 28*, 3–13.

Park, R. E. (1915). The city: suggestions for the investigation of human behavior in the city environment. *American Journal of Sociology, 20*, 577–612.

Rabinow, P. (2004). *Anthropos today. Reflections on modern equipment*. Princeton: Princeton University Press.

Rittel, H. W. J., & Webber, M. M. (1973). Dilemmas in a general theory of planning. *Policy Sciences, 4*, 155–169.

Rockström, J., Steffen, W., Noone, K., Persson, A., Chapin, F. S., Lambin, E. F., et al. (2009). A safe operating space for humanity. *Nature, 461*, 472–475.

Sahlins, M. D., & America, V. O. (1964). *Culture and environment: the study of cultural ecology*, Voice of America, U.S. Information Agency.

Scott, J. C. (1998). *Seeing like a state: how certain schemes to improve the human condition have failed*. New Haven: Yale University Press.

Seto, K. C., Reenberg, A., Boone, C. G., Fragkias, M., Haase, D., Langanke, T., et al. (2012). Urban land teleconnections and sustainability. *Proceedings of the national academy of sciences of the United States of America, 109*, 7687–7692.

Smith, P., Gregory, P. J., Van Vuuren, D., Obersteiner, M., Havlík, P., Rounsevell, M., et al. (2010). Competition for land. *Philosophical Transactions of the Royal Society of London B: Biological Sciences, 365*, 2941–2957.

Star, S. L., & Griesemer, J. R. (1989). Institutional ecology, translations and boundary objects—amateurs and professionals in Berkeleys-Museum-of-Vertebrate-Zoology, 1907–39. *Social Studies of Science, 19*, 387–420.

Steffen, W., Richardson, K., Rockström, J., Cornell, S. E., Fetzer, I., Bennett, E. M., Biggs, R., Carpenter, S. R., De Vries, W., De Wit, C. A., Folke, C., Gerten, D., Heinke, J., Mace, G. M., Persson, L. M., Ramanathan, V., Reyers, B., & Sörlin, S. (2015). Planetary boundaries: Guiding human development on a changing planet. *Science,* 347.

Turner, B. L., Lambin, E. F., & Reenberg, A. (2007). The emergence of land change science for global environmental change and sustainability. *Proceedings of the National Academy of Sciences, 104*, 20666–20671.

United Nations (2015). World Population Prospects. The 2015 Revision. New York: United Nations Department of Economic and Social Affairs.

Verburg, P. H., Crossman, N., Ellis, E. C., Heinimann, A., Hostert, P., Mertz, O., Nagendra, H., Sikor, T., Erb, K. -H., Golubiewski, N., Grau, R., Grove, M., Konaté, S., Meyfroidt, P., Parker, D. C., Chowdhury, R. R., Shibata, H., Thomson, A., & Zhen, L., Land system science and sustainable development of the earth system: a global land project perspective. *Anthropocene* (in press).

Verburg, P. H., Mertz, O., Erb, K.-H., Haberl, H., & Wu, W. (2013). Land system change and food security: towards multi-scale land system solutions. *Current Opinion in Environmental Sustainability, 5*, 494–502.

WBGU (2011). *Welt im Wandel. Gesellschaftsvertrag für eine Große Transformation,* Berlin, Wissenschaftlicher Beirat der Bundesregierung Globale Umweltveränderung.

Wetzel, D. J. (2013). *Soziologie des Wettbewerbs. Eine kultur- und wirtschaftssoziologische Analyse der Marktgesellschaft,* Jena, VS Verlag für Sozialwissenschaften.

Young, O. R., Lambin, E. F., Alcock, F., Haberl, H., Karlsson, S. I., Mcconnell, W. J., et al. (2006). A portfolio approach to analyzing complex human-environment interactions: institutions and land change. *Ecology and Society, 11*, 15.

Part I
Going Beyond Distal Drivers in Land Use Competition

Chapter 2
Conceptualizing Distal Drivers in Land Use Competition

Jörg Niewöhner, Jonas Ø. Nielsen, Ignacio Gasparri, Yaqing Gou, Mads Hauge, Neha Joshi, Anke Schaffartzik, Frank Sejersen, Karen C. Seto and Chris Shughrue

Abstract This introductory chapter explores the notion of 'distal drivers' in land use competition. Research has moved beyond proximate causes of land cover and land use change to focus on the underlying drivers of these dynamics. We discuss the framework of telecoupling within human–environment systems as a first step to

J. Niewöhner (✉)
Institute of European Ethnology and Integrative Research Institute THESys, Humboldt-Universität zu Berlin, Berlin, Germany
e-mail: joerg.niewoehner@staff.hu-berlin.de

J.Ø. Nielsen
Department of Geography & IRI THESys, Humboldt-Universität zu Berlin, Berlin, Germany
e-mail: jonas.ostergaard.nielsen@hu-berlin.de

I. Gasparri
Instituto de Ecología Regional, CONICET-Universidad Nacional de Tucumán, Tucumán, Argentina
e-mail: ignacio.gasparri@gmail.com

Y. Gou
School of Geosciences, University of Edinburgh, Edinburgh, UK
e-mail: Y.Gou@ed.ac.uk

M. Hauge · N. Joshi
Department of Geosciences and Natural Resource Management, Geography Section, University of Copenhagen, Copenhagen, Denmark
e-mail: mmh@ign.ku.dk

N. Joshi
e-mail: npjo@ign.ku.dk

A. Schaffartzik
Institute of Social Ecology, Alpen-Adria Universität Klagenfurt-Wien-Graz, Vienna, Austria
e-mail: anke.schaffartzik@aau.at

F. Sejersen
Department of Cross-Cultural and Regional Studies, University of Copenhagen, Copenhagen, Denmark

K.C. Seto · C. Shughrue
Yale School of Forestry and Environmental Studies, Yale University, New Haven, Connecticut, USA

J. Niewöhner et al. (eds.), *Land Use Competition*,
Human-Environment Interactions, DOI 10.1007/978-3-319-33628-2_2

come to terms with the increasingly distal nature of driving forces behind land use practices. We then expand the notion of distal as mainly a measure of Euclidian space to include temporal, social, and institutional dimensions. This understanding of distal widens our analytical scope for the analysis of land use competition as a distributed process to consider the role of knowledge and power, technology, and different temporalities within a relational or systemic analysis of practices of land use competition. We conclude by pointing toward the historical and social contingency of land use competition and by acknowledging that this contingency requires a methodological–analytical approach to dynamics that goes beyond linear cause–effect relationships. A critical component of future research will be a better understanding of different types of feedback processes reaching from biophysical feedback loops to feedback produced by individual or institutional reflexivity.

Keywords Telecoupling · Social space · Systemic effects · Competition as process · Power/knowledge

2.1 Distal Drivers—A Conceptual Challenge

Land as a biophysical entity always has a specific presence in Euclidian space. Its extent can be measured in areal and topographic terms. Its position can be located relative to a global grid of longitude and latitude. Land use, therefore, necessarily possesses a local component. Land is always used some*where*. Yet the notion of 'land use' complicates the picture. Some*one* is doing the using. For a long time, this has been straightforward: An individual or a group of people is using land for their own purpose, e.g., to live and produce something on it, most commonly food.

Today, land use at all spatial levels is influenced by long-distance flows of raw materials, energy, products, people, information, and capital creating a need for novel theoretical and methodological approaches to the analysis of causal relations in land systems. An increasing spatial decoupling of drivers and outcomes in current land use patterns is challenging the straightforward relationship of local land use and consumption most evident in subsistence agriculture; think, for example, of a tropical forest being protected from agricultural use through the international 'Reducing Emissions from Deforestation and Forest Degradation' (REDD+) program. This forest might simultaneously be used by the international group of participating states as a carbon sink to slow down global climate change, it may serve local people to sustain their livelihoods in multiple ways, and it might also be of particular symbolic or spiritual significance for yet another group of people.[1]

[1] We emphasize that this symbolic dimension is by no means restricted to indigenous groups. Debates in Germany in the 1980s about 'Waldsterben,' i.e., the dying of forests due to increasingly acidic rain, made abundantly clear how important forests are in German discourse and symbolism. Not least the concept of the 'risk society' has been developed by German sociologist Ulrich Beck under the impression of the debates about dying forests.

Moreover, the REDD+ protection of a forest area may also serve the purpose of protecting biodiversity. Hence, many nonhuman actors also use the forest.

Thus, the same piece of land can be subjected to multiple uses at the same time. Land use may involve agents of different kinds, such as human and nonhuman. What may be considered 'use' and a 'legitimate use' depends on epistemic and ethical perspectives and varies with the observer. And the reasons why a piece of land is used in a particular way may not be found locally.

Land use science research has grappled with the spatiotemporally distributed nature of land use for decades. When the analysis of remote sensing data was producing land cover maps documenting changes over time at different scales, the question 'why' these changes were occurring loomed large immediately. To address this new set of questions, a distinction between proximate causes and distal or underlying drivers has been established (Lambin and Geist 2006). Others have spoken of direct and indirect effects on land use transitions (Marcotullio 2014) or of pressures and drivers (Young et al. 2006). Proximate causes are considered direct and local, distal drivers underlying or root. 'The proximate causes of land use change explain how and why local land cover and ecosystem processes are modified directly by humans, while underlying causes explain the broader context and fundamental forces underpinning these local actions. In general, proximate causes operate at the local level (individual farms, households, or communities) and underlying causes originate from regional (districts, provinces, or country) or even global levels, though complex interplays between these levels of organization are common. As a result, underlying causes also tend to be complex, formed by interactions of social, political, economic, demographic, technological, cultural, and biophysical variables' (Guyer et al. 2007).

Research on proximate causes has long been primarily a challenge in terms of data availability, quality, and analysis. It has set out from the initial questions 'what is happening where?' and 'who does what on a specific piece of land?' Contributions from physical geography, and particularly the remote sensing community, have provided an understanding of land cover dynamics that enables the research community to now ask questions about underlying and distal drivers of land use change. Understanding these drivers requires the development and deployment of suitable (remote) sensor technology and increasingly fine-grained temporal and spatial analytical techniques to increase coverage and analytical resolution. Yet, a rather more complex difficulty lies in the integration of biophysical land cover and land use data with qualitative, often disaggregated 'ground-based' data. The chapters by Joshi and Gou in this section demonstrate the challenges that arise when land use changes rapidly and when categories such as 'degraded forest' struggle to discriminate sufficiently enough and meaningfully between actual land use practices. A thick analysis of such processes that are as much biophysical as they are social requires empirical data on the natural and cultural dynamics, on land cover and ecosystem characteristics as well as markets, regulation, social and moral orders, and cultural and environmental history. And it requires multiple disciplinary perspectives and conceptual advances to relate them to each other. Established conceptual distinctions between nature and culture are unlikely to be up to the task (Palsson et al. 2013).

Research on underlying and often distal drivers faces another problem. It is often not feasible to reconstruct—let alone understand or predict—all underlying factors and their interactions that have shaped a particular decision or event and thus 'caused' a particular land use. Given the important context of global social and environmental change, it is often necessary to consider the manifold interactions between individual land use decisions and thus discern trends, feedbacks, and emergent effects that only manifest at levels of analysis above individual decisions or events. Research on distal drivers in global land use change thus needs to select and prioritize from an overwhelmingly complex intermingling of relevant factors. This is a process that is informed by empirical data, but cannot in itself be determined by it. Prioritization needs to be theoretically and conceptually shaped and legitimated.

2.2 The Telecoupling Framework

The concept of 'teleconnection and telecoupling' offers one such conceptual framework with which to analyze distal drivers of land use change under conditions of global social and environmental change (Friis et al. 2015; Seto et al. 2012). As set out originally, the framework starts from a systems-theoretical perspective and assumes that human and natural systems are coupled. 'Systems are defined as a set of human and natural components interacting to form a whole' (Liu et al. 2014). Telecoupling occurs when 'causes generate a telecoupling between a minimum of two coupled human and natural systems, which produce effects that are evident in one or more of the systems. A telecoupling is produced by agents that facilitate or hinder the flows of material/energy or information among the systems' (ibid.: 122). Analyzing processes of telecoupling relies on five components: coupled human natural systems, flows, agents, causes, and effects.

2.2.1 Systems

Systems are divided into sending, receiving, and spillover systems. Spillover systems are systems that are affected by telecoupling processes, although they have until that moment not been connected or not been known to be connected to the sender–receiver coupling. Telecoupling thus focuses attention on effects that emerge anew or that have thus far not been visible through the dominant lenses of knowledge production and governance (Eakin et al. 2014).

2.2.2 Flows

Flows forge the connections between systems. They mediate the coupling. 'Flows exchange material, energy, or information between the systems. Material or energy includes biophysical and socioeconomics entities (goods, food, natural resources, organisms, and carbon), whereas information includes knowledge and agreements (trade agreements, land titles, and agricultural techniques)' (Liu et al. 2014). Flows can also be thought of as manifestations or traces of processes that underlie them, and so measurement of flows can be thought of as an indirect measurement of effects from coupling processes.

2.2.3 Agents

As coupled systems comprise human and natural components, the concept of the agent has been introduced. The perhaps more obvious 'actor' is in the social sciences commonly reserved for human agents; i.e., it is tied to concepts of intentionality, motivation, and behavior as used in psychology and economics and as rooted in much of Western philosophy (Palsson et al. 2013). Agent on the other hand marks a less anthropocentric and more inclusive 'site of agency' and thus incorporates not only human individuals and groups of people, but also herds of animals or government agencies. The boundaries of the category 'agent' are not clearly defined within the telecoupling framework.

2.2.4 Causes and Effects

'Causes are factors that generate dynamics (emergence, change in strength) of a telecoupling. Most telecouplings have multiple causes: cultural, economic, political, technological, or ecological. These are interacting categories' (Liu et al. 2014). The environmental and socioeconomic consequences and impacts of a telecoupling process are analyzed as effects. They are structurally categorized into types: indirect effects, feedbacks, cascading, and legacy effects with often nonlinear and time lagged dynamics (Liu et al. 2014). In addition, system dynamics are understood as hierarchical reaching from flows between multiple systems (high), to a focus on the facilitating agents and cause–effect relationships (intermediate) to particular characteristics of interest (low).

It is important to note that today virtually all land systems are telecoupled (Eakin et al. 2014). Flows of trade, migration, transnational land deals, and financial capital or species invasions are enveloping the globe. In some ways, this is not new. Ever since antiquity, human settlements have been shaped by local social and environmental conditions as well as regional and global flows. The local has never really been local but always been shaped by constant exchange processes. Yet, the exchange of people, information, and goods today unfolds with unprecedented speed, intensity, and extent. The degree and quality of interdependence has increased with the liberalization of trade and financial markets, the fall of many socialist governments, and the rapid unfolding of information and communication technologies all happening since the late 1970s and at full speed really only since 1989. The resource use of 7.3 billion people has produced a situation of manifold localized resource scarcities and raised matters of distribution with a renewed urgency. 'Feedbacks within systems are now tighter, more rapid, and multiscalar; the potential for rapid acceleration to systemic transformation (thresholds) or crisis arising from multiple systems interactions is higher' (ibid.: 145). Lastly, different forms of industrialization and urbanization at an unprecedented scale are taking place across the globe, leading to new settlement patterns and changing lifestyles and in conjunction largely driving more energy and resource intensive forms of dwelling (Creutzig et al. 2015). And, last but not least, the scientific community is producing new knowledge about these global interdependencies contributing to what one might call a 'planetary moment' in human and natural history: Never before have the planetary dimensions of human action been specified more clearly than today.

Telecoupling is meant to function as an umbrella concept under which to discuss the effects of these macrochanges on concrete transitions in land use change globally. Yet, it has been noted that this will not be achieved in any simple additive or mechanical fashion of factors within a single analytical framework. Rather, there exists a 'need to integrate epistemologies, methodologies, and analytical approaches to expand on place-based land use tradition through a focus on new networks and system interactions involved in land change' (Lambin and Geist 2006). The current global configuration might be conducive to such a pathway as ethical concerns about global imbalances start to play a greater role in decision-making processes across scales of governance. New forms of social contracts, moralities, and empowerment arise and sustainability and resource limitations are now increasingly important motivations for action.

The telecoupling framework is being developed (Friis et al. 2015) in order to be able to deal adequately with the complexities of processes shaping land systems under conditions of global change as reflected in changes in communication and trading patterns, land governance structures, and policy regimes as well as processes of transition that more often than not do not follow linear paths toward modernity. Overall, scholars have diagnosed an increasing spatial decoupling of local land uses from the most important driving forces (Seto et al. 2012). The relative simplicity of the telecoupling framework, which offers much heuristic value, reveals limitations, when interacting forces of land use change are situated within vastly different ecological, economic, political, and cultural contexts.

It has therefore been suggested to carefully widen the disciplinary knowledge base upon which the telecoupling framework rests. Friis and her colleagues have suggested learning from

- economic geography to differentiate the analyses of networks of actors and their power relations through approaches to global production networks and value chains;
- socioeconomic metabolism studies to add material—energetic components and temporality to established economic analyses of flows by drawing on ecological economics, ecological anthropology, industrial, and social ecology;
- political ecology to address questions of power and agency in processes of land systems transitioning with particular attention on human–environment relations;
- cultural anthropology to better understand scaling issues, the demarcation of system boundaries, and 'unexpected' couplings through the thick analysis systems of exchange within which virtually all land systems are embedded (Friis et al. 2015).

This volume adds further observations that may contribute to the broadening of land system science's knowledge base. This section in particular discusses the challenges to the notion of 'distal' drivers in relation to land use competition. In the following, we introduce and explore analytical dimensions of the notion of 'distal drivers' in land use competition that go beyond a largely Euclidian understanding of proximate and distal to also consider social, institutional, and epistemological aspects.

2.3 Analytical Dimensions of 'Distal Drivers' in Land Use Competition

2.3.1 Going Beyond Proximate and Distal: Competition as a Distributed Process

Urbanization, transnational markets, and the increasing importance of internationally coordinated land protection/conservation activities have shifted analytical attention from the immediate pressures on land use toward the more complex interactions of underlying drivers. Supply and use chains, international flows of governance, certification schemes, the transnational mobility of people, goods, services and knowledge, the systemic effects in ecological systems, increasingly networked forms of social movements, and resistance are processes that operate in complicated spatial, temporal, and institutional constellations that are difficult to contain within coherent systems thinking.

Hence, the notion of the distal driver has become central to the analysis of land use change and competition (e.g., Seto et al. 2012). While basic economic understandings of competition may frame competition as that between two actors over a resource that only one can attain, it is clear from the above that land use competition is a highly distributed process. It is distributed across space and time, across agents, and across flows that operate in very different modes and according to different logics.

In such a constellation, the distinction between proximate and distal loses significance. A proximate pressure, e.g., the decision of a local farmer to plant crop X instead of Y, may be direct and local relative to the plot of land under investigation. Yet, the farmer's decision between competing alternatives is only one step in a distributed process. This process may involve evaluating past success of other crops grown on this land, checking prices on markets in different locations, considering the institutional (legal, cultural, and ethical) setting and implications of this process, mediating positions within a social network, and pondering knowledge about global change and the implications of this decision for the local social—ecological system or even the global climate. This means that individual land use decisions are embedded in multiple social, political, historical, and ecological contexts and as such each decision is conditional upon more or less distal drivers. When it comes to competition over land, it is therefore important to understand that decisions regarding land use are mediated in many different ways in which the (Euclidian) spatial component is integrated into a broader spatial and temporal set of processes. As outlined in the overall introduction to this book, we understand competition over land use in the same way. Competition must be understood as a relational concept that focuses our analytical attention on the properties and qualities of relations between drivers in land use change. Adding to this, the following sections explore the increasingly distributed and processual nature of competition.

Moving away from an understanding of land use competition as a single decision-making moment and toward seeing it as a mediated process requires paying analytical attention to the contingent nature of decision-making processes. Consider transnational agricultural markets as an example. Their dynamics are crucial in formatting local land use. Econometric approaches have made valuable inroads into connecting a quantitative understanding of demand and supply, quality of produce, and buyer–seller relationships to land use patterns (Garrett et al. 2013). Others, often qualitative approaches, particularly from human geography, anthropology, and science and technology studies, are beginning to emphasize the role of historically developed connections between countries and regions, the role of personal trust between producers and traders, or the role of knowledge about market dynamics and local contexts of production and consumption (Ouma et al. 2013). The competition between different land use decisions is thus a complex process set within a social history and embedded within concrete social interaction, economic and technical infrastructures as well as institutional framings and cultural and moral orders. Add to this the complex dynamics of biophysical connections and flows at different spatial and temporal scales and it becomes clear that a social-ecological understanding of land use competition needs a broad portfolio of theories and methods to capture the processual nature and multiple embeddedness of land use competition (Young et al. 2006).

2.3.2 Space: Material and Social Distance

While largely meant in the sense of 'underlying' in the literature, the term 'distal' in distal drivers is firmly rooted in a Euclidian understanding of space. This is not accidental but reflects the understanding of space prevalent in the land use science literature. Space is commonly understood in material or biophysical terms. Distance is hence measured in Euclidian terms. A driver of land use change may be out of sight and operate 'at a distance.' A prominent example might be the changing patterns of consumption in urban centers driving patterns of agricultural production in increasingly faraway hinterlands. These may be increasingly meat-based diets in industrializing and urbanizing economies or post-material trends toward organic produce from community-supported agriculture in the urban centers of the Global North.

Physical distance may play a role in these cases. Seto and colleagues have already differentiated distance and the patterning of exchange in a spatially explicit model of urban land teleconnections (Seto et al. 2012). They focus on the distribution of senders and receivers within what they suggest ought to be conceived of as an urban–rural continuum. This is a useful framework as it begins to shift from actors of urban–rural teleconnections to the spatially explicit analysis of telecoupling processes. We suggest a further expansion of this thinking here: Physical distance remains a relevant measure, e.g., where issues of space-dependent transaction costs are concerned. Yet, the key message of this type of analysis lies in preparing the analyst for the unexpected, i.e., for the fact that drivers of local land use may be operating at a distance or in Seto's case in multiple sites all at different distances. This is really a functional understanding of distance that only happens to take physical shape, because of geographical thought styles and mapping methodologies. It is not difficult to imagine the same distribution of drivers, but on the basis of social rather than physical distance.

Social space and distance cannot be measured in Euclidian terms. What it is, how it operates, and how it may be studied depends very much on the analyst's social theoretical stance (Niewöhner 2014b). Most social scientists will agree that within groups of people living together forms of social order emerge. These shape to a considerable degree how individuals are positioned relative to each other and how they may interact. What kind of social and moral order emerges within a group of people, how it is continuously legitimated and (de)stabilized, how and by whom it might be changed, and how it operates in everyday life depends on a multitude of aspects reaching from rather stable structural and institutional elements often captured in socioeconomic terms to much more fluid individual motivations and actions, as well as various forms of interaction between these dimensions. Most social scientists today thus consider emerging social orders as comprising dispositional and situational elements. Processes of social ordering unfold within and thus reproduce social space (e.g., Bourdieu 1984). Social space positions agents toward each other and thus produces social differentiation. Such processes of differentiation within social space may be analyzed, for example, with an emphasis on individual and group-based social and cultural capital, a focus on power, knowledge, and access to political and economic resources or shared strategies of identification and belonging. They result in groups of people, where the notion of group is understood in relational terms, i.e., as distance within social space from others, rather than as homogeneous and stable entities with a fixed culture or identity.

Distances can be mapped in social space as they can in physical space. And the two forms of space interact. Segregation in cities, for example, can be understood as differences in social position and forms of capital inscribing themselves into material form through the size of plots of lands and houses, architectural styles, etc. The structure of the physical space will then act to manifest and reproduce distinctions in social space.

This brief discussion of physical and social space illustrates that 'distance' can take different forms. Hence, one can easily imagine teleconnections between different contexts and actors that are socially distant but physically close. Particularly in the highly dense and socioeconomically differentiated urban centers of the world, effects on land use competition may arise in close physical proximity to each other, but at a vast distance in social terms as they relate to entirely different social worlds. In contrast, diasporic networks of people that have considerable impact on the distribution of economic means in many contexts may be physically distant, but socially close. Many cases of land use competition develop in the context of sharp increases in the value of land. Market integration, the rise of biofuels, or technological advances enabling agricultural production are typical examples. Investigating such cases, it is important to consider the entanglement of physical and social space as emerging markets change existing social spaces. The chapter by Hauge in this section is a case in point. Detailing how middlemen rise to relative prominence in South Vietnam thereby changing established local social orders and the distribution of agency within a local social-ecological system, Hauge shows how a particular land use change is embedded in a historically, politically, and socially mediated process of competition between two types of rice cultivation and marketing.

It is one of the key challenges for an interdisciplinary land systems science to better understand the mutual constitution of physical and social space or, more broadly speaking, the manifold entanglement of nature and culture. For the emerging environmental social sciences and humanities, this means first and foremost coming to terms with the biophysical environment as a social phenomenon (Palsson et al. 2013). For the natural sciences, it requires an appreciation of the social and historical contingency and complexity of 'the social' as a unit of analysis *sui generis*.

2.3.3 Systemic and Relational Perspectives

The notion of distal drivers as outlined above is rooted within systems thinking: Human and natural components interact in coupled social-ecological systems. Systems thinking enables a view from outside or above, identifying system boundaries, components, and linkages (Young et al. 2006). The point of systems thinking is not to identify the independent effects of individual factors, but rather to focus on synergies and interactions between factors (Lambin and Geist 2006). The focus is on processes more than on nodes. Agent-based modeling in social-ecological systems analysis reflects this desire. Here, nodes or actors are assigned simple rules to then analyze the patterning that arises from these agents in interaction.

We expand on this systemic approach here to suggest what in anthropology is referred to as a relational perspective (Beck 2008; Niewöhner 2014a; Strathern 1991) and what science and technology studies calls relational materialism or actor–network theory (Law and Callon 1992; Law and Mol 2002; Latour 2005). Relational thinking shifts the basic analytical unit from the human actor of rational choice theories to the concrete practices of making connections between agents. It tries to better understand the actual 'how' of making connections within practices of land use competition, whereas the systemic approach focuses on the 'what' of links within systems. We do not argue that the relational perspective is superior to the systemic approach. Rather we think of the two as complementary and mutually enlightening in analysis. Where systems thinking has its strength in quantifying flows and analyzing the topography of social-ecological systems, relational thinking comes into its own by analyzing the quality of relations, the adequacy of system boundaries, and the topology of patterns of land use competition in practice (Marres 2012). In this way, analytical perspectives may be reversed with one starting from actors and systems, the other from relations and practices.

2.3.4 Patterns of Practice in Land Use Competition

Lambin and Geist rightly point out that the search for independent drivers of land use change might be mistaken. Instead, 'different patterns or modes may represent the interactions between the various causes of land change.' The two authors distinguish four modes of interaction between drivers: (1) one cause dominating all others; (2) a connection of causes in causal chains, i.e., in sequential order; (3) independent but synchronous action of several factors on one variable; and (4) interventions in synergetic factor combinations producing feedbacks and reciprocal action (Lambin and Geist 2006: 62ff.).

This systematic is driven by a top-down view onto a network of factors or a system. Following on from our suggestion of a relational perspective, we suggest another understanding of the notion of mode or patterning here that is useful to further differentiate our understanding of connections among drivers. We suggest paying attention to the different patterns of practice within which land use competition unfolds. Social and cultural anthropology has long argued that human practices are characterized by particular patterns and that participating in these patterns brings order to how people act in particular group- and context-specific ways (Roepstorff et al. 2010). This view rests on the basic insight that action is always embedded in three heuristically different environments: normative orders (institutions, markets, and moral values), social dynamics, and material conditions (cf. Parsons 1937). The idea of patterns of practice emphasizes that actors, who want to solve practical problems of land use competition, must always do so interactively with others in concrete social-ecological situations (Alexander 1988). These interactions form patterns of practice and develop their own routines and implicit logics.

It is of course not only anthropology that has developed this perspective. The concept of 'languages of valuation' in ecological economics (Martinez-Alier 2008) is making a very similar point. Here, the emphasis is on the different values that people may assign to a piece of land. It is the attempt to broaden the economic or monetary logic often in operation in ecosystem services into a set of languages of valuation within which those affected by land use change may express their interests or stakes.

Institutional analysis in turn explores the idea of rules in institutionally formatted processes of competition, particularly around matters of land use and resource use (e.g., Ostrom et al. 1993). Such rules may be formally codified as in laws and regulations. Yet, they may also be conventionalized or routinized at the level of everyday social-ecological practice.

Knowledge plays an important role in shaping such routines and conventions. For example, the widespread mismatch between environmental information accessible to local populations and macrolevel institutions has been noted (Lambin and Geist 2006). A better understanding is needed of how institutional arrangements operate across scales and how this affects people involved in and affected by land use competition. Restoring a sense of agency to local people is an important claim in many land use conflicts (Poteete and Ostrom 2004). Yet, it would be a mistake to analytically treat 'local people' as if they were disconnected from the rest of the world. Not least anthropological research has shown to what degree and how people are embedded in global flows of people, goods, and information (e.g., Hannerz 1993; Strathern 1992). While it is clear that powerful structural asymmetries with regard to agency and knowledge exist between groups of actors, any simple global regimes versus local people framework is bound to miss important elements of land use competition dynamics.

We thus need to pay attention to how these patterns of practice in land use competition are actually constituted. In particular, it is important to better understand the entanglement between ecological, economic, political, and social agents. This is not least a methodological challenge as most methods—e.g., econometrics, governance analysis, ecological analyses, or ethnographic methods—have coevolved with their objects of study and thus operate in rather domain-specific ways.

2.3.5 Power and Knowledge Structure What Counts as Legitimate

Environmental social sciences keep emphasizing the importance of power and hegemony in processes of land use competition. The land sciences have so far not developed an explicit focus on the role of power, albeit for different reasons: The physical sciences are primarily interested in the states and dynamics of the material world and thus do not have an immediate need for a theory of (sociopolitical) power. For many economists, power in the context of competition refers to the risk of concentration of market power in monopolistic structures. Political power is thus understood as the regulating framework within which market dynamics are analyzed.

For the social sciences, power is a key variable in analyzing social processes such as competition. Conceptually, power comes in two forms. One is largely congruent with the commonsensical meaning of the word and refers to the power to enforce one's will against the will of others in a social situation (e.g., Weber 1922/2002). The exercise of power may be agreed upon or coercive, and it may be legitimate or not. In democratic systems, power is given to elected representatives. In the analysis of political constellations that are set up to govern land use competition, this notion of power is conventionally used. Actors have the power to enforce agreed upon rules. This usually applies to territorial jurisdictions such as nation states or provinces. Within the trans- and particular the international realm, the role of these powers is weakened due to the absence of institutions with the power to enforce rules and impose sanctions. Institutional analysis, as well as analyses of governance regimes, usually operates implicitly or explicitly with such a notion of power.

A second notion of power plays a much lesser role in the land science community. This notion originates with the French philosopher Michel Foucault (Foucault 1972). Foucault ties the notion of power not to individual people or institutions, but to discursive practices and formations. A discursive formation comprises not only the language, codes, conventions, and habits, but also the material artifacts of a particular time and society. Discourse shapes the orders of truth; i.e., what may count in any given society as true and real. Power then operates within these discursive formations through the micropolitics of everyday interaction. In land use competition, ecosystem services may serve as a good example (e.g., Daily and Matson 2008). The valuation of ecosystems through particular mechanisms is a major driver in land use competition. It brings such diverse issues as natural resources, biodiversity, or indigenous knowledge onto a common denominator—the 'service'—to further process toward a consensus on the value of a particular piece of land or ecosystem. The rapid spreading of the concept of ecosystem service assessment throughout the world can be understood as driven by a discourse that makes it seem true and legitimate that nature ought to be valued in order to trade, develop, or protect it. This process is not simply driven by a few powerful global actors—though these, of course, exist. It can occur, because scientific and market expertise within a larger capitalist market mode of governance align to make the valuation of nature in a particular way seem meaningful while alternatives appear to be naïve or simply not thinkable (Chan et al. 2012).

This is not about judging the meaningfulness of the concept of ecosystem services. It is an analytical perspective that investigates the conditions that lead to such a concept becoming the dominant lens through which to understand the value of land. Scientific knowledge plays a key role here. Much of ecosystem service assessment relies on scientific data about land and land use. The same is true for many other concepts such as metabolic flows, biodiversity, or planetary boundaries.

Scientific knowledge and its circulation globally has become a key driver of land use and land use competition. In its most obvious form: Global climate change would not exist in its current form if it were not for global climate science. This is not only saying that the way we know and problematize our planet is heavily shaped by scientific knowledge, particularly as it is impossible to experience *global* climate change directly and without the help of global climate science, but also pointing out that scientific knowledge about global climate change shapes our responses and interventions. Scientific knowledge circulates globally and this knowledge increasingly drives land use decisions. For the global land science community, this means being 'reflexive,' i.e., constantly questioning the data produced, the methods used, and the categories developed as well as their respective wider societal and ecological consequences. Gou (Chap. 6) in this volume contributes to this debate by asking about the role of radar versus optical sensors in detecting land degradation and the largely unquestioned normativity of the label 'degraded.' Reflexivity means being aware of the looping effects between scientific knowledge, the categories produced on the basis of that knowledge, and the effects these categories have in the real world (Hacking 2006, see also Chap. 1). The role of technology in this process, e.g., sensor technology development in remote sensing, is vital.

2.3.6 Technology Makes Connections

Technology has become an important driver in land use competition. First and foremost, information and communication technologies increase the rate, extent, and speed of information exchange. For example, in the context of market access, being able to communicate instantly over vast distances is changing social relations and systems of exchange. Research on globalization shows how a world that has always been networked through trade routes and colonizing powers is now connected much more intimately. Diasporic cultures develop in close connection with people in the home countries, migration routes are rapidly being communicated across vast regions, social movements connect and learn in global networks: information, knowledge, goods, and people flow differently through such global scapes (Appadurai 1991).

Agricultural technology also develops and spreads quickly through global agricultural businesses and information exchange. As Gasparri shows in this volume

(Chap. 4), the arrival of new, more salinity tolerant soy plants in the Chaco region rapidly turned what was a dry forest largely inhabited and managed through local pueblo systems into one of the hotspots for global soy production, thus triggering massive land use change. Importantly, technology holds the potential to make unforeseen connections and create new trajectories. Such surprising connections have already been discussed in the context of the telecoupling concept. Spillover effects and the coupling of systems so far unconnected is a major point of analytical attention. The role of technology in helping to establish these connections outside of known trajectories and systems of governance must not be underestimated.

2.3.7 Recognizing Different Temporalities in Land Use Competition

The concept of underlying drivers of land use competition points to the increasing functional distance of local land use from the highly mediated processes of global competition and decision-making. The role of different temporalities is crucial in this context. We suggest to distinguish three forms of temporality.

History

Patterns of land use and that of land use competition have environmental and sociocultural histories. As particularly Hauge, Schaffartzik, and Gasparri (Chaps. 3, 4 and 5) demonstrate in this section, land use patterns have a legacy in environmental, political, and social contexts. History can be considered part of a discourse (see above, Foucault) that structures to some degree what may count as feasible and legitimate. It also helps to understand social and political relations, traditions and routines of land use, belief systems, and the potential for conflict. History, however, is not only the consecutive succession of past events, e.g., of political regimes or property systems. Rather, in many regions of the world, several historically formed systems of land governance exist at the same time. One may speak of a palimpsest insofar as several different systems of governance may ‘shine’ through what is currently considered the official rule and code of conduct. Layers of governance may relate back to a colonial past, to times of occupations, or to different settlement periods or political systems. This means that land use competition does not unfold according to one official land governance system. Instead, actors may find ways of legitimately drawing on older systems in order to stake their claim. Oftentimes, the analytical distinction between formal and informal is introduced to clearly distinguish an official way of how things should be done and other, somehow vernacular ways of how things are actually done. This perspective is deeply embedded in a modern Euro-American understanding of nation states with their heavily rationalized bureaucratic structures. For many regions of the world, what is readily lumped together as ‘informal’ constitutes the norm and understanding the logics of such historically layered systems of governance is paramount to understanding land use competition.

It is also worth noting that most historians today will not conceptualize history only as a representation *of* the past. History is always also a representation *for* something; i.e., it is written today in a particular context, by particular people with a particular goal in mind. Hence, the idea of history as a single linear string of events is problematic. Multiple histories usually emerge on the basis of different sources and different readings of archival sources. The question who tells the history of a particular region in what way is an important one. Historical narratives offer frameworks to understand and legitimate developments in the present and they help to make possible futures and expectations more or less plausible (Brown and Michael 2003). Hence, contested histories, particularly in regions of the world where written archives are nonexistent or have been produced by colonial powers for very particular purposes, need to be analyzed with great care rather than taken for granted as a given narrative.

Potentiality

Many commentators of social-ecological development have pointed out that modernity and its almost universal underlying capitalist market logic entails the promise of a better future: political and social systems that engender reliability in social expectations, secure or higher income, better health care, safer and cleaner living environments, more extensive education, and partaking in the world of consumer goods. These are the promises and expectations built into the discourse of global development and market integration. They may remain implicit or they may be discussed explicitly often when people are faced with their expectations not being met over extended periods of time.

An extensive literature on aspirations, hope, and expectations demonstrates how powerful such factors can be a driving force in the present (Kornienko 2014). In the context of land use competition, it is important to be analytically aware of and sensitive to these imagined or materialized social potential(itie)s (Sejersen 2015) that come with certain prospects of land use development. Sejersen has shown in the context of industrial development in Greenland that decisions to allow the building of large industrial sites in the country are not only driven by typically rationalized decision-making based on actual cost-benefit analyses. Rather the potentiality of a development plays an important role. What narratives about a different future may be attached to a certain piece of land, a landscape, or a proposed site of development? These are powerful factors that do not work predominantly at the level of individual imagination, but rather operate as social imaginaries that help to form collectives and alliances around particular visions of the future.

Response Time

History and potentiality are rooted within the same temporality of linear progression—or be it multiple and contested. Response time on the other hand introduces a different temporality altogether. The concept of response time originates in physics and describes the characteristic time it takes for a state variable to respond to a perturbation. It has been used in nonlinear systems theory to identify variables that have strong two-way interactions (Werner and McNamara 2007). It has been proposed that variables describing human and natural systems have characteristic response times.

This is not only a technical measure that may interest modelers. Understanding land use competition as a mediated process as discussed above means being aware of differences and mismatches in response times between social and physical processes. For example, a particular derivative market operated online and globally may (be able to) respond within seconds or less to news affecting a region's agricultural productivity, e.g., political unrest or new knowledge about developments in regional climate. Changes in the agricultural system, on the other hand, feedback more slowly onto the economic market. This mismatch in timescale might be thought of as an inherent instability in this system where the market system drives irreversible physical change. It highlights the role of social institutions and governance schemes that evolve on the timescale of the natural system and thus stabilize this temporal asymmetry, which in turn may alter investments and business models on the ground-raising issues about social inequality and environmental damage.

Joshi's chapter in this section demonstrates how variations in the gold price change land use quite dramatically on the ground with massive implications of increased environmental degradation (Chap. 7). In her case study, local miners are able to respond quickly to changing market prices, but regulation and enforcement of sanctions are slow to react to the volatility of the market and rapidly moving actors. The result is a dynamic in land use and land use competition that means that those people lose out who require a more stable social-ecological system (Brazil nut harvesters in Joshi's case). Hence, response time within systems and particularly differential response times in coupled systems create very real governance problems. This also demonstrates how the characteristic timescale of coupling can be used to set meaningful analytical boundaries. If the prevalence of coupling language implies an intractably interconnected world, the notion that not all coupling is created equal—that processes acting on different time scales may have linear interactions over short time frames, as in a market driving deforestation—can be exploited to decide which processes are important to the system over a particular time frame. Particularly, the time lags inherent in many environmental developments are problematic in this sense, because land use competition has difficulty in pricing in externalities that have not yet materialized and that are highly uncertain.

2.4 Conclusion: The Contingency of Land Use Competition

We have shown here how an extended notion of distal drivers may address concerns of power, knowledge, and technology. Such an understanding of distal drivers, we suggest, lies at the heart of land use competition framed not as an individual decision event, but rather as a distributed process. Processes of competition are distributed across geographical, institutional, and social scales, they incorporate different understandings of space, they involve multiple histories and futures, and they need to account for different temporalities, in particular the idea of differential response times within a social-ecological system.

We suggest a relational understanding of land use competition with the concept of 'distal drivers' asking important questions about the nature—quantity and quality—of the relations that are being forged or discarded through competitive processes. An ontologically very heterogeneous set of agents comes to light in this analysis: nature, human actors including society and institutions, knowledge, social and moral orders, and technological artifacts. They all contribute to the specificity of distal drivers in particular processes of land use competition. One might say: They situate land use competition (Haraway 1988), with 'situate' not meaning 'locate' but rather 'embed', within broader social-ecological systems and their contexts (Granovetter 1985).

This type of embedding is a historically and socially highly contingent process: It may unfold along a particular path, but that path is not the necessary path. This has important consequences for our epistemological approaches to studying land use competition. In contingent processes, straightforward cause and effect relationships are hard to specify. This alerts us, firstly, to the important role of uncertainty and ambiguity in social-ecological systems. Secondly, it urges us to carefully consider system boundaries and the range within which our knowledge claims may be considered valid. For example, the price of a good on the market may predict crop choice for the next-growing season for a particular case with a high degree of accuracy and certainty. Hence, operating within the narrow constraints of one growing season, one product and a fixed set of actors may allow us to usefully treat land use competition as a linear set of functions. Yet, extending the system boundaries to include distal drivers, as we have done in this chapter, frames land use competition as a complex set of processes that interact in nonlinear fashion and are virtually impossible to predict. The analysis of the dynamics of contingent processes is in itself contingent, i.e., heavily influenced by methodological and analytical choices. While the plausibility of the respective approaches may be assessed according to their own terms, their legitimacy needs to be considered in a wider scientific, political, and moral context.

The contingency of land use competition thus alerts us to the situatedness of our own knowledge production: situated in a particular time and culture, in a particular set of methods, and a particular disciplinary set of concepts and theories. Distal drivers are too heterogeneous and too complexly related to be adequately known through a single epistemological framework. Neither geoscience, geography, or economics, nor anthropology or philosophy holds the all-encompassing methodological framework to deal with all facets of distal drivers in land use competition. Hence, research on distal drivers will require humility and epistemological openness to translate between the very distal thought styles concerned with distal drivers of land use competition.

References

Alexander, J. C. (1988). *Action and its environments. Toward a new synthesis*. New York: Columbia University Press.

Appadurai, A. (1991). Global ethnoscapes: Notes and queries for a transnational anthropology. In R. G. Fox (Ed.), *Recapturing anthropology: Working in the present*. School of American Research: Santa Fe.

Beck, S. (2008). Natur|Kultur. Überlegungen zu einer relationalen Anthropologie. *Zeitschrift Für Volkskunde, 104*, 161–199.

Bourdieu, P. (1984). *Distinction: A social critique of the judgement of taste*. Harvard University Press.

Brown, N., & Michael, M. (2003). A sociology of expectations: retrospecting prospects and prospecting retrospects. *Technology Analysis and Strategic Management, 15*, 3–18.

Chan, K. M. A., Guerry, A. D., Balvanera, P., Klain, S., Satterfield, T., Basurto, X., et al. (2012). Where are cultural and social in ecosystem services? A framework for constructive engagement. *BioScience, 62*, 744–756.

Creutzig, F., Baiocchi, G., Bierkandt, R., Pichler, P.-P., & Seto, K. C. (2015). Global typology of urban energy use and potentials for an urbanization mitigation wedge. *Proceedings of the National Academy of Sciences, 112*, 6283–6288.

Daily, G. C., & Matson, P. A. (2008). Ecosystem services: From theory to implementation. *Proceedings of the National Academy of Sciences of the United States of America, 105*, 9455–9456.

Eakin, H., Defries, R., Kerr, S., Lambin, E. F., Liu, J., Marcotullio, P. J. et al. (2014). Significance of telecoupling for exploration of land use change. In: S. Kc, & A. Reenberg (Eds.), *Rethinking global land use in an urban era*. Cambridge: MIT Press.

Foucault, M. (1972). Truth and Power. In C. Gordon (Ed.), *Power/knowledge: selected interviews and other writings* (pp. 1972–1977). New York: Pantheon Books.

Friis, C., Nielsen, J. Ø., Otero, I., Haberl, H., Niewöhner, J., & Hostert, P. (2015). From teleconnection to telecoupling: taking stock of an emerging framework in land system science. *Journal of Land Use Science*, 1–23.

Garrett, R. D., Rueda, X., & Lambin, E. F. (2013). Globalization's unexpected impact on soybean production in South America: linkages between preferences for non-genetically modified crops, eco-certifications, and land use. *Environmental Research Letters*, 8.

Granovetter, M. (1985). Economic action and social structure: the problem of embeddedness. *American Journal of Sociology, 91*, 481–510.

Guyer, J., Lambin, E., Cliggett, L., Walker, P., Amanor, K., Bassett, T., et al. (2007). Temporal heterogeneity in the study of African land use. *Human Ecology, 35*, 3–17.

Hacking, I. (2006). *Kinds of people: Moving targets*. London: British Academy Lecture.

Hannerz, U. (1993). When culture is everywhere: Reflections on a favorite concept. *Ethnos, 58*, 1–11.

Haraway, D. (1988). Situated knowledges: the science question in feminism and the privilege of partial perspective. *Feminist Studies, 14*, 575–599.

Kornienko, K. (2014). Waiting, hope, democracy, and space: how expectations and socio-economic rights shape two South African Urban Informal Communities. *Journal of Asian and African Studies*.

Lambin, E. F., & Geist, H. (2006). *Land-use and land-cover change: Local processes and global impacts*. Berlin/New York: Springer.

Latour, B. (2005). *Reassembling the social: An introduction to actor-network-theory*. Oxford: Oxford University Press.

Law, J., & Callon, M. (1992). *The life and death of an aircraft: a network analysis of technical change*.

Law, J., & Mol, A. (2002). Complexities: An introduction. In J. Law & A. Mol (Eds.), *Complexities: Social studies of knowledge practices*. Durham, NC: Duke University Press.

Liu, J., Hall, V., Moran, E., Nagendra, H., Swaffield, S. R., & Ii, B. L. T. (2014). Applications of the telecoupling framework to land-change science. In S. Kc, & A. Reenberg (Eds.), *Rethinking global land use in an urban era*. Cambridge: MIT Press.

Marcotullio, P. J. (2014). Globalization, economic flows, and land-use transitions. In K. C. Seto & A. Reenberg (Eds.), *Rethinking global land use in an urban era*. Cambridge, MA: MIT Press.

Marres, N. (2012). On some uses and abuses of topology in the social analysis of technology (or the problem with smart meters). *Theory Culture and Society, 29*, 288–310.

Martinez-Alier, J. (2008). Languages of valuation. *Economic and Political Weekly, 43*, 28–32.

Niewöhner, J. (2014a). Ökologien der Stadt. Zur Ethnografie bio- und geopolitischer Praxis. *Zeitschrift für Volkskunde, 110*, 185–214.

Niewöhner, J. (2014b). Raum aus anthropologischer Perspektive. In J. Oßenbrügge & A. Vogelpohl (Eds.), *Theorien in der Raum- und Stadtforschung – Eine Einführung*. Westfälisches Dampfboot: Münster.

Ostrom, E., Schroeder, L., & Wynne, S. (1993). Analyzing the performance of alternative institutional arrangements for sustaining rural infrastructure in developing countries. *Journal of Public Administration Research and Theory: J-PART, 3*, 11–45.

Ouma, S., Boeckler, M., & Lindner, P. (2013). Extending the margins of marketization: frontier regions and the making of agro-export markets in northern Ghana. *Geoforum, 48*, 225–235.

Palsson, G., Szerszynski, B., Sörlin, S., Marks, J., Avril, B., Crumley, C., et al. (2013). Reconceptualizing the 'Anthropos' in the Anthropocene: integrating the social sciences and humanities in global environmental change research. *Environmental Science and Policy, 28*, 3–13.

Parsons, T. (1937). *The structure of social action*. New York: Free Press.

Poteete, A. R., & Ostrom, E. (2004). Heterogeneity, group size and collective action: the role of institutions in forest management. *Development and Change, 35*, 435–461.

Roepstorff, A., Niewöhner, J., & Beck, S. (2010). Enculturing brains through patterned practices. *Neural Networks, 23*, 1051–1059.

Sejersen, F. (2015). *Rethinking greenland and the arctic in the era of climate change: new northern horizons*. Routledge, Taylor & Francis Group.

Seto, K. C., Reenberg, A., Boone, C. G., Fragkias, M., Haase, D., Langanke, T., et al. (2012). Urban land teleconnections and sustainability. *Proceedings of the National Academy of Sciences of the United States of America, 109*, 7687–7692.

Strathern, M. (1991). *Partial connections*. Savage, Md, Rowman & Littlefield Publishers.

Strathern, M. (1992). Parts and wholes. Refiguring relationships in a post-plural world. In A. Kuper, (Ed.), *Conceptualizing society*. London/New York: Routledge.

Weber, M. (1922/2002). *Wirtschaft und Gesellschaft - Grundriß der verstehenden Soziologie*. Tübingen, Mohr.

Werner, B. T., & McNamara, D. E. (2007). Dynamics of coupled human-landscape systems. *Geomorphology, 91*, 393–407.

Young, O. R., Lambin, E. F., Alcock, F., Haberl, H., Karlsson, S. I., Mcconnell, W. J., et al. (2006). A portfolio approach to analyzing complex human-environment interactions: Institutions and land change. *Ecology and Society, 11*, 15.

Chapter 3
At a Distance from the Territory: Distal Drivers in the (Re)territorialization of Oil Palm Plantations in Indonesia

Anke Schaffartzik, Alina Brad, Melanie Pichler and Christina Plank

Abstract Exponential growth in oil palm land and palm oil production in Indonesia currently makes the country the world's largest producer of this vegetable oil. Throughout its tumultuous political past from the 1960s until today, conditions were created which enabled the expansion of oil palm plantations. Under President Suharto's "New Order," territorialization processes were used to bring land and people under the control of an increasingly powerful central government. Plantations were instrumental in this regard and additionally formed the basis for the production of palm oil as an important export commodity in the opening of the Indonesian economy. In the wake of the 1997/1998 Asian economic crisis and with the end of Suharto's rule, Indonesia entered a period of reform marked by decentralization and reterritorialization processes. Oil palm plantations continued to grow as foreign investment and plantation ownership by private businesses became increasingly relevant. Throughout both of these periods, the land-use decisions, which fostered the expansion of oil palm plantations, were not made by the people on the ground, but at a spatial, temporal, and functional "distance." While distal drivers strongly shaped the development of land use, the consequences of these changes had little impact on the drivers. As plantations expanded farther into territories already claimed for other forms of land use (e.g., rainforest, subsistence agriculture land, indigenous land), they were increasingly likely to confront competing claims to land. Under the perceived greater political freedom of the reform period, the competing claims more often triggered conflicts, which, however, had limited bearing on these distal drivers.

Keywords Land-use practices · Globalization · Governance · Conflict · Palm oil

A. Schaffartzik (✉)
Institute of Social Ecology, Alpen-Adria Universität Klagenfurt-Wien-Graz, Vienna, Austria
e-mail: anke.schaffartzik@aau.at

A. Brad · M. Pichler · C. Plank
Department of Political Science, University of Vienna, Vienna, Austria

J. Niewöhner et al. (eds.), *Land Use Competition*,
Human-Environment Interactions, DOI 10.1007/978-3-319-33628-2_3

3.1 Conceptualizing Drivers of Plantation Expansion

Palm oil is consumed as an oil and as an ingredient in processed foods across the globe on a daily basis. At an average of more than 2 kg per person and year, it is second to the consumption of only one other type of vegetable oil—that of soybeans (FAO 2014). Palm oil is not used for food alone: Currently, more than 50 % of global palm oil is used for industrial purposes, mainly in soaps and cosmetics, but also increasingly as a feedstock for the production of agro-biodiesel.[1] The latter use is especially relevant in Southeast Asia (Zhou and Thomson 2009). Palm plantations are considered the fastest growing monoculture in the world (Gerber 2011) and almost half of global growth between the early 1960s and 2010 occurred in Indonesia (6 million ha). In the two decades between 1991 and 2011 alone, oil palm plantations expanded from an area corresponding to less than 0.5 % to more than 3 % of Indonesia's land area (FAO 2014).

The drivers of this expansion originate outside of the oil palm plantations and the realm of local control: From the archipelago's history under Dutch Colonial rule to the influence of international organizations, especially during the second half of the twentieth century, to increasing rates of foreign investment in the present-day Indonesian plantation sector, the growth in palm oil production has always been enabled and facilitated through changes at various spatial and temporal scales. Production is currently also more strongly tied to satisfying foreign rather than domestic final demand: In 2011, more than ¾ of the palm oil produced in Indonesia were exported (FAO 2014). The plantation expansion is also linked to environmental impacts, which extend into other countries and into the future: It constitutes a significant risk to prevailing ecosystems and biodiversity (Koh and Wilcove 2008; Fitzherbert et al. 2008) and has a potentially significant impact on climate change through associated deforestation and drainage of peat lands (Fargione et al. 2008; Reijnders and Huijbregts 2008; Germer and Sauerborn 2008).

Although the unbroken exponential growth of palm oil production in Indonesia may suggest otherwise, plantation expansion is often heavily contested. The associated land-use changes infringe upon other types of land use and exclude groups of people from the land. The dispossession of people from their livelihood resource has often occurred forcefully and has been associated with social conflict (Colchester et al. 2006; Marti 2008; Colchester and Chao 2013; Larsen et al. 2014).

From the first phase of plantation expansion beginning in the 1960s to the continuing exponential growth of palm oil production into the twenty-first century, Indonesia has undergone tremendous political and economic change. Even in tumultuous times, the expansion of oil palm plantations was, in itself highly dynamic, a constant. As will be argued in this chapter, oil palm plantation

[1]The term "agro-biofuels" (and hence agro-biodiesel) is used within this chapter in order to underline the focus on large-scale agricultural production of biofuel feedstock.

expansion can even be considered to have been instrumental to the ongoing political and economic change. The political and economic developments in Indonesia which are directly linked to the expansion of oil palm plantations extend from a military coup in 1965 and the subsequent centralization and authoritarian rule of President Suharto under the "New Order" regime (orde baru) to the decentralization process during the reform period (reformasi) following the Asian crisis and the fall of Suharto in 1998.

In this chapter, the expansion of palm oil production in Indonesia is discussed in terms of the distal drivers by which it was shaped during the politically and economically distinct periods of New Order and Reform, respectively. In the DPSIR[2] model proposed for environmental assessments by the European Environment Agency in the mid-1990s, "drivers" (D) are the causes of a "pressure" (P) which may change the "state" (S) of an environmental system, and thus have an "impact" (I), and possibly solicit a "response" (R) (EEA 1995). This model affords the advantage of linking environmental impacts to their drivers and thus potentially to the root of the problem. Early on, applications of the DPSIR framework distinguished direct and indirect pressures on the environment: pressures, which were directly exerted, and pressures, which indirectly led to proximate pressures. A similar distinction was later made with regard to drivers, differentiating between immediate drivers, which were proximate to the pressures they caused, and underlying drivers, which could cause the former. Building on this distinction, it was more recently proposed that in assessing environmental impacts and finding an effective response to environmental pressures, it is necessary to consider not only "proximate" drivers which can be directly linked to a pressure on the environmental system but to additionally take into account so-called distal drivers (Geist and Lambin 2002; Geist 2005, xxxi). These drivers not only exert pressures across a distance, but they may also (indirectly) be affected by these environmental pressures (e.g., through feedback loops) or by the changes in states which they cause (e.g., if a threshold is crossed) (Liu et al. 2013; Turner et al. 2007). In the following analysis, drivers were also considered if they were temporally distal, i.e., if changes that took place in the past continue to shape current social and environmental pressure, and/or functionally distal, i.e., if they served an objective which was not land-use related but had "side effects" on land use. The three types of distal drivers are not mutually exclusive but in fact often overlap. In Indonesia, the "distance"—both in the literal, spatial, sense of the word and its figurative, temporal and functional meaning—between drivers and their socioecological impact was decisive in enabling the continuation of large-scale land-use change, in spite of intense protest and resistance from indigenous and peasant as well as environmental movements. The "distance" removed underlying causes from the sphere of influence of local actors and kept the (local) consequences of the distal drivers from directly affecting the more influential actors.

[2]Driving force, Pressure, State, Impact, Response (EEA 1995).

3.2 (Re)terriorialization of Indonesia's Plantations

The analysis of distal drivers of palm oil expansion in Indonesia and the ensuing land-use competition benefits from the conceptualization of the processes of centralization under the New Order regime and of decentralization during the reform period as territorialization and reterritorialization (for a detailed discussion of these processes, please see Brad et al. 2015).

Territorialization is an important means of state control over land (Vandergeest and Peluso 1995; Peluso and Lund 2011) and can be understood as a "process by which states attempt to control people and their actions by drawing boundaries around a geographic space, excluding some categories of individuals from this space, and proscribing or prescribing specific activities within these boundaries" (Vandergeest 1996: 159). Territorialization processes enable the accumulation of capital from land-based resources on which capitalist societies depend (Smith 1984). The control over land and the subsequent accumulation process, referred to as primitive accumulation (ursprüngliche Akkumulation) by Marx (1867), has been analyzed in contemporary politics as "accumulation by dispossession" (Harvey 2005). This enclosure and subsequent control of land for accumulation purposes are mediated by institutional arrangements, legal processes, and state regulations. The framework of analysis spanned by the territorialization concept allows for the consideration of drivers which are spatially or temporally distal (as opposed to proximate) and of drivers, which are functionally distal in the sense that the objectives behind them are not directly land-use related. The focus on territorialization during New Order provides a structural framework, which allows for the consideration of specific distal drivers as they emerge from an analysis of land-use competition focused on underlying power relations.

Globalization processes have intensified sub- and supranational reconfigurations of plantation land control which imply processes of reterritorialization (Peluso 2005; Sassen 2005; Peluso and Lund 2011; Corson 2011). During the reform period in Indonesia, including greater liberalization of the economy after the Asian crisis and the fall of Suharto, existing centralized territorialization processes were challenged. In the decentralization process, power to control land and make land-use decisions was transferred to district governments at the subnational scale and also to international banks and foreign investors—especially Malaysian and Singaporean companies—at the supranational scale. Reterritorialization strategies can be contradictory and contested while constructing "the place-based and territorial preconditions for accelerated global capital circulation [...] on multiple spatial scales" (Brenner 1998: 3). Both territorialization and reterritorialization processes are closely related to enclosure and dispossession processes, enabling the continuous accumulation of capital by public and/or private actors. By examining the processes of reterritorialization during the reform period, the shifts in relevant distal drivers to Indonesia's foreign trade and international involvement in Indonesia can be considered. Competing alternative claims to land often respond to specific territorialization processes in the context of Indonesia's palm oil expansion. Examples for such alternative claims include the enclosure of land for conservation purposes or the reclaiming of customary lands by indigenous groups.

3.2.1 Centralization and the Influence of Jakarta on Plantation Expansion

In 1965, the "Guided Democracy" under Indonesia's first president, Sukarno, was brought to an end by a bloody military coup. Changes, which had been set in motion under Sukarno, were incompatible with the control over land that the New Order regime of General Suharto sought to establish. Sukarno's vision of an "Indonesian socialism" had involved the nationalization of land held privately by individuals or international companies and the redistribution of this land to landless peasants. The expression of this political will was the Basic Agrarian Law (BAL) which also introduced customary rights (adat) as the legal basis of agrarian law (Republik Indonesia 1960a, Article 5). A government regulation was passed which restricted the maximum land area that could be managed by an individual or a family to between 5 and 20 hectares, depending on population density and land quality (Republik Indonesia 1960b), seeking to avoid high concentrations in landownership. The BAL did, however, allow for the continuation (and thus for the subsequent expansion) of plantation agriculture by introducing a cultivation/exploitation rights title (HGU—Hak Guna Usaha). Companies could own more than the maximum permitted land area if they were granted such an HGU which could then remain valid for up to 60 years (Republik Indonesia 1960a, Article 29). The HGU itself is an interesting case-in-point illustrating the influence of temporally distal drivers in Indonesia's oil palm expansion: Under Sukarno, colonial plantation land was excluded from the BAL land reform program and the HGU allowed for the continuation of this particular land-use structure. Following the West Papua conflict, in 1957, Dutch-owned plantation land was nationalized and martial law was implemented, giving the military the power to control much of the highly lucrative plantation land. The HGU must thus also be understood as a benefit offered to the military elites. It would additionally prove to be a temporally distal future driver in the continued expansion of oil palm plantations.

Despite the selective continuation of plantation agriculture in the course of agrarian reform and land distribution, the break of Suharto's New Order with the land reform legacy was pronounced: In the wake of the coup, the reform was stopped and all radical peasant movements were banned. An estimated half a million members and alleged sympathizers of the Indonesian Communist Party (PKI—Partai Komunist Indonesia) and related peasant organizations (e.g., the Indonesian Peasant Front) were killed in this short period. The, often violent, removal of people from the land and the categorical exclusion of large population segments from land-use decisions enabled a specific process of centralized territorialization and initiated agricultural expansion within a capitalist production system (Farid 2005; Peluso et al. 2008).

In 1967, Suharto issued the Basic Forestry Law (BFL) in order to establish the national government's control over the Outer Islands' forests, declaring more than 70 % of the Indonesian land area as "state forest land" and transferring control over all of this land to the Directorate General of Forestry. This centralization of power in Jakarta meant that in order to have influence on land use in Indonesia—the majority of the land area of which was (and is) covered by forests—it was necessary to have influence with the decision makers in the capital city. In contrast to the provisions for the recognition of customary adat land claims and thus for the local control over land under the BAL, the BFL assigned control over forest land to the national government (which shared some of its decision-making power with district and provincial governments) and excluded indigenous groups from any decision making: All land not covered by centrally documented property rights was declared to be state land and as such was subject to a zoning system (Barr et al. 2006). Thereby, over 2/3 of state forestland was approved for production (approximately 93 Mha). This allowed for the establishment of commercial logging as a major pillar of the Indonesian economy and facilitated the expansion of plantations into fallows, grasslands, scrub, and secondary forests (Colchester et al. 2006). The zoning system not only ignored the particular usage rights of an entire segment of the population, but also failed to consider the (ecological) functions of the existing land cover (e.g., of primary forests). The BFL thus provided the legal preconditions for the dispossession of people and for the expansion of both commercial logging and plantation agriculture.

The control of the central government in Jakarta over the development of oil palm plantations on the other islands[3] and, by extension, over the rural population was additionally asserted through the instrument of contract farming: Those smallholders whose claims to land were recognized in spite of the provisions of the BFL were integrated into a core-smallholder scheme of plantation agriculture. In the PIR (Perkebunan Inti Rakyat) scheme, a core plantation (inti) was associated with a number of smallholder (rakyat/people) plots of approximately two hectares per family, in a core-to-smallholder ratio of about 20:80 or 30:70. This arrangement provided the state with an important lever for its processes of territorialization and enclosure. The core plantation was charged with the management of all smallholder plots for which it provided access to national and international markets (Zen et al. 2005). The often state-owned core plantation company was also the focal point of the state's control over the agricultural system. The Indonesian government assisted

[3]Until 1980, oil palm plantations were essentially confined to three provinces in Sumatra where they followed in the trodden paths of colonial rubber plantations: In 1975, North Sumatra contributed 94 % of total Indonesian palm oil production, followed by Aceh (5 %) and Lampung (1 %) (Kementerian Pertanian Republik Indonesia 2013).

this company in gaining access to land, developing infrastructure, and provided subsidized capital for plantation development. In terms of territorialization, the contract-farming scheme brought newly enclosed land under state control. Local claims to land were subsumed under plantation agriculture.

During the period of centralization under Suharto's New Order, the political strategy aimed at bringing both land and rural population under state control must be considered to have been a dominant driver of the expansion of oil palm plantations. Simultaneously, the high growth rate of approximately 6 % per year of oil palm land meant that output of oil palm fruit and thus of palm oil was becoming increasingly relevant. During the first period of this expansion, however, income from Indonesia's agricultural sector had continuously fallen from over 50 % of the country's GDP in 1965 to approximately 36 % in 1976 (Pitt 1980). The opportunities for agricultural growth fundamentally changed during the last quarter of the twentieth century. With the oil crises of the early 1970s, Indonesia significantly increased its exports of fossil fuels which accounted for 80 % of Indonesia's physical exports by 1980 (Schaffartzik et al. 2014) and oil export revenues rose markedly. Along with long-term loans procured by the Indonesian government, the increasing budget volume allowed for greater financial investments in agricultural development. The main driver behind this particular investment decision was the "green revolution"[4] promoted by international financial institutions (for Indonesia most importantly the World Bank and the Asian Development Bank) from the late 1960s onward and implemented in Indonesia with the central goal of decreasing import and aid dependence for rice as the most important food crop. The higher outputs of agricultural commodities were, however, following requirements made by the loan-granting banks, almost exclusively dedicated to export: While only 2 % of Indonesia's physical exports were of biomass origin in 1960, this share rose continuously to 10 % by 1990. The relevance of the export market opened the production of Indonesian palm oil to new driving forces originating from the country's trade partners. Until the late 1980s, 80–90 % of Indonesia's annual palm oil production was exported, and most important destinations for these exports were European: In 1986, 62 % of total palm oil exports went to the Netherlands, Germany, the UK, and Italy. By the late 1990s, the Netherlands, an important port of entry into the European market, continued to receive the largest share of imports (22 %), but was followed by the growing Asian markets of India (20 %), Malaysia (17 %), and China (10 %) (FAO 2014). This changing trade pattern illustrates the impact of the economic development in the Asian region on production in Indonesia. The country's export orientation coincided with a gradual opening of the economy to foreign investment and trade. While the country was heavily dependent on foreign aid following independence, foreign direct investment (FDI) was increasingly sought instead. FDI multiplied by a factor of 42.6 between 1981 and

[4]The "green revolution" was essentially aimed at establishing a high-input high-output agricultural system in which higher yields were to be achieved through the use of fertilizers, pesticides and herbicides, agricultural technology, and new crops types along with efficient management techniques.

1997 from 126 million (or 0.2 % of GDP) to 5.4 billion constant 2005 US$ (more than 2 % of GDP; The World Bank 2013).

The production of palm oil for a growing export market and the expansion of oil palm plantations were mutually linked. On the one hand, high levels of domestic and foreign demand provided an incentive for higher production levels. On the other hand, higher yields (attained through mechanization of agriculture following the "green revolution," see above) and further extensification of palm oil production made the maintenance of the (foreign) sales market a necessity. The expansion of plantations continued on the island of Sumatra while territorialization processes of the Outer Islands facilitated investments by national and foreign private-owned conglomerates there. This was accompanied by an extension of the contract-farming scheme. From the mid-1980s on, the PIR plantation model was merged with the Indonesian resettlement program (transmigrasi) to form PIR-trans. Peasants and landless people from the densely populated islands of Java and Bali were moved to the Outer Islands of Kalimantan, Sumatra, and Sulawesi where they were provided with land for oil palm development under the contract-farming scheme, enabling central control over the resource-rich, fertile, and only sparsely populated areas of the country (McCarthy et al. 2012).

In the expansion of oil palm plantations, a complex array of distal drivers is at play. These tend to simultaneously feature elements of a spatial, a temporal, and a functional disconnect from the pressures and practices on the ground. With the BFL, the central government in Jakarta made far-reaching decisions on how land was to be used, ensuring that vast forest areas would be used in an (economically) productive manner. Likewise, contract-farming schemes brought the expanding oil palm plantations under central control while both government spending and foreign investment were decisive in determining the rate of plantation expansion. The forced migration under transmigrasi not only provided labor in regions of increasing agricultural activity, but also gave the government influence over local communities in which inhabitants diverse in terms of languages, culture, and beliefs were assimilated under a national cultural identity. Although none of these factors were specifically geared toward palm oil production, they formed a framework, which enabled the expansion of oil palm plantations as Suharto's government strove for greater centralization.

3.2.2 Decentralization and the New Influence of District Governments

In the wake of the Asian financial crisis and the fall of Suharto's authoritarian regime in 1998, a reform process (reformasi) was initiated by Indonesian and international actors. These reforms were directly relevant for the continued expansion of oil palm plantations: They transferred important decision-making power with regard to land use from central state institutions to the district level.[5] After more than three decades of a highly centralized system of land control and territorialization, the transfer of political, administrative, and fiscal decision-making power to the district governments (Barr et al. 2006) constituted a significant change in the drivers of land use and oil palm expansion. District governments gained the right to release plantation permits (Ministry of Agriculture 2007, Article 17), which began to provide an important source of palm oil-related income for the districts during a time when most revenues generated from palm oil production (e.g., export taxes) remained with the central government.[6] Next to the district government's right to stipulate land division and ownership, the decentralization process also led to regional plantation regulations. These governed such important issues as the compensation payments for local landowners and endorsed the given form of plantation structure. The district head (bupati) thereby became an important stakeholder in plantation development (Gillespie 2011). Many members of district governments were simultaneously shareholders in agribusiness companies or land brokers and thus additionally influential in controlling access to land.[7]

These reterritorialization processes were accompanied by a fundamental change in the contract-farming schemes established under Suharto. The state subsidies for the PIR and PIR-trans schemes severely declined with the beginning of the Asian financial crisis in 1997. During the reform period, the Indonesian government therefore sought to attract private investment to support the plantation sector. The Indonesian government liberalized the plantation sector and introduced so-called partnership models (kemitraan) to replace the PIR and PIR-trans schemes. Under these partnership models, the core company owned approximately 80 % of the jointly managed plantation area, and only 20 % was allocated to smallholders (McCarthy et al. 2012). The companies were no longer dependent on state support for their endeavors as they were entitled to negotiate directly with local landowners for the integration of their land into the plantation area (Republik Indonesia 2004, Article 9). Once the plantation had been established, these smallholders were reimbursed with dividends of the plantation's profit (Li 2011). Between 2001 and 2010, the number of smallholders tied to plantation companies as contract farmers increased by 140 % (Badan Pusat Statistik 2012).

During the period of reformasi, the increasing use of agro-fuels both directly and indirectly triggered a rise in demand for Indonesian palm oil: In particular within the Southeast Asia region, this oil is one of the most important feedstocks in the production of agro-biodiesel; internationally, the growing use of other vegetable

[5]The country's political program of government decentralization was mainly triggered by regional expressions of discontent (e.g., separatist movements in resource-rich provinces like Aceh and West Papua) and the pressure of the International Monetary Fund and the World Bank for structural adjustments following the Asian crisis (McCarthy 2004).

[6]Interview with Sawit Watch, Bogor, December 3, 2013.

[7]Interview with an activist, Bogor, December 5, 2013.

oils (rape seed, soybean, sunflower) for agro-biodiesel production necessitated a substitute for these oils in human nutrition and other industrial uses. From climate protection to resource security, a number of policies implemented in other countries acted as significant drivers of Indonesian palm oil production.

The introduction of the partnership plantation model coincided with the expansion of oil palm plantations further into the Outer Islands, driven both by the need to expand production for export and by the decentralization process. Due to the latter, the district governments could now secure income for their districts through plantation concessions, taxes, and foreign direct investments. The districts also competed for these forms of income and granted increasing numbers of concessions in attempts to outbid other districts. As a result, oil palm plantations increasingly expanded into economically less developed regions and districts. While in 1975, almost 90 % of oil palm plantation land was located in one province (North Sumatra), in 2011, the largest share of oil palm plantation area (21 %) was in the Riau Province (on the island of Sumatra) and 31 % of all oil palm plantation area was on the island of Kalimantan (Kementerian Pertanian Republik Indonesia 2013). Overall, out of the 6 million hectares on which oil palm plantations in Indonesia expanded between 1962 and 2012, more than 70 %, or 4.3 million ha, were planted during the last 14 years alone (FAO 2014).

Although the political and institutional changes in Indonesia during the reform period were pronounced, they did not alter the course of oil palm expansion. The system of plantation agriculture as inherited from the colonial period and extended during New Order continued to act as a temporally distal driver. The decentralization process led to increased competition among the district governments for foreign investment and government subsidies and drove plantation expansion in a functionally distal manner. The involvement of foreign investors and the high foreign demand for Indonesian palm oil exports introduced a new array of spatially distal drivers in which the global surge in demand for vegetable oil for the purposes of biodiesel production played a considerable part.

3.3 Competing Claims to Land as a Response to Plantation Expansion

The more strongly the impacts of oil palm expansion (e.g., deforestation, dispossession of people from their land and subsequent livelihood loss, loss of biodiversity) made themselves noticed on the ground, the more they—and by extension the distal drivers which had led to these impacts—solicited responses. These were and are articulated as competing claims to land, which bore high conflict potential, if they were, as was commonly the case, not resolved but suppressed. The exclusion of certain groups of people from land-use decisions was a central element of territorialization processes and was upheld during the reform period, fostering struggles to (re)claim land for other people and/or other uses. Whether the "distance" between the drivers of plantation expansion and their socioecological impact was spatial, temporal, and/or functional, it made it difficult for local actors to address these drivers and thus to tackle the problems they were facing at the root. Spatially distal drivers mostly remained unaffected by local change, the window of opportunity to impact temporally "distal" drivers lay in the past, and arguments which allowed for the inclusion of functionally "distal" drivers in a local case were difficult to formulate. In conjunction, this distalness kept the (local) consequences of the distal drivers from directly affecting the more influential actors. Claims to land were successful only if they were based on bridging the disconnect between drivers and impact or if they addressed the distal drivers at their point of origin. The major types of competing claims to land in the context of plantation expansion in Indonesia can be distinguished by the group(s) which make(s) the respective claim and by the argument on which they base their claim.

3.3.1 Customary Land Rights

Customary land tenure as an alternative mode of local land control has been largely ignored since the colonial period. Although the Indonesian Constitution as well as the Basic Agrarian Law (BAL) generally recognized customary rights and could act as temporally distal drivers in favor of this claim, several laws and regulations—especially the Basic Forestry Law (BFL, see Sect. 3.2.2)—do not. While resistance to large-scale dispossession was violently suppressed under Suharto, the greater political freedom of the reform period was conducive to the reassertion of the claim to customary land rights (Benda-Beckmann and Benda-Beckmann 2010). A recent success for this claim was achieved by overcoming both the spatial and functional distance inherent in the expansion of plantations despite competing land-use forms: Intense lobbying with decision makers at both the district and the national level on the part of the Alliance of Indigenous Peoples (AMAN—Aliansi Masyarakat Adat Nusantara)[8] and other civil society organizations led to a landmark ruling of the Constitutional Court in 2013 according to which forest land covered by customary rights was to be excluded from state forest land. In making this claim, the organizations were able to address it at the same abstract level of land functions, which had been introduced with the BFL and remained dominant in governing issues of access to land. The implementation of this ruling will require reterritorialization of forestland and district and central authorities as well as business will face competing claims to "their" land.

[8]For more information see: http://www.aman.or.id/.

3.3.2 Social Justice

Agrarian movements such as the Indonesian Peasant Union (SPI—Serikat Petani Indonesia) and the Consortium for Agrarian Reform (KPA—Konsorsium Pembaruan Agraria) seek to obtain redistribution of land to peasants and landless poor in order to achieve greater social justice and food sovereignty (also see Sect. 3.3 of this volume). Like the claim to customary land rights, these claims are subject to a temporally distal driver potentially in their favor as they are legally based on the Basic Agrarian Law, which grants every Indonesian citizen the right to possess land for her or his livelihood (Republik Indonesia 1960a). While agrarian movements were violently suppressed under New Order, the reform period reopened some political maneuvering room and brought agrarian reform initiatives back to the agenda.[9] Next to lobbying in state institutions, the claim has also been pursued via land occupation and settlement projects of landless farmers. Both these forms of addressing the claim partially create a feedback between the actions of the district and the central government and their local impacts. In 2005, the largest peasant organization in Java, the Sundanese Peasant Union (SPP—Serikat Petani Pasundan), occupied more than 15,000 ha of state plantations and forest land, calling for an alternative territorialization in the Gurat District in West Java, 44 % of the area of which is under control of plantation companies and the State Forest Corporation (SFC) (Peluso et al. 2008). In many cases, the peasants continue to cultivate this land, and even return after eviction, although they have no recognized legal claim to it.

3.3.3 Economic Growth

The redistribution of land in favor of individual land titles and counteracting what is deemed to be excessive state control is requested by international financial institutions such as the World Bank and USAID and also by the National Land Agency (BPN—Badan Pertanahan Nasional) and is subject to a set of drivers that can be located at the national and the supranational level. This claim is made on the basis of the perceived need for economic growth to which a land market with individual and registered property rights and the possibility to transfer these rights is considered to be conducive (e.g., World Bank 2014). While it is not functionally necessary to expand palm oil production in order to foster economic growth, in conjunction with other factors (see Sect. 3.2.2), this claim nonetheless contributes to such a development. As part of the Land Management and Policy Development Project, funded by the World Bank and administered by the BPN, approximately 1 million land titles were issued between 2004 and 2009 (Waren and Lucas 2013; USAID 2010). The agrarian movement (see Sect. 3.3.2) had intensively lobbied

[9]Interview with KPA, Jakarta, July 12, 2011.

with the central government in order to include a more comprehensive agrarian reform and redistribution of land within this project. Under the claim to land for economic growth, however, the Ministry of Forestry and the Ministry of Agriculture refused to participate in the redistribution process, using the project instead to legalize land titles for people already using state land without any significant land redistribution (Rachman 2011).

3.3.4 Environmental Protection

International organizations like Conservation International or the World Wildlife Fund (WWF) call for the establishment of nature conservation reserves and wildlife corridors in Indonesia (Maddox et al. 2007) in order to protect the local and the global environment. In order to achieve the latter, it is argued that the deforestation associated with plantation expansion must be stopped. The interests driving such claims may thus originate across all scales from local to national to supranational. The implementation of environmental claims relies on a specific form of territorialization, which excludes people with competing claims (e.g., subsistence agriculture, hunting, and gathering) from the area, often by violent means. Nature reserves may depend on the police, the military, or private security companies to enforce boundaries and limit access. Conflict has arisen, for example, with regard to the Hutan Harapan Rainforest in the district of Jambi, Sumatra, which the conservation company PT REKI holds as private land property in order to promote ecosystem restoration and gain carbon credit (Hein and Faust 2014).[10] The same land is claimed by members of the peasant union SPI (see above) for subsistence agriculture and livelihood purposes.[11]

3.4 Can the Distance Be Bridged?

Despite the changes in Indonesia's political and economic orientation from the late 1960s until today, land use for oil palm plantations continuously expanded. Territorialization under New Order and reterritorialization during the reform period were instrumental in enabling this expansion in the face of competing claims to land by segments of the Indonesian population as well as by international organizations and institutions. Even though an intermittent attempt was made to further the development of secondary production and services in the Indonesian economy, the country continues to rely heavily on the extraction of raw materials and the export

[10]For another example of land-use decisions in which carbon credit plays a role, see Gasparri (Chap. 4).

[11]Interview with SPI, Jakarta, January 25, 2012.

of primary commodities. This pathway is largely enabled through state control over land and thus over access to resources.

As the expansion of oil palm plantations continues to threaten rural livelihoods and cause irreversible environmental damage, an understanding of the distal drivers, which result in these socioecological impacts, is a prerequisite in making the issue addressable in an effective manner. The existing plantation infrastructure on some of the islands, which the newly independent Indonesia inherited from the Dutch Colonial rulers, shaped the initial expansion of oil palm plantations. In a similar way, contract-farming schemes shape today's large-scale agriculture in predefining that an agricultural unit consists of a central, managing "core" with decision-making power and smallholder plots subject to those decisions. The politically defined mandate of keeping the multiethnic Indonesian population in the 15th largest country in the world, dispersed on over 18,000 islands, under some form of central control must be considered as a strong impetus in upholding the system of plantation agriculture. While the expansion of plantation agriculture into rainforests, subsistence agriculture land, and indigenous territory can be seen as a driver of the increasing pressure resulting from competing claims to land, it must also be understood as a response to the perceived potential for conflict. The expansion of oil palm plantations has brought some degree of state control over the Outer Islands as well as over the rural population through contract-farming schemes.

The growing oil palm plantation area was a prerequisite to Indonesia's exponentially increasing palm oil production, which is almost exclusively dedicated to export. The integration of the Indonesian economy in the global market introduces new distal drivers, which exert pressure not only on the plantation agriculture but also on the entire political and economic system. In early 2014, for example, India raised its import tariff on vegetable oil in order to protect domestic producers. Indonesia, with its new focus on the Asian export market (see Sect. 3.2.2), was forced to lower export tariffs in order to protect this important market (Dutta and Supriatna 2014).

Rather than being taken for granted, the political and economic conditions of Indonesia's plantation expansion must be understood within the context of the exerted power over both nature and the people, which they represent. When claims to land were successfully made, they were based, at least in part, on bridging the disconnect between distal drivers and their corresponding impact. This occurred through intensive lobbying efforts at the point of origin of distal drivers, by the World Bank and the agrarian movement, for example, with the Indonesian government. Global environmental impacts, such as climate change, constitute an example of how distal drivers may be relinked to their, in this case global consequences. In raising awareness and support for the protection of Indonesian rainforest, non- and intergovernmental organizations have profited from partially closing the knowledge gap between consumption of palm oil in high-income countries and the local socioenvironmental impacts: Public media campaigns and certification schemes for palm oil products have been highly important in this regard. Certification schemes translate socioecological into economic valuation. While they thus constitute an example of bridging the disconnect between production conditions and consumption, they simultaneously illustrate that the "distance" between drivers and their impacts allows the more influential actors to define the terms on which the claims of local actors will be heard.

References

Badan Pusat Statistik. (2012). Statistik Kelapa Sawit Indonesia 2012 [Indonesian Palm Oil Statistics]. Jakarta: Badan Pusat Statistik [Statistics Indonesia].

Barr, C., Resosudarmo, I., Dermawan, A., & McCarthy, J. (2006). *Decentralization of forest administration in Indonesia: Implications for forest sustainability, economic development, and community livelihoods*. Bogor: CIFOR.

Brad, A., Schaffartzik, A., Pichler, M., & Plank, C. (2015). Contested territorialization and biophysical expansion of oil palm plantations in Indonesia. *Geoforum, 64*, 100–111. doi:10.1016/j.geoforum.2015.06.007.

Brenner, N. (1998). Global cities, glocal states: global city formation and state territorial restructuring in contemporary Europe. *Review of International Political Economy, 5*(1), 1–37. doi:10.1080/096922998347633.

Colchester, M., & Chao, S. (2013). Conflict or consent? The oil palm sector at a crossroads. Forest Peoples Programme, Sawit Watch and TUK Indonesia.

Colchester, M, Jiwan, N., Sirait, M., Firdaus, A. Y, Surambo, A., & Pane, N. (2006). Promised Land: palm oil and land acquisition in Indonesia: implications for local communities and indigenous peoples. Bogor: Forest Peoples Programme (FPP)/Sawit Watch.

Corson, C. (2011). Territorialization, enclosure and neoliberalism: non-state influence in struggles over Madagascar's forests. *Journal of Peasant Studies, 38*(4), 703–726. doi:10.1080/03066150.2011.607696.

Dutta, R., & Supriatna, Y. (2014). *Indonesia may cut palm oil export tax to offset India tariff rise*. Reuters.

EEA. (1995). *General strategy for integrated environmental assessment at the European Environment Agency*. Copenhagen: European Environment Agency.

FAO. (2014). FAOSTAT Database. Rome: Food and Agriculture Organization of the United Nations (FAO). http://faostat.fao.org/.

Fargione, J., Hill, J., Tilman, D., Polasky, S., & Hawthorne, P. (2008). Land clearing and the biofuel carbon debt. *Science, 319*(5867), 1235–1238. doi:10.1126/science.1152747.

Farid, H. (2005). Indonesia's original sin: mass killings and capitalist expansion, 1965–66. *Inter-Asia Cultural Studies, 6*(1), 3–16. doi:10.1080/1462394042000326879.

Fitzherbert, E. B., Struebig, M. J., Morel, A., Danielsen, F., Brühl, C. A., Donald, P. F., & Phalan, B. (2008). How will oil palm expansion affect biodiversity? *Trends in Ecology and Evolution, 23*(10), 538–545. doi:10.1016/j.tree.2008.06.012.

Geist, H. J. (2005). *Our Earth's changing land: an encyclopedia of land-use and land-cover change*. Greenwood Publishing Group.

Geist, H. J., & Lambin, E. F. (2002). Proximate causes and underlying driving forces of tropical deforestation tropical forests are disappearing as the result of many pressures, both local and regional, acting in various combinations in different geographical locations. *BioScience, 52* (2), 143–150. doi:10.1641/0006-3568(2002)052[0143:PCAUDF]2.0.CO;2.

Gerber, J.-F. (2011). Conflicts over industrial tree plantations in the South: who, how and why? *Global Environmental Change, 21*(1), 165–176. doi:10.1016/j.gloenvcha.2010.09.005.

Germer, J., & Sauerborn, J. (2008). Estimation of the impact of oil palm plantation establishment on greenhouse gas balance. *Environment, Development and Sustainability, 10*(6), 697–716. doi:10.1007/s10668-006-9080-1.

Gillespie, P. (2011). How does legislation affect oil palm smallholders in the Sanggau District of Kalimantan, Indonesia?. SSRN Scholarly Paper ID 2005590. Rochester, NY: *Social Science Research Network*. http://papers.ssrn.com/abstract=2005590.

Harvey, D. (2005). *The new imperialism*. New York: Oxford University Press.

Hein, J., & Faust, H. (2014). Conservation, REDD+ and the struggle for land in Jambi, Indonesia. *Pacific Geographies, 41*, 20–25.

Jianguo, L., Hull, V., Batistella, M., DeFries, R., Dietz, T., Fu, F., et al. (2013). Framing sustainability in a telecoupled World. *Ecology and Society, 18*(2). doi:10.5751/ES-05873-180226.

Kementerian Pertanian Republik Indonesia. (2013). Basis Data Statistik Pertanian [Agricultural Statistical Database]. Jakarta: Kementerian Pertanian Republik Indonesia [Ministry of Agriculture, Republic of Indonesia].

Larsen, R., Jiwan, N., Rompas, A., Jenito, J., Osbeck, M., & Tarigan, A. (2014). Towards 'hybrid accountability' in EU biofuels policy? Community grievances and competing water claims in the central kalimantan oil palm sector. *Geoforum, 54*(July), 295–305. doi:10.1016/j.geoforum.2013.09.010.

Koh, L. P., & Wilcove, D. S. (2008). Is oil palm agriculture really destroying tropical biodiversity? *Conservation Letters, 1*(2), 60–64. doi:10.1111/j.1755-263X.2008.00011.x.

Li, T. M. (2011). Centering labor in the land grab debate. *Journal of Peasant Studies, 38*(2), 281–298. doi:10.1080/03066150.2011.559009.

Liu, J., Hull, V., Batistella, M., DeFries, R., Dietz, T., Fu, F. et al. (2013). Framing sustainability in a telecoupled world. *Ecology and Society, 18*. doi:10.5751/ES-05873-180226.

Murray, L. T. (2011). Centering labor in the land grab debate. *Journal of Peasant Studies, 38*(2), 281–298. doi:10.1080/03066150.2011.559009.

Maddox, T., Priatna, D., Gemita, E., & Salampessy, A. (2007). The conservation of tigers and other wildlife in oil palm plantations: Jambi Province, Sumatra, Indonesia. Vol. 7. ZSL *Conservation Report*. London: Zoological Society of London (ZSL).

Marti, S. (2008). Losing ground. The human rights impacts of oil palm plantation expansion in Indonesia. *Friends of the Earth*, LifeMosaic, Sawit Watch.

Marx, K. (1867). *Capital* (Vol. I). London: Penguin.

McCarthy, J. F. (2004). Changing to gray: decentralization and the emergence of volatile socio-legal configurations in Central Kalimantan, Indonesia. *World Development, 32*(7), 1199–1223. doi:10.1016/j.worlddev.2004.02.002.

McCarthy, J. F., Gillespie, P., & Zen, Z. (2012). Swimming upstream: local indonesian production networks in 'globalized' palm oil production. *World Development, 40*(3), 555–569. doi:10.1016/j.worlddev.2011.07.012.

Ministry of Agriculture. (2007). Regulation of Minister of Agriculture Number: 26/Permentan/OT.140/2/2007 Regarding Plantation Business Permit Guidelines. Jakarta.

Peluso, N. L. (2005). Seeing property in land use: local territorializations in West Kalimantan, Indonesia. *Geografisk Tidsskrift-Danish Journal of Geography, 105*(1), 1–15. doi:10.1080/00167223.2005.10649522.

Peluso, N. L., Afiff, S., & Rachman, N. F. (2008). Claiming the grounds for reform: agrarian and environmental movements in Indonesia. *Journal of Agrarian Change, 8*(2–3), 377–407. doi:10.1111/j.1471-0366.2008.00174.x.

Peluso, N. L., & Lund, C. (2011). New Frontiers of land control: introduction. *Journal of Peasant Studies, 38*(4), 667–681. doi:10.1080/03066150.2011.607692.

Pitt, M. (1980). Alternative trade strategies and employment in Indonesia. In *Trade and Employment in Developing Countries, Individual Studies* (Vol. 1, pp. 181–238). University of Chicago Press, Chicago.

Rachman, N. F. (2011). *The resurgence of land reform policy and agrarian movements in Indonesia*. Berkeley: University of California.

Reijnders, L., & Huijbregts, M. A. J. (2008). Palm oil and the emission of carbon-based greenhouse gases. *Journal of Cleaner Production, 16*(4), 477–482. doi:10.1016/j.jclepro.2006.07.054.

Republik Indonesia. (1960a). Undang-Undang Republik Indonesia No. 5 Tahun 1960 Tentang Peraturan Dasar Pokok-Pokok Agraria. Jakarta.

Republik Indonesia (1960b). Peraturan Pemerintah Pengganti Undang-Undang Nomor 56 Tahun 1960 Tentang Penetapan Luas Tanah Pertanian. Jakarta.

Republik Indonesia. (2004). Undang-Undang Nomor 18 Tahun 2004. Tentang Perkebunan. Jakarta.

Sassen, S. (2005). When national territory is home to the global: old borders to novel borderings. *New Political Economy, 10*(4), 523–541.

Schaffartzik, A., Mayer, A., Gingrich, S., Eisenmenger, N., Loy, C., & Krausmann, F. (2014). The global metabolic transition: regional patterns and trends of global material flows, 1950–2010. *Global Environmental Change, 26*(May), 87–97. doi:10.1016/j.gloenvcha.2014.03.013.

Smith, N., (1984). *Uneven development: Nature, capital and the production of space*. Basil Blackwell: Oxford.

The World Bank. (2013). *World development indicators 2013*. Washington, D.C.: World Bank Publications.

Turner, B. L., Lambin, E. F., & Reenberg, A. (2007). The emergence of land change science for global environmental change and sustainability. *Proceedings of the National Academy of Sciences, 104*(52), 20666–20671. doi:10.1073/pnas.0704119104.

USAID. (2010). USAID country profile: property rights & resource governance: Indonesia. United States Agency for International Development. http://usaidlandtenure.net/sites/default/files/country-profiles/full-reports/USAID_Land_Tenure_Indonesia_Profile_0.pdf.

Vandergeest, P. (1996). Mapping nature: territorialization of forest rights in Thailand. *Society and Natural Resources, 9*(2), 159–175. doi:10.1080/08941929609380962.

Vandergeest, P., & Peluso, N. L. (1995). Territorialization and State Power in Thailand. *Theory and Society, 24*(3), 385–426. doi:10.1007/BF00993352.

von Benda-Beckmann, F., & von Benda-Beckmann, K. (2010). Multiple embeddedness and systemic implications: struggles over natural resources in Minangkabau since the Reformasi. *Asian Journal of Social Science, 38*(2), 172–186.

Waren, C., & Lucas, A. (2013). Indonesia's land titling program (LAP)—the market solution? In C. Waren & A. Lucas (Eds.), *Land for the people: the state and agrarian conflict in Indonesia* (pp. 93–113). Athens, USA: Ohio University Press/Swallow Press.

World Bank. (2014). Land policy: sector results profile. securing land tenure rights to reduce poverty and promote growth. Washington, D.C.: *The World Bank Group*. http://www.worldbank.org/en/results/2013/04/15/land-policy-results-profile.

Zen, Z., Barlow, C., & Gondowarsito, R. (2005). Oil palm in indonesian socio-economic improvement a review of options. *Working Papers in Trade and Development*. Canberra: Research School of Pacific and Asian Studies, ANU.

Zhou, A., & Thomson, E. (2009). The Development of Biofuels in Asia. *Applied Energy, Bio-fuels in Asia*, 86, Supplement 1 (November), S11–20. doi:10.1016/j.apenergy.2009.04.028.

Chapter 4
The Transformation of Land-Use Competition in the Argentinean Dry Chaco Between 1975 and 2015

Nestor Ignacio Gasparri

Abstract The Dry Chaco in Argentina is among the most dynamic deforestation frontiers in South America. Land-use competition in this region today mainly relates to trade-offs between on the one hand ecosystem services important for local communities (e.g., fuelwood, forage, hunting, subsistence farming) and on the other side global demands for both agricultural commodities (e.g., soybean, beef) and conservation (e.g., of carbon stocks or biodiversity). Over the last four decades, land-use competition in the Dry Chaco has shifted from the local/market mode to the national/politic mode and recently to a global scale under a combination of political and market mode. Different actors and sectors try to shape land-use competition by shifting it onto a scale and into a mode more favorable for their own objectives. On the one hand, the agribusiness sector and provincial governments alike try to conserve land-use competition playing out at local-to-regional scale, with little regulation, and under market forces. On the other hand, local communities (indigenous communities and traditional small-scale farmer cattle ranchers) as well as regional NGOs try to shift land-use competition to the national and even global level. Associated with the upscaling of the competition process, new actors have emerged and become incorporated into land-use competition processes in the Dry Chaco. The national government takes on the role of a mediator to resolve conflicts, but also to create new framework conditions and legislation (most importantly a National Forest Law) for regulating land-use competition. The global community joins land-use competition by adding new options for land use (e.g., carbon stocks, conservation) and by market mechanisms that feedback on producers (e.g., sustainable or green labels). Distal drivers related to agricultural commodity trade, initially, promoted asymmetries in favor of the agribusiness sector. However, in the long run, distal drivers may also act to partially counterbalance the original asymmetries and to result in more balanced outcomes between the often conflicting

N.I. Gasparri (✉)
Instituto de Ecología Regional, Universidad Nacional de Tucumán y CONICET, CC34, Yerba Buena 4107, Tucumán, Argentina
e-mail: ignacio.gasparri@gmail.com

J. Niewöhner et al. (eds.), *Land Use Competition*,
Human-Environment Interactions, DOI 10.1007/978-3-319-33628-2_4

aspiration of the actors involved in land-use competition in the Dry Chaco of Argentina.

Keywords Ecosystem services • Soybean • Scaling • Conflict • Governance

4.1 Introduction

The Dry Chaco ecoregion, stretching into Bolivia, Paraguay, and Argentina (Fig. 4.1 and Chap. 13), has been among the most dynamic deforestation frontiers in South America during the last decade (Hansen et al. 2013). The Argentine part of the Dry Chaco has experienced particularly high deforestation rates since the 1970s, with a strong acceleration during the 1990s and the 2000s especially in the north Argentine Dry Chaco (Fig. 4.1) (Gasparri et al. 2013). The main driver of deforestation has been the booming soybean economy, oriented toward the global market. In addition to the direct conversion of forests due to soybean expansion, displacement of cattle ranching to the Dry Chaco due to soybean expansion elsewhere leads to additional deforestation for pastures and beef production (Gasparri and Le Polain 2015; Gasparri et al. 2013). The wave of deforestation to produce soybean and beef in the Dry Chaco is thus an interesting example of a deep structural change in a land system and strongly intensifying competition for land mainly driven by what we frame here as distal drivers.

The recent agribusiness expansion in the Dry Chaco has resulted in strong tensions and conflicts with traditional local communities that use the natural resources as common pools in the region, indigenous people, and the conservation sector. In particular, land-use competition in the Dry Chaco unfolds around the trade-offs between local ecosystem services (ES, e.g., forage, hunting, fuelwood) versus agriculture production for global market (e.g., soybean). However, other ecosystem services—for example, carbon management for global climate regulation or conservation of emblematic species (e.g., *phantera onza*)—are also attributed to and demanded from the region and thus also compete with traditional land uses and local provision of ES in the region.

The rising unrest in the region attracted attention from actors at national and global scale. Associated with the increasing competition for land in the region, national and global actors were incorporated in the process and gradually created a context for land-use planning as a way to solve discrepancies in expectations over how to use land in the Chaco region. Thus, the Dry Chaco region in three decades shifted from little land-use planning to trans-scalar land-use planning (Rudel and Meyfroidt 2014). In this work, I use the telecoupling framework (Liu et al. 2013; Eakin et al. 2014; Friis et al. 2015) to analyze the linkages among actors and ecosystems in the Dry Chaco, as these linkages tend to unfold over large, and increasing, distances. These linkages could be fluxes of material and capital, as in the case of the soybean trade, but also fluxes related to technology or information. An important characteristic of these linkages is also feedback mechanisms, for

Fig. 4.1 The dry Chaco in South America and the north Argentine Dry Chaco at the center of land-use competition analyzed in this chapter

example, zoning regulations, policy implementation (e.g., REDD +), or consumer pressure via value chains (e.g., green labels) in response to growing concern about environmental depletion in the region. Using the telecoupling lens, I will analyze land-use competition in Argentina's Dry Chaco, considering land-change dynamic, key actors, and drivers of this amplifying competition for land.

4.2 Key Actors of Land-Use Competition in the Dry Chaco

A number of key groups of actors are involved in, or experience the outcomes of, the recently increasing competition for land in Argentina's Dry Chaco. In this section, I will briefly describe these actor groups and their claims to land in order to provide context for the assessment of changes in the land-use competition described in following sections (see also Chap. 13).

Agribusiness actors have been the main proponents of the recent wave of land-use change in the Dry Chaco of Argentina. The soybean production in South America oriented toward exports is an archetypal telecoupled system (Liu et al. 2013). The production sector includes medium and large agribusiness companies, relying heavily on technology and capital. These agribusiness companies also have access to complex networks of agricultural services (e.g., agrochemical application, machinery, and transport logistic) and commercialization. Agribusiness companies operating in the Dry Chaco of Argentina commonly combine soybean production and cattle ranching (Gasparri and Le Polain 2015) and are heavily integrated into the global market through channels of commercialization and the vegetable oil industry in Argentina. The linkages to distal drivers through information, technology, and trade promote strong asymmetries within processes of land-use competition in favor of the sector promoting agriculture and cattle production in contrast with the interests of indigenous communities and traditional rural populations.

The local population in the Dry Chaco is a second key group of actors engaging and being affected by the increasing competition for land. Traditional communities are mainly Spanish descendants (criollos) and rely heavily on the local provisioning of ecosystem services. Many criollos live in small homesteads inside natural areas (puestos) devoted to extensive cattle ranching. The puesto system consists of sheds for livestock and a watering source. Some of the puestos have a long history of more than four generations and evolved into small communities, sometimes with state infrastructure (e.g., schools). The livestock (mainly cows and lambs) feed on the litter fall and herbs from the forest understory (especially during the dry season) and return to the puesto regularly for water. The criollos living in the puestos depend on a range of services provided by the forest, including wood for fuel, building construction, and in some cases charcoal for selling on the market. Hunting for food is also common, especially for peccaries and other small mammals (Altrichter 2006). Historically, puestos were developed on fiscal and large private areas (commonly with absent owners), often in areas with unresolved land property rights. Forests for forage, hunting, and wood were historically used by local people as a common pool resource (Altrichter and Basurto 2008).

A third important group of actors are indigenous people. The Dry Chaco is the area with the largest indigenous population in Argentina, with the main indigenous groups being the Quom and Wichís. The subsistence and culture of these groups is based on hunting and gathering, and traditionally, these people lived a nomadic lifestyle. Today, indigenous groups still use a wide range of forest resources and widely practice hunting, fishing, and fuelwood collection in a traditional way. Additionally, indigenous people depend on particular forest resources such as fiber (chaguar) and wood (palo santo) to generate crafted objects to be sold in cities to tourists. Finally, the forest areas are highly valued as a source of cultural identity among indigenous groups (Palmer 2005). Indigenous communities live on lands with a variety of land tenure settings, including in some cases with formal titles. However, indigenous communities use the resources as communal resources or sometimes in a territory system affecting different sectors through the year—in both cases not restricting access to those holding formal land titles.

The NGOs in the region traditionally were more oriented to social and agrarian development than conservation. The Dry Chaco in Argentina is an area with a high proportion of poor rural populations. Common problematics attended to by the NGOs were water accessibility, subsistence agriculture and cattle health, and welfare. However, the advances of deforestation created new constellations and challenges. Many in the local population working with and living of natural resources identified common interests with the nature conservation community. Likewise, the region's traditional NGOs gradually focused more on nature conservation and created synergies with national organizations such as Greenpeace or Fundación Vida Silvestre (WWF in Argentina).

4.3 Three Periods of Land-Use Change and Increasing Competition for Land in the Argentine Dry Chaco

The evolution of land-use competition in the Argentinean Dry Chaco can be summarized in three main time periods that also demonstrate how competition gradually adds scales (from local to global) as well as transitions in the mode and type of arguments made. In the first period, land-use competition mainly occurred at local-to-regional scale and within a market mode. This period started early in the 1970s with the introduction of soybean as an alternative crop, triggering agricultural expansion and deforestation (Gasparri and Grau 2009). During the 1990s, deforestation accelerated as demand for soybean surged, access to global markets increased, and the introduction of the new GM soybean strains provided opportunities for agribusiness companies to expand cultivation (Grau et al. 2005a, b). At the same time, social and environmental conflicts intensified locally raising concerns taken up at national scale.

The second period, starting around 2000, was characterized by an upscaling of the problem to national–international level with social demands from criollos and conservation groups directed at national government to assume a role of control and regulation over the deforestation process. During the second period, despite the continuing importance of market forces, a policy mode with negotiation and implementation of regulation rose to prominence, mediated to a large degree by national and provincial governments. The second period was also characterized by the increasing availability of scientific data and knowledge about the deforestation process and the environmental and economic trade-offs involved in it. The second period ended with a new National Forest Law passed in 2007 to regulate and curb deforestation and create a new framework for land-use competition.

In the third period since 2007, land-use competition has continued to be strongly influenced by global market forces, but increasingly also by global concerns about deforestation. Land-use competition during this period added mechanisms of negotiation at global scale with market mechanisms to promote more sustainable practices (e.g., RTRS-Round Table on Responsible Soy) and at national scale with deforestation regulation (zoning forest areas for the new National Forest Law). As deforestation progressed and information on its massive extent became available (Hansen et al. 2013), interest by the international scientific community and the general public increased (e.g., Rolling Stone 2014). Likewise, global actors became engaged in land-use competition in the Dry Chaco, most importantly conservation organizations such as the World Wildlife Fund or The Nature Conservancy. Finally, new global conservation initiatives were connected to payments for ecosystem services, such as REDD +. In the following sections, each of these three periods is described in detail.

4.3.1 Period I: Local Competition (1970–2000)

Technology plays an important role in land-use competition, providing new land-use options. In the case of soybean cultivation, agronomic research in early 1970s (Gasparri and Grau 2009) and later in 1997, with the GM soybean introduction in Argentina, expanded the areas suitable for soybean substantially. This meant soybean cultivation was suddenly possible in areas of the Dry Chaco where traditionally only cattle ranching had been suitable (annual rainfall below 800 mm/year), and thus, the competition for land had been low (Grau et al. 2005a, b; Zak et al. 2008).

The initial phase of land competition connected to the expansion of soybean occurred "face to face" between the agribusiness sector on the one hand and the traditional land users in the Chaco (i.e., puestos and indigenous communities) on the other. This competition was characterized by large asymmetries in relation to capital, access to technology, information, as well as in the lobby capacity at the governmental scale. During this period, land-use competition was largely resolved within a market mode with widespread land transactions and land concession from public lands into the private sector in order to incorporate these lands into agricultural production. These conditions promoted a rapid deforestation process, triggered by well-capitalized farmers and agribusinesses companies. This led to a process of "accumulation by dispossession" in the region (Cáceres 2015), resulting in the widespread exclusion of traditional users from access to the forest areas by companies that reinforced the private control over land and natural resources. During this period, the government did not implement any policies to regulate land-use change or deforestation, but limited itself to regulating agriculture through taxes in the 1970s and 1980s. During the 1990s, an even more liberal policy regime with few state interventions into the agricultural market was implemented (Gasparri and Grau 2009). However, the liberalization of policy increased land-use conflicts around the deforestation process. These began to mobilize actors at national scale, especially conservation NGOs that incorporated deforestation into their national priority agenda setting up a new constellation for land-use competition.

4.3.2 Period II: Upscaling of Competition to the National Level (2000–2007)

Starting from 2000, national policy-making left the path of economic liberalization in favor of a larger degree of intervention and regulation. In addition, the year 2002 represented an inflection point in the Argentinean economy following an economic crisis that ended a strong devaluation of the national currency, which in turn had marked impacts on deforestation rates (Gasparri et al. 2013). Finally, the year 2002 was a tipping point, because much technical and scientific information about the extent and pace of deforestation became available, with a strong impact on public perceptions of agricultural expansion and the soybean boom in the Chaco. In particular, the first national map of native forest in Argentina became available in December 2002 (SAyDS 2004), and for the first time, Argentina had a reliable estimate of its forest area. In 2004, the national government presented the first estimation of deforestation rates for the period 1998–2002 and for the period 2002–2004 for the most dynamic regions of the country (UMSEF 2014). These reports were widely publicized and used by NGOs and regional development institutions to support arguments about the necessity and urgency of regulating the agricultural expansion and deforestation process. At the same time, scientific papers published in international scientific journals unraveled the patterns, drivers, and outcomes of deforestation in the Dry Chaco, starting in 2004 (Zak et al. 2004), but soon followed by a wide range of contributions (e.g., Grau et al. 2005a, b; Boletta et al. 2006).

During this period, the national conservation NGOs from Argentina incorporated deforestation in the Chaco on their priority agenda. In 2005, Fundacion Vida Silvestre (the Argentine representation of WWF) in collaboration with other NGOs and The Nature Conservancy carried out a regional assessment for the entire Great Chaco, identifying priority areas for conservation (FVS et al. 2005). Greenpeace Argentina started a national forest campaign active until today (Greenpeace 2014) and brought the deforestation issue further into public discussion via the participation of famous artists and intellectual personalities from Argentina.

A defining moment for land-use competition in the Dry Chaco during the early 2000s was the conflict around the Pizarro Provincial Reserve. In the Anta sector (a very dynamic deforestation frontier), the government of Salta Province decided to change the status of a provincial protected area (i.e., the Pizarro Provincial Reserve) to lease to the private sector with the objective to convert it into agricultural land. The Pizarro Reserve includes areas traditionally used by a Wichi community and is also of conservational concern as it marks an important transitional forest from the Yungas to the Dry Chaco forests. The case of a provincial government trying to change the status of a protected area with traditional use and conservation value (arguments for creating this provincial reserve) was a tipping point discussion surrounding deforestation and conservation issues. The case of the Pizarro Reserve became emblematic and moved concerns about deforestation to the national level, ending with national government intervening and the creation of new federal protected areas (Reserva Natural Pizarro; Hufty 2008). The Pizarro Reserve conflict also reinforced the alliance between indigenous communities, criollos, and the conservation community. NGOs with a long-term presence in the region coordinated in the Red Agroforestal Chaco Argentina (REDAF) shared information related to deforestation and developed a common position regarding the agribusiness expansion and its environmental and social conflicts in the Chaco (REDAF 2014).

One of the main demands of conservation and local development organizations was the political recognition of the necessity of a regional planning process in order to mitigate trade-offs between agricultural expansion, the needs of local communities, and conservation goals. Likewise, clearer and stronger policy interventions from the government at all scales were commonly called for. For example, the legislation in Salta Province included a requirement for an environmental impact assessment (EIA) prior to deforesting a property, but restricted to the particular land plot to be deforested. As such, this regulation clearly overlooked the regional, aggregated impacts of deforestation, which was problematic in the eyes of conservation and development organizations. For example, in the eastern part of Salta Province, indigenous communities and criollos appealed to the National Supreme Court of Justice for legal protection, which then requested the government of Salta Province to develop a "Cumulative Environmental Effects Assessment" and the establishment of a moratorium for deforestation until this assessment was carried out (Paruelo et al. 2011; Seghezzo et al. 2011).

As result of the social interest over the deforestation process at national scale, the new National Forest Law of Argentina (i.e., Ley N° 26.331) was developed, discussed, and finally passed in the National Congress in December 2007. Implemented in 2009, the new legislation created the framework for a new phase of land-use competition in the Chaco, centered on negotiation and regulation. In the National Forest Law, the federal government requested provincial administrations to prepare zoning plans for all natural forests. These plans were required to indicate sectors available for conversion into other land uses (*green zones*), sectors designated to sustainable forest management, but where deforestation is not allowed (*yellow zones*), and sectors with conservation value where deforestation is forbidden and forest management has restrictions (*red zones*). Additionally, the National Forest Law made it mandatory to develop this zoning plan in a participatory process and to engage in negotiations with society at large and all actors with interests in forest or land in the Chaco (e.g., the agribusiness sector, conservation organizations, local community; Collazo et al. 2013).

4.3.3 Period III: Moving to the Global Scale (The 2007–2014)

Three main developments further distinguish competition for land before and after 2007. First, land-use competition has entered a new phase with the Chaco increasingly being integrated into and influenced by land-use dynamics in the South American soybean and cattle ranching region as a whole, as well as global concerns about the deterioration of ecosystem services (e.g., carbon stocks) and biodiversity. Other prominent South American regions implicated in these extended processes of land-use competition include the Cerrado region in Brazil and the Chiquitania region in Bolivia. In particular, many South American agribusiness companies have invested in different countries and production activities, increasing opportunities for transnational land-use leakage and displacements (Gasparri and le Polain 2015).

Second, the increase in global concerns about the situation in the Dry Chaco promotes the development of feedbacks between global and local actors, discourses, and concerns. For example, the severe phase of deforestation in the Dry Chaco increased the coverage in the international media (e.g., Rolling Stone 2014; New York Times 2012; BBC 2011) and moved the issue up on the agenda of environmental organizations (e.g., WRI 2014). The linkages between the deforestation in the Dry Chaco and the consumption of soybean (or meat from animals fed by soybean) are becoming increasingly visible in public and political discourses. They are becoming a matter of concern for end consumers as well as national governments in countries importing soybean from Argentina, resulting in feedback mechanisms through supply chains operating in the Dry Chaco. For example, a number of certification schemes have been developed to steer soybean production into more sustainable and responsible modes, such as the Round Table on Responsible Soy (RTRS 2014). This initiative started in 2006 and promotes principles and criteria (P&C) (approved in 2010) covering different aspects of land-use practices such as the use of agrochemicals, natural area conversion, and local communities' rights. These P&C have resulted from negotiations between the soybean producers, industry, and civil society at large, not only in Argentina (RTRS 2014). This initiative thus clearly represents an example of a telecoupling (see Chap. 2), promoting a feedback mechanism between widely separated areas of production and consumption, in order to promote more sustainable practices in the production region. The first soybean farms producing under RTRS standards were accredited in 2011 in Argentina and Brazil.

Third, payment for ecosystem services schemes (PES) emerged since 2007 as another distant driver affecting land-use competition in the Dry Chaco. Most importantly, initiatives aiming at reducing carbon emission from deforestation and degradation or promoting forest restoration became important (UN-REDD+ 2014). Argentina was not traditionally part of negotiations about emissions related to deforestation, in part because Argentina was considered "temperate" with respect to its climate and thus a country without "tropical" deforestation. However, partially supported by the situation in the Dry Chaco, Argentina was incorporated into the REDD + process as a partner country in 2009, and in 2014, the UN-REDD + program approved the preparation of national capacities for implementing a REDD + program in Argentina (UN-REDD+ 2014).

The global initiatives such as REDD + and RTRS represent global governance initiatives that add land-use options (carbon for global climate regulation) or variations in the mode of production (responsible soybean) in the Dry Chaco. These global governance initiatives are the result of concerns from the international community (e.g., soybean consumers or climate change community) over developments in the region and demonstrate how processes of land-use competition clearly shift from relatively local face-to-face interactions to incorporating increasingly distal drivers. These distal drivers are represented within processes of land-use competition through new actors, new knowledge, and new interests and demands. Most importantly perhaps, these new distal drivers in the Dry Chaco have transformed forest areas, so far considered predominantly for agricultural use and dwelling purposes, into potential carbon stock for global climate regulation. This has now been established as a plausible land use for the coming years in the region. Also significant, soybean production under RTRS standards represents a land-use option that is not new in itself, but that carries new requirements, including the restriction for expansion in areas cleared before 2008 and the fulfillment of social and environmental standards.

4.4 Strategies for Dealing with Emerging Land-Use Competition in the Dry Chaco

The transformation of the mode and scale of land-use competition in the Dry Chaco of Argentina is the result of the constant negotiation among, and incorporation of, different actors and evermore distant drivers into a trans-scalar land-use planning. Figure 4.2 shows a schematic representation of the dimension of land-use competition (scale and modes), the position of different actors, and their intention or resistance to changing the scale and modes of competition. Three main strategies in relation to the transition in the scale and modes of land-use competition can be distinguished: (a) the "resistance strategy" of the agribusiness sector and to some degree the provincial government, (b) the "transitioning strategy" that represents the intention of some actors (specially local communities) to change the mode of the competition process, exerting pressure on actors at larger scales and organizational

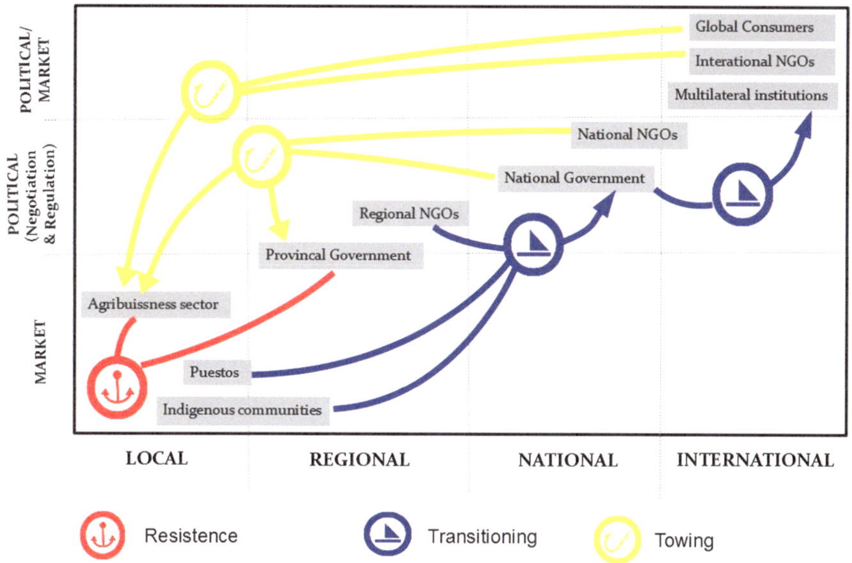

Fig. 4.2 Schematic representation of actors engaged in land-use competition related to the soybean economy in the Argentine Dry Chaco

levels (e.g., national governments) to participate in and mediate the process, and (c) the "towing strategy" of higher-level actors trying to incorporate the agribusiness production sector and local governments into regional planning, negotiations, and sustainable land management practices.

The local conservation sectors, criollos and indigenous communities, face land-use competition at local scale, but try to transition the process to higher levels (national and international), by creating alliances and by highlighting the situation and conflicts in the Argentine Chaco to the national and international community. Additionally, a principal strategy has been to demand the national government to intervene in the process and to create regulation to limit and regulate the decisions and actions taken by the agribusiness production sector and the provincial governments. Hence, these local actors tend to evade competition in a pure market mode to avoid structural asymmetries of power largely related to resource access with respect to capital, political contacts, and agricultural technology (see Chap. 13). They shift upscale and into a political mode, where their ethical and legal claims stand a better chance of gaining legitimacy and value.

The production sector (especially agribusiness) pursues a range of strategies. In spite of the participation of producer organizations in global negotiations (e.g., RTRS), the agribusiness sector tried to anchor land-use competition in a market-oriented mode and at local scale (i.e., at property level) where the agriculture and cattle producting actors have clear advantages over other sectors. The resistance of the production sector to negotiations on national or regional scale has become particularly clear during the preparation of the National Forest Law and the associated forest zoning maps. Some strategies to resist the changes in land-use competition, as documented for the Salta Province (Seghezzo et al. 2011), included: (a) resistance to the implementation of the National Forest Law based on the argument that deforestation is linked to progress and rural development, (b) pressure on the provincial government to issue deforestation permissions right before the National Forest Law was implemented (Leake and De Ecónomo 2008), (c) lobbying modifying the zoning map after it had been approved in a participatory process to incorporate particular request of the agribusiness sector, and (d) active participation in the zoning discussions, pushing for a fast resolution and for covering the "necessity" of meeting new land demands of the agricultural production sector.

Despite the regional zoning logic inherent in the National Forest Law, the agriculture sector and provincial governments have tried to keep land-use planning at farm level. Under the current zoning regime, some farms are completely inside yellow zones where deforestation is forbidden. After the implementation of the National Forest Law, the agricultural sector has started to heavily push for downscaling the application of the forest law to the farm level. Some provinces, such as Formosa and Santiago del Estero, have already included some mechanisms to permit deforestation in regulated areas, adjusting the proportion in each farm to license deforestation according to the provincial forest zoning map. Moreover, the productive sector argues that each farm could be zoned in the three forest zones. For example, the Salta provincial government supports the strategy of the production sector to move land-use planning to farm level arguing that the zoning map is only for orientation and that the status of every property can be revised (La Gaceta 2014). This tendency to move land-use planning to farm level undermines the utility of the zoning regime at the regional scale.

On the other hand, provincial governments traditionally have strong links with the local productive sector and many of their actions are targeted at retaining influence on land-use planning and regulation at the regional (province) scale. Provinces have thus been reluctant to discuss issues of natural resource use and land-use planning with national government based in the federal structure of Argentina. In addition, for the discussion about the zoning of forest areas, provincial governments provided little information in relation to conservation value and ecosystem service provision from the relevant areas (Gautreau et al. 2014). As a result, the forest zoning was implemented at provincial scale without an ecoregional vision and coordination between provinces. This fragmented planning processes, separate for each province, thus resulted in diverse criteria being applied and in inconsistencies in the forest classification across provincial administrative boundaries (Collazo et al. 2013). This situation implies a high risk of unintended conservation outcomes affecting connectivity at regional scale (Piquer-Rodriguez et al. 2014).

The government at national level responded to calls from the conservation sector and the public at large by creating mechanism of negotiation and regulation to solve conflicts. Examples of this type of intervention are the actions of the National Supreme Court of Justice in the conflict in eastern Salta, the creation of the National Reserve Pizarro (Hufty 2008) and the implementation of the new National Forest Law. In general, national government seeks to draw provincial governments and the agribusiness sector into a national negotiation process, where conservation organizations and local communities are also present and have a voice, with the general hope for a more balanced outcome of land-use competition in the Dry Chaco. Additionally, the national government presents to the global community its policies for regulating deforestation in different multilateral institutions and tries to promote synergies between global initiatives and national regulation. Some examples include the UN-REDD + initiative related to forest zoning or UN Global Environmental Facility projects related to forest ecosystem payments in the Chaco region (e.g., projects GEF#3623#5044#5338; GEF 2014).

Finally, the global and international actors (e.g., European soybean consumers' association and international conservation NGOs) also push forward a change in the mode of land-use competition in the Dry Chaco. These actors promote a mix of negotiations and market incentives creating initiatives as the RTRS and conservation projects in the region.

4.5 Conclusion

The Argentinean Dry Chaco is an example of a region where distal drivers enforce land-use competition in major ways. These distal drivers generate new land-use options in the region (e.g., soybean production early during the 1970s, more recently payments for carbon protection for global climate regulation). This affects land-use competition in the Chaco in major ways. While Chap. 13 in this volume has highlighted the important role of new actors representing these new drivers, this case study has emphasized the role of technology for creating new types of land-use competition as well as the intricate shifts in the mode of competition between liberal markets on the one hand, and on the other different forms of regulation operating across scales.

In relation to distal drivers of land-use competition, I described here how the differential access to global markets and transnational value chains, capital, information, and technology has led actors to pursue different strategies in navigating and shaping processes of land-use competition in their favor. Actors from the agribusiness production sector tried to retain control over land-use competition at the farm level and under liberalized market conditions. Local communities tried to push land-use competition to higher levels of policy-making, awareness, and regulation (e.g., national level) and promoted alliances with national and global actors. As a result, the gradual incorporation of national and international actors shifted the scale at which land-use competition is negotiated and plays out from the local market mode to the national-policy and global policy-market modes. This analysis reaches an ambivalent conclusion: As Baumann et al. also conclude (Chap. 13), distal drivers and telecoupling mechanisms create asymmetries between competing actors. Yet transnational flows of scientific data and knowledge as well as the cross-scalar global networking of public and political concerns also create opportunities to counteract asymmetries by introducing new actors, legitimate marginalized concerns in different ways, and broadening access to political decision-making processes by establishing new forums for negotiation.

Acknowledgments I am grateful to Tobias Kuemmerle, Sofia Nanni, and Yann le Polain, as well as to the editors for their highly valuable comments and suggestions that contributed to improving this article. We also would like to thank to Humboldt-University, Berlin and the Argentine Fondo Nacional de Ciencia y Técnologia (FONCyT, PICTO 2011 N° 0098), for their support.

References

Altrichter, M. (2006). Wildlife in the life of local people of the semi-arid Argentine Chaco. *Biodiversity and Conservation, 15*, 2719–2736.

Altrichter, M., & Basurto, X. (2008). Effects of Land Privatisation on the Use of Common-pool Resources of Varying Mobility in the Argentine Chaco. *Conservation and Society, 6*, 154.

BBC. (2011). Forest fears as paraguay's Chaco region sees land rush. August 16, 2011. http://www.bbc.com/news/world-latin-america-14032060

Boletta, P. E., Ravelo, A. C., Planchuelo, A. M., & Grilli, M. (2006). Assessing deforestation in the argentine chaco. *Forest Ecology and Management, 228*, 108–114.

Cáceres, D. M. (2015). Accumulation by dispossession and socio-environmental conflicts caused by the expansion of agribusiness in argentina. *Journal of Agrarian Change, 15,* 116–147. doi:10.1111/joac.12057.

Collazo, M., Panizza, A. & Paruelo, J. M. (2013). Ordenamiento Territorial de Bosques Nativos: Resultados de la Zonificación realizada por provincias del Norte argentino. *Ecología Austral, 23,* 97–107.

Eakin, H., DeFries, R., Kerr, S., Lambin, E. F., Liu, j., Marcotullio, P. J., et al. (2014). Significance of telecoupling for exploration of land-use change. In: Seto, K. C. & Reenberg, A. (ed.) *Rethinking Global Land Use in an Urban Era*. Strüngmann Forum Reports, 14, pp. 141–162. MIT Press. ISBN: 9780262026901.

Friis, C., Nielsen, J. Ø., Otero, I., Haberl, H., Niewöhner, J., & Hostert, P. (2015). From teleconnection to telecoupling: taking stock of an emerging framework in land system science. *Journal of Land Use Science*. doi:10.1080/1747423X.2015.1096423.

FVS (Fundación Vida Silvestre Argentina), TNC (The Nature Conservancy), Fundación DeSdel Chaco & Wildlife Conservation Society-Bolivia.(2005). *Evaluación ecoregional del Gran Chaco Americano* - 1a ed. Buenos Aires. p. 24

Gasparri, N. I., & Grau, H. R. (2009). Deforestation and fragmentation of Chaco dry forest in NW Argentina (1972–2007). *Forest Ecology and Management, 258*, 913–921.

Gasparri, N. I., Grau, H. R., Gutiérrez, Angonese, J. (2013). Linkages between soybean and neotropical deforestation: Coupling and transient decoupling dynamics in a multi-decadal analysis. *Global Environmental Change, 23,* 1605–1614.

Gasparri, N. I., Polain, Le, & de Waroux, Y. (2015). The coupling of South American soybean and cattle production frontiers: new challenges for conservation policy and land change science. *Conservation Letters, 8,* 290–298. doi:10.1111/conl.12121

Gautreau, P., Langbehn, L. & Ruoso, L. (2014). Movilización de información en el Ordenamiento Territorial de Bosques Nativos de Argentina, In: Terceras Jornadas Nacionales de Investigación y Docencia En Geografía Argentina. (pp. 29). May 2014, Argentina: Tandil.

Greenpeace, (2014). Campaña de Bosques. http://www.greenpeace.org/argentina/es/campanas/bosques/

GEF (Global Environmental Facilities) (2014). GEF-projects http://www.thegef.org/gef/project_list.

Grau, H. R., Gasparri, N. I., & Aide, T. M. (2005a). Agriculture expansion and deforestation in seasonally dry forests of north-west Argentina. *Environmental Conservation, 32*, 140–148.

Grau, H. R., Aide, T. M., & Gasparri, N. I. (2005b). Globalization and soybean expansion into semiarid ecosystems of argentina. *Ambio, 34*, 265–266.

Hansen, M. C., Potapov, P. V, Moore, R., Hancher, M., Turubanova, S. Tyukavina, A, (2013). High-resolution global maps of 21st-century forest cover change. *Science, 342*, 850–853.

Hufty, M. (2008). Pizarro protected area: A political ecology perspective on land use, soybeans and Argentina's nature conservation policy, In: Galvin, Haller, T. (ed.) People, protected areas and global change: participatory conservation in Latin America, Africa, Asia and Europe. pp. 145–174.

La Gaceta. (2014). Suspenderán la recategorización de áreas para desmontes. Edition 08/29/2014. http://www.lagacetasalta.com.ar/nota/2846/politica/suspenderan-recategorizacion-areas-para-desmontes.html.

Leake, A., & De Ecónomo, M. (2008). *La deforestación de Salta (2004–2007).* Fundación Asociana. Argentina: Salta.

Liu, J., Hull, V., Batistella, M., Defries, R., Dietz, T., Fu, F., et al. (2013). Framing sustainability in a telecoupled world. *Ecology and Society, 18*, 26.

New York Times. (2012). Vast tracts in Paraguay Forest being replaced by ranchers. March 24, 2012. http://www.nytimes.com/2012/03/25/world/americas/paraguays-chaco-forest-being-cleared-by-ranchers.html.

Palmer, J. H. (2005). *La buena voluntad Wichi: Una espiritualidad* indígena. Ed.: Grupo de Trabajo Ruta 81. Buenos Aires. Argentina. p. 386. ISBN 987-22489-0-7.

Paruelo, J., Verón, S., & Volante, J. (2011). Elementos conceptuales y metodológicos para la Evaluación de Impactos Ambientales Acumulativos (EIAAc) en bosques subtropicales. El caso del este de Salta. *Ecología Austral, 21*, 163–178.

Piquer-Rodríguez, M., Torella, S., Gavier-Pizarro, G., Volante, J., Somma, D., Ginzburg, R., & Kuemmerle, T. (2015). Effects of past and future land conversions on forest connectivity in the Argentine Chaco. *Landscape Ecology, 30*, 817–833.

Redaf. (2014). Red agroforestal chaco argentina. http://redaf.org.ar/.

RTRS-Round Table on Responsible Soy. (2014). http://www.responsiblesoy.org.

Rudel, T., & Meyfroidt, P. (2014). Organizing anarchy: The food security–biodiversity–climate crisis and the genesis of rural land use planning in the developing world. *Land Use Policy, 36*, 239–247.

Rolling Stones. (2014). Green Going Gone: The Tragic Deforestation of the Chaco. July 28, 2014. http://www.rollingstone.com/culture/news/green-going-gone-the-tragic-deforestation-of-the-chaco-20140728.

SAyDS. (2004). Atlas de los Bosques Nativos Argentinos. Buenos Aires (Argentina): Secretaria de Ambiente y Desarrollo Sustentable. p. 243.

Seghezzo, L., Volante, J. N., Paruelo, J. M., Somma, D. J., Buliubasich, E. C., Rodriguez, H. E., et al. (2011). Native forests and agriculture in salta (Argentina): Conflicting visions of development. *Journal of Environment & Development, 20*, 251–277.

UMSEF -Unidad de manejo del sistema de evaluación forestal. (2014). Perdida de bosque nativo y tasa de deforestación. http://www.ambiente.gov.ar/?idarticulo=311.

UN-REDD +. (2014). The United Nations collaborative programme on reducing emission form deforestation and forest degradation in developing countries. http://www.un-redd.org/.

WRI (World Resources Institute). (2014). Five overlooked deforestation hotspots. http://www.wri.org/blog/2014/03/5-overlooked-deforestation-hotspots.

Zak, M. R., Cabido, M., & Hodgson, J. (2004). Do subtropical seasonal forests in the Gran Chaco, Argentina, have a future? *Biological Conservation, 120*, 589–598.

Zak, M. R., Cabido, M., Cáceres, D., & Díaz, S. (2008). What drives accelerated land cover change in central Argentina? Synergistic consequences of climatic, socioeconomic, and technological factors. *Environmental Management, 42*, 181–189.

Chapter 5
Mind the GAP: Vietnamese Rice Farmers and Distal Markets

Mads Martinus Hauge

Abstract The region of the Mekong River Delta is the most important rice-producing region in Vietnam. Changes in global rice market demands combined with the Vietnamese state's interest in the economic development of its rural areas are currently influencing agricultural production practices in the region. With a lead farmer in the settlement of Truong Thang, Tam, as the key figure, this chapter investigates what happens when local rice farmers are encouraged to adopt a new production model and internationally recognized "good agricultural practices" (GAPs) in order to become part of a global high-value rice production network. The aim of this chapter was to emphasize some of the local processes of change that are part of a larger regional transformation process with the potential to change the dynamics of the global rice market. For the farmers, the production of GAP-certified high-value rice intended for export markets is radically different from what they are used to. It necessitates a new kind of knowledge, a new organization of production, new time cycles, a new distribution of responsibility, and the formation of new relations between people in their settlements. The analysis of this chapter is based on empirical fieldwork conducted in the Truong Thang settlement in Can Tho Province of the Mekong River Delta region. Based on this work, the key argument of the chapter is that the conversion of production practices that is currently taking place is not a simple case of local adaption to an international demand. Rather, it is a complex process of reconfiguration of societal and territorial embeddedness as the conversion of production practices is necessarily entangled with settlement and household spaces. The conversion of practices essentially relies on local processes of negotiation, on the entrepreneurship of individuals, and on the willingness and ability of farmers to internalize new values, i.e., perform the change.

Keywords Global production networks · Land-use practices · Drivers · Governance · High-value rice

M.M. Hauge (✉)
Department of Geosciences and Natural Resource Management, Geography Section, University of Copenhagen, Copenhagen, Denmark
e-mail: mmh@ign.ku.dk

J. Niewöhner et al. (eds.), *Land Use Competition*,
Human-Environment Interactions, DOI 10.1007/978-3-319-33628-2_5

5.1 Introduction

"Your ears will hear and your eyes will see," explains Tam, the lead farmer of a group of rice-producing households in the settlement of Truong Thang. With these words he underlines that only by means of demonstration can his fellow farmers be inspired to change their mind-sets and start converting their rice production practices. Five years of his hard work has recently resulted in the first yield of high-value jasmine rice from 45 farming households covering a total of 66 ha of land. During these years, Tam has become the embodiment of distant market impulses, acting as a persuader, negotiator, mediator, educator, and role model in a process that has the potential to transform the way in which farmers in Truong Thang farm rice. However, there is no reason for Tam to rest on his laurels, not yet. The 45 households are only in the first phase of the transformation from low- to high-value rice production. The first crop was hugely important, as it demonstrated some of the potential of a so-called Small Farmers, Large Fields (SFLF) model that the authorities in the Can Tho Province are encouraging in all of its rural districts. The aim of this first crop was twofold, according to Tam: first, to establish a coherent production area and get the farmers to work together and grow the same high-value rice variety; second, to have farmers experience that the rice-export company with whom they entered into a contract was keeping its word of coming back to buy up the harvest. When I spoke to Tam, farmers were only partially introduced to some of the more challenging requirements related to the production of high-value rice destined for more demanding markets in terms of quality and food safety. A central requirement among these is adherence to the industry-defined production standard of GlobalGAP. "You have to do the simplest things first, you cannot jump to the difficult things, otherwise the farmers cannot and will not do it," Tam explains, emphasizing that the new production practices are radically different and must be introduced gradually to influence his fellow farmers and make the transformation successful.

This chapter examines how farmers in the settlement of Truong Thang in Can Tho Province, southern Vietnam, driven by the potentiality to tap into a high-value rice market, have started to convert their production processes and land use from low- to high-value rice. A central point of this chapter is that the conversion is not only made possible by extra-regional demands and state mediation, but essentially relies on local processes of negotiation and on the entrepreneurship of individuals and the willingness of farmers to convert their practices. The resonance this conversion has with distal markets is negotiated and mediated by a range of national and regional institutions and individual actors on its path to the Truong Thang settlement. The conversion of production practices therefore relies on the individual strategies of the actors involved and on their ability to cooperate in making the transformation possible. In the literature on global production networks (GPN), this relationship between the local (regional) and the global is known as a "strategic coupling," that is, "the dynamic processes through which actors in regions coordinate, mediate and arbitrage strategic interests between local actors and their counterparts in the global economy" (Yeung 2009: 332). This is essentially a process of "molding" local assets into compliance with the demands of a global network, thus stressing that global forces always are played out locally.

A strategic coupling is conceptualized to exhibit three attributes: It is strategic because it happens through the active interventions and intentional actions of the parties involved; it is time–space contingent as it is non-permanent and subject to change (Yeung 2009; Coe and Hess 2011); and it transcends territorial boundaries and geographical scales "as actors from different spatial sites (states, regions, and localities) converge and their practices radiate out to diverse geographical scales—some global and some highly local" (Yeung 2009: 333). The development of the concept of strategic coupling has primarily been informed by the experiences of Asian "tiger economies" that have inserted themselves into technologically intensive and service-orientated GPNs (see, for instance, Lee 2009; Yang et al. 2009; Yang 2009, 2014; Yeung 2009; Neilson et al. 2014; Yeung 2014; Lee et al. 2014). However, in this chapter, the concept is applied to an agricultural resource-based, regional development process involving the inclusion and molding of the practices of small-scale production units: farming households.

In order to discuss some of the challenges connected to the strategic coupling, I employ the GPN framework's concept of embeddedness. The concept promotes the perspective that each individual point in the production network is embedded in social and spatial arrangements. The precise nature and articulation of GPNs and their integration into specific locations are deeply influenced by the concrete sociopolitical, institutional, and cultural contexts within which they are embedded, produced, and reproduced (Dicken 2011). However, the social and spatial arrangements, which constrain and facilitate a strategic coupling, can be porous and subject to change. The GPN approach draws on the work of Karl Polanyi, Mark Granovetter, and Paul DiMaggio, among others, and operates with three overlapping dimensions of embeddedness (Hess 2004): the societal embeddedness of actors, that is, how their cultural backgrounds and institutions influence and shape the actions of individuals and collectives; network embeddedness, which refers to the anchorage of actors in networks in which they are or become integrated; and territorial embeddedness, which takes into account the fact that all economic activities are carried out in space and anchored in particular places (Hess and Coe 2006). This chapter illustrates how working toward inclusion in high-value rice GPNs embeds farmers in the logic of this GPN (network embeddedness), thus challenging and coshaping existing norms and institutions in Truong Thang (territorial and societal embeddedness).

Historical contextualization is pivotal to explaining how the Truong Thang farmers act when confronted with a new model of high-yield and high-quality rice production and why they have come to stand on the threshold of a production methods' conversion. In the following section, I therefore present the post-1975 history of rice production and land management in the Mekong River Delta (MRD). Having established the historical context, the chapter will go on to analyze the change that is currently taking place in the Truong Thang settlement. The chapter uses empirical data collected during fieldwork in the Truong Thang settlement, December 2012 to March 2013, and March 2014. Group interviews, a household survey, and in-depth semi-structured interviews with selected households, individuals, authorities, and company representatives were the main methods applied.

5.2 Historical Contextualization: The Molding of Local Assets Over the Past Four Decades

The strategic coupling that is emerging between the MRD region and high-value rice-export markets builds on several decades of local assets' molding. Rice farmers in Truong Thang, like most farmers in all of the MRD, have been subject to several transformations affecting their rice production practices and thus their livelihoods: regulatory and ideological transformations in land use; infrastructural developments which have accommodated a revolution in the production of rice; and, more recently, the government's attempts to overcome some of the barriers that exist in the rice industry in Vietnam. The following unfolds these transformations.

5.2.1 Regulatory and Ideological Transformations

The area that today constitutes the settlement of Truong Thang was subject to heavy fighting during the Second Indochina War (the "Vietnam War"). Consequently, most of the residents fled to seek refuge in neighboring districts less affected by war. When the war ended in 1975, its former residents bit by bit returned to their land. Shortly afterward, Hanoi-initiated policies began to influence life in Truong Thang. At its 4th National Convention in 1976, the Vietnamese Communist Party (VCP) adopted Resolution 4, which urged all party leaders in the provinces of southern Vietnam to begin a gradual collectivization of their agricultural production. Furthermore, from 1976 onward, the VCP also initiated a land reform in the south with the aim of appropriating and redistributing the so-called excessive land from the upper middle class to the poor and landless (Raymond 2008; Gorman 2014). Land was distributed based on the number of household members: 0.1–0.15 ha/adult and 0.08–0.1 ha/child under sixteen and adult over sixty (De 2005). In total, more than 500,000 ha, or 25 % of the total agricultural area of the MRD, were redistributed (Gorman 2014).

These reforms never crystalized into the prosperous agricultural production they were intended to. By the early 1980s, Vietnam and its planned economy were in crisis, epitomized by food shortages in the north, which placed an estimated 9.3 million people in danger of famine (Gorman 2014). The agricultural sector in the south, traditionally a region with an agricultural surplus, had missed its centrally planned targets. The total amount of rice grain available in the country dropped, which led Vietnam to increase its imports. Observers of Vietnam's agricultural development generally agree that it was the structure of the sector, and especially the collectivization policies, that caused the food availability problems that the country was experiencing (Scott 2009; Akram-Lodhi 2005; De 2005). Key problems included (1) a lack of incentives to work diligently, (2) a lack of upkeep of collective property, and (3) administrative burdens. Also, the state's procurement prices that farmers received for their output was low, and some farmers sought to avoid them by selling outputs on the informal markets, where prices were reported to have been up to ten times higher.

As early as in 1981, the VCP reacted by implementing a series of reforms. Again, restructuration was the means, but this time decollectivization and commercialization of the agricultural sector nation-wide were the aims. Directive 100 (1981), Resolution 10 (1988), and the land law (1993, but amended several times since) were the pillars of the restructuring. Following Directive 100, farmers became the masters of three production stages (planting, caring, and harvesting), while other stages (land use, crop choice, land preparation, irrigation, and input supply) were still under the control of the authorities (De 2005). Resolution 10 aimed to further deregulate agricultural management. Entitling farmers to the land they had owned prior to 1975 with land-use rights lasting ten, fifteen, or twenty years on the basis of renewable leases was one result of this process (De 2005). The new land law followed in 1993, resulting in the transformation of land ownership from the collective back into private hands. This restructuring eventually commercialized agriculture in Vietnam and gave farmers the right to make all managerial decisions regarding their land (De 2005). Farmers were now also allowed to own machinery and draft animals, while private traders were allowed to trade in the inputs and outputs of farm products.

These reforms were part of the more general reform process known as Doi Moi ("renovation"), with which the VCP has aimed to create a socialist-oriented market economy in Vietnam. During the reform process, the country has slowly opened its economy to international trade. Since then, Vietnam has experienced rapid changes in the agricultural sector with the intensification of, among other things, rice production. As part of this development, the MRD region has been able to reestablish its position as the country's "rice bowl" (Nguyen et al. 2007). Between 1995 and 2009, the MRD increased its paddy production by almost 60 %, and the delta now produces 54 % of total annual national production and 90 % of total annual national rice exports. The region is thus the main contributing region in ensuring Vietnam's shift from being a rice-importing country to becoming one of the largest rice exporters in the world (Garschagen et al. 2012).

5.2.2 Infrastructural Developments and Green Revolution Influence

The agricultural sector in the MRD has been subject to additional changes regarding both the availability of water for irrigation and the adoption of new production methods. Shortly after the unification of Vietnam in 1975, the ability to maintain a year-round rice production was gradually facilitated by the government through the development of a network of canals in the Truong Thang. The canals allowed water to reach fields it had not done before, making it possible for vessels to penetrate the MRD region's rural hinterland and transport the farmer's produce to processing facilities. Agricultural extension officers and engineers employed by the local authorities provided the overall plan, while the local residents supplied labor for the construction of the canals.

Controlling water decoupled the agricultural production from dependence on seasons. Previously, only one crop a year was the norm. The farmers would plant Lua Mua (seasonal rice) at the start of the rainy season and harvest it around Tet (the Vietnamese New Year). In the 1960s, a new rice variety, locally called the Than Nong ("miracle rice"), was introduced in the delta. However, it was only after the war, and especially after the Doi Moi reforms, that the region really began to experience the benefits of Green Revolution technologies such as mechanical inputs, fertilizers, pesticides, and continuous further refinements to high-yielding rice varieties. Cooperation between the International Rice Research Institute (IRRI), regional research institutes, and local universities in the delta has played a significant role in developing new generations of high-yielding, disease-resistant rice varieties adapted to the natural conditions of the delta (Ut and Kajisa 2006). Since the 1990s, through public media and local agricultural extension officers, the government has also been active in promoting a more efficient use of pesticide and fertilizer inputs in rice production.

Water management and the introduction of technologies have enabled farmers to first double (1980s) and later triple (1990s) rice cropping. Today, most farmers in the area plant IR 50404, a rice variety which is a direct descendant of the Than Nong miracle rice and known for its high yields but low quality and low price on the global market. Much of the production in the MRD is exported, and the farmers in Truong Thang settlement are thus already part of production networks, which extend beyond the borders of Vietnam. However, earnings are low due to, on the one hand, world market price of the produced rice variety, and, on the other hand, ingrained structural deficiencies of the Vietnamese rice production sector. The latter impacts on the former, as the ability to secure a better market position, are hampered by structural deficiencies, which according to the government render the sector, in its current form, unsuitable to secure a sustainable economic development of rural areas (Can Tho People's Committee Report 2013; FAO 2013).

5.2.3 Small Farmers, Large Fields Model: A Government-Initiated Attempt to Overcome Structural Deficiencies of the Rice Sector

The main problems, as identified by the government, are the role of the middlemen, the size of the production units (the farms), and the variety of agricultural practices that goes into rice production. These will be sketched out in the following.

Currently, the rice production network is reliant on a vast number of small traders who roam the MRD region's rivers and canals to connect the production of rice with processing and exporting companies. These "middlemen," firstly, undermine exporters' ability to control the quality and purity of the rice, as they mix together harvests of different quality and origin. Secondly, the fact that they also operate with very short time horizons, as they want to earn as much as possible in the here and now, makes them unreliable buyers of the farmers' produce. Farmers often experience that middlemen break agreements and never show up to collect their produce or refuse to pay the agreed price. The farmers themselves have small and fragmented plots on which they make decisions regarding production practices and rice varieties. Small and fragmented plots create a number of disadvantages for the individual farmers, but also for exporting companies. For the individual farmer, it increases the costs of production and thereby hampers the ability to invest in future productions by acquiring new production technologies and machines, while it decreases the bargaining power vis-à-vis buyers. For exporters, it means that the produce that is brought to them by the middlemen is undifferentiated and the inability to control the purity and quality of rice makes it impossible for them to establish themselves as well-reputed brands in international markets (Ut and Kajisa 2006). In order to address these deficiencies, the government has been trying to encourage a new model called the "SFLF." This model and its implementation form one of the outcomes of Resolution 26 of 2008 (called Tam Nông, meaning agriculture, farmers, and rural areas), whose principle goal is to "build up a comprehensively developed agricultural sector in a modern and stable manner with large-scale commodity production, high yield, good quality, better efficiency, high competitiveness" (FAO 2013: 15). Resolution 26 came with the realization that, although the development of rural areas in Vietnam over the period of implementation of Doi Moi reforms has been quite comprehensive, its achievements are still far from fulfilling Vietnam's potential. The government sees the introduction of the SFLF model as a means through which the standard of living of its rural population can be raised and untapped trade potentials exploited.

The ideas of the SFLF model are as follows: (1) to get the many land-possessing households of a particular area to use the same production procedures in all phases of the production (technologies, production management, harvesting, preservation, and sale of produce); (2) to gain access to profitable world markets by introducing internationally recognized principles of GAP, which aim to provide safety for consumers and producers, be environmentally friendly, and enable the tracking of products; and (3) to create direct links between producers and exporters with connections to the world market (Can Tho People's Committee Report 2013). In the model, exporters will have to join the farmers and share the risks and the benefits. This is expected to create a stable supply chain and raise the confidence of the end buyers (importers from other countries). The overall ambition is an enhanced quality of rice, a more positive perception of rice produced in Vietnam, and more profitable prices for all actors involved.

Several government institutions are involved in implementing the model. Each province's departments of Agriculture and Rural Development (at the provincial, district, and communal levels) are in charge of the actual implementation. They facilitate the initial introduction of the idea to both farmers and exporters, provide platforms of dialogue between the respective actors to help overcoming potential mistrust, and advise farmers on how to grow according to the GAP principles through its agricultural extension officers. Importantly, the local government oversees the signing of contracts between farmer groups and exporters, and it mediates between them should disputes arise. The central government also creates incentives for exporting companies to enter into contracts with farmers directly by, for example, providing export licenses (Vietnam Food Association, Ministry of Industry and Trade) and cheap state loans (Ministry of Planning and Investment) for building storage and drying facilities[1] dependent on this activity.

5.3 The Current Situation: Transformations Under Distal Influences

Forty years of developments in the agricultural sector, and especially initiatives launched during the last decade, are contributing to the emergence of a strategic coupling where local assets are being molded and matched to the demands of a more profitable distal market. The coupling is highly dependent on local forces. These forces include the individuals' entrepreneurship, the farmers' interests in cooperating with each other and with the rice buyers, and generally the farmers' willingness to internalize new values. In the following, I trace the resonance of the global high-value rice market at the level of the farmers' rice production practices in the Truong Thang and explore the ways in which it is negotiated at household and settlement level.

5.3.1 The Global High-Value Rice Market: A Potentiality with Challenges

Significant changes in demand and consumption patterns over the last decades in North America and Europe have made accessing high-value rice production networks potentially interesting for Vietnamese farmers. Different ideas, on the one hand, about the purity and danger of food, what a proper meal is, what the proper ways to treat farmland and livestock are, and, on the other hand, the responsibility of governments to protect producers and consumers from food risks, fluctuate and change over time (Dicken 2011). For many consumers, food has become a statement of lifestyle, and the increased affluence has stimulated a desire for greater

[1]Under the Decree 109/2010/ND-CP of November 4, 2010, ownership of storage and drying facilities has become a requirement for acquiring an export license.

choice in food products. While white rice was previously a uniform product, today consumers all over the world expect to find jasmine, basmati, organic, etc., rice on the retailers' shelves.

The agri-food business under, for example, EU legislation is charged with the legal responsibility for ensuring that certain food safety requirements are met (Graffham 2006). However, it is up to the industry itself to create systems that enable compliance. One of these systems is the GlobalGAP standard, a so-called industry-driven, collective, business-to-business standard (Tallontire 2007) in which actors of the industry can participate in the standard-setting process. Most of these participants are, however, retailers and suppliers/producers from Western countries (GlobalGAP 2014). The standard meticulously defines good agricultural practices (GAP) for the production of safe agricultural products of good quality (for rice production farmers need to comply with almost 200 principles). Adherence to the standard is in principle voluntary, but in practice, it is mandatory for agricultural exports to much of the Western market.

Since its introduction in 1997, the standard has been adapted to take country-specific circumstances into account. One example of this is that in many developing countries, production takes place among small-scale farmers with limited financial capabilities. Therefore, it has become an option for small-scale farmers to become certified as a group under the GlobalGAP Standard (GlobalGAP 2014). This entails a collective distribution of some of the costs of adhering to the standard (such as the construction of secure warehousing for inputs) and paying for a third party to take care of the certification. A significant attribute of group certification, however, is that each individual farmer becomes dependent on the (in-) activities of fellow farmers, as these (in-)activities can have direct consequences on the ability of the group to adhere to the standard and thus gain access to the market. The GlobalGAP standard therefore functions as a mechanism that pushes the responsibility for providing safe food onto producers around the globe. For retailers, the standard is a technology that enables them to govern from a distance (Foucault 2007; Dean 1999) and to organize their supply chains. It does this by providing a language and a clearly defined set of requirements aimed at constituting a shared frame of reference for all the actors involved. For the Vietnamese farmers, the standard is a minimum requirement for accessing high-value GPNs. The standard enters and embeds the Truong Thang settlement along with the SFLF model as elements that need to be negotiated, adopted, and internalized to fulfill the strategic coupling potentiality.

5.3.2 Mobilizing Tam

This brings us back to Truong Thang and the local lead farmer, Tam. Tam is perhaps the key figure in the process of converting production practices in Truong Thang. Tam is proud of establishing a group of farmers and linking them with an export company and speaks with authority as he explains how the local conversion of production practices has taken place. Five years ago, he was chosen by the settlement and the commune's Department of Rural and Agricultural Development to participate in a class held at the University of Can Tho. In the class, he was informed about the benefits of the SFLF model and about how, in combination with adherence to the GlobalGAP standard's principles of agricultural production, the integration of the model would provide access to a higher paying segment of the world market. As part of the class, Tam also visited areas in the province where the SFLF model had been implemented.

Upon returning to the village, Tam and the settlement leadership, in cooperation with the commune's Department of Agriculture and Rural Development, started to implement the SFLF model by searching for an export company willing to engage with Truong Thang farmers. The incentives that the government had provided for the exporting companies (see above) made this an easy task. The district's Department of Agriculture and Rural Development also facilitated the initiation process by organizing a meeting in which exporting companies active in the district were invited to register the amount and locations of hectares they would be interested in contracting. In this match-making process, Tam soon found a suitable company. Together, they drafted an initial contract distributing and stating the rights and responsibilities of both signing parties, which Tam could present to the other farmers in the settlement.

5.3.3 Mobilizing the Farmers

Tam turned out to be a perfect change agent, as several farmers described him as a highly skilled and trustworthy farmer with the ability to spearhead such a significant change. However, two key challenges needed to be tackled in order for his efforts to be successful. First, he needed to establish a reasonably sized group of farmers in order for it to be attractive to the exporter. Here, Tam's own reputation was of the essence. Past experiences with the erratic behavior of buyers of rice, as well as of forced collectivization after unification, made it difficult to persuade farmers to join forces and commit to just one buyer. In order to negate this, Tam organized an introductory meeting in which he physically and symbolically stood next to the company representative, in order to vouch for the company. At the meeting, the proposal was met with skepticism by the farmers. Therefore, the meeting was followed by visits to each individual household. Together with an agricultural extension officer and representatives of the settlement leadership, Tam walked around the settlement with the aim of persuading households to join the group. Eventually, 45 households did. The local authorities played a pivotal role in this process of recruitment. They oversaw the drafting and signing of the contract and guaranteed to act as mediators if disputes should arise between the signing parties. However, formally establishing the group and signing the contract with the company were just the initial step. The vital part was what followed, namely Tam's support to his fellow farmers at every step of the production of the first crop to be sold to the company. Part of his support strategy was to use his own land as a demonstration field, in which all of the GlobalGAP Standard's requirements were followed. He also visited the households involved on a daily basis to give advice if they were interested in receiving them from him. Tam's strategy reflects the fact that he can only inspire—not force—the farmers to change their cultivation practices: In the end, it is their own choice.

5.3.4 *Reconfiguring Social and Spatial Arrangements in the Settlement*

The production of jasmine rice intended for export markets is radically different from what farmers in the settlement are used to. It necessitates a new kind of knowledge, a new organization of production, new time cycles, a new distribution of responsibility, and the formation of new relations between people in their settlements. The territorial and societal embeddedness of the households involved are thus starting to be reconfigured as the group members commit to cooperating. This reconfiguration is manifested not only physically but also organizationally and discursively, as it entails a remolding of existing consciousness of place for the involved farmers. Interviews with farmers revealed that several aspects regarding the conversion of production practices caused friction, as their evolving perception of their role in the settlement was sometimes in contrast to their traditional possibilities and restrains. This friction highlights the fact that land-use competition in this case involves not only competition between diverse forms of land uses, but also between different perceptions of place. Of these frictions (1) losing flexibility, (2) adapting to new production practices, and (3) trusting the exporter and the SFLF model were of greatest importance.

For the individual farmer, converting production practices to meet the GlobalGAP standards and the SFLF model makes his or her daily life less flexible. Inclusion in global high-value rice production networks entails that the individual farmer's practices are structured by the farming group's decisions and the contracting company's requirements. As part of the SFLF model, responsibility for rice production, thus, shifts from the individual farmer to the group. The conversion of production practices, therefore, requires individual farming households to enter into new forms of societal and territorial embeddedness, as they cannot partake without aligning their practices with those of the group. They will be held accountable if they deviate in any way from the group's expectations. The friction between individual independence and the group's social control is especially embodied in the journal that the farmers are or will be expected to keep of all agricultural activities, as this poses several challenges for the farmers (in the first SFLF-produced crop, only ten households were given this task, but in the future, all will have to carry it out). Journal keeping is one of the basic requirements of the GlobalGAP standard, put in place to enable transparency and traceability in the production network. In keeping the journal, the farmers are not only challenged by high levels of illiteracy (mostly in the older generation); the record also quickly comes to constitute the constant presence of their fellow farmers and the exporting company in their individual households. Often, the individual farmer does not understand why it is necessary to note everything down, and the journal becomes a specific object in which the discomfort ingrained in restructuring individual practices and senses of place is materialized.

After the first SFLF-produced crop in the spring of 2014, several farmers had still not come to terms with the new production principles, nor with them being dependent on the group. One farmer referred enviously to farming households, which had not joined the group, as "freedom farmers." Another said that he did not want to join the group, but had basically been forced to do so. His fields were surrounded by those of group members, and therefore, he had had to join their cycle of irrigation etc.

Adapting to the new production practices also requires a whole new way of understanding the production process itself. The aim of the production is no longer just to produce rice (calories) in as great a volume as possible. Now, the aim is to ensure that the produce is healthy and that, for instance, excessive usage of chemical inputs does not harm end consumers. Several farmers confessed in interviews that in their production of the first crop, they had hardly changed their practices and basically did not follow the instructions that Tam and the agricultural extension officer had given them. They had, for instance, refrained from buying the more expensive inputs that Tam had advised them to do, because "all pesticides come from the same company, so it does not matter" (male farmer). Also, they did not adhere to advice on when and how to use the inputs: "If there is pest we will spray, and if the leaves start to turn yellow or red we will just use fertilizers, and then they become green again. Farmers will just do as they like." (female farmer).

Adjusting the production process necessitates an internalization of the values and practices of the GlobalGAP originating from a foreign market concerned with foreign ideas about nutrition and health. This is not an easy task. During interviews, Tam shared his conclusion that it will take time to make this adjustment and that it hinges on the financial success of the first few crops. If the farmers can see that their extra efforts are rewarded, they will be more ready to adapt, and Tam can begin to introduce the farmers to more of the GlobalGAP standard's requirements.

The building of trust is vital if changes to production practices are going to materialize. It is only when the farmers trust in one another, in the SFLF model, and in the relationship with the exporting company and perceive the latter as beneficial to themselves and the group that they will wholeheartedly begin to change their practices. The field study clearly pointed to the fact that a change agent such as Tam is of crucial importance. Tam used all his powers of persuasion to convince his neighboring farmers to join the group, and he constantly needed to maneuver in order to maintain their faith in him. A vital element is the negotiation of the price of the produce from the first crop. According to the contract, the exporter is obliged to pay 100 VND/kg above the market price, and Tam carefully contacted every person knowledgeable about the market price and, for the first time, used the Internet as a source of information about pricing. This he did in order to ensure that group members could neither find any reason to question the price nor Tam's role in defining it, nor the export company's intentions. Interviews with the farmers testified to Tam's success in this matter, as they all declared that they were impressed and encouraged by the fact that the exporter kept its promises and returned to the settlement at the time of the harvest to pick up the produce at the agreed price.

5.4 Concluding Remarks

The chapter has illustrated some of the virtues attached to a resource-based strategic coupling. This form of strategic coupling relies on the active enrollment and mobilization of farmers scattered around the delta. Various institutions and individuals are cooperating to enable Vietnamese rice farmers to acquire access to global high-value markets. Central to this strategic coupling process are the ways in which the SFLF model and internationally recognized production standards of GlobalGAP are being negotiated and possibly adopted and internalized by farmers. In Truong Thang, the actions of a key trusted entrepreneurial farmer, Tam, are of vital importance for the process to be successful. Tam essentially mediates the requirements of the GPN and the logic of the SFLF model to his community and by doing so plays a central role in the successful conversion of local production practices. The conversion challenges and requires a reconfiguration of the societal and territorial embeddedness of individual farmers in the group. The transformation therefore manifests itself not only physically but also organizationally and discursively, bringing with it a remolding of the existing consciousness of place on the part of the households involved. Land-use competition in this case therefore involves competition between not only different forms of land uses and production practices but also between different perceptions of place, of how farmers see themselves as part of a community and as farmers. The transformation of production practices and the territorial and societal embeddedness of farmers may only be happening slowly, but past experiences of boom crops have shown that in large areas, Vietnamese farmers, when convinced, can convert their production practices promptly and industriously. This was, for instance, the case for the robusta coffee bean and the Pangasius fish industries (see Agergaard et al. 2009; Hansen and Trifković 2014). The dynamics of the world market in these industries changed dramatically almost the moment Vietnam entered the arena. The same may well be in the pipeline for high-value rice.

References

Agergaard, J., Fold, N., & Gough, K. V. (2009). Global–local interactions: Socioeconomic and spatial dynamics in Vietnam's coffee frontier. *Geographical Journal, 175*(2), 133–145.

Akram-Lodhi, A. H. (2005). Vietnam's agriculture: Processes of rich peasant accumulation and mechanisms of social differentiation. *Journal of Agrarian Change, 5*(1), 73–116.

Coe, N. M., & Hess, M. (2011). Local and regional development: A global production network approach. *Handbook of local and regional development* (pp. 128–138). London: Routledge.
Can Tho People' Committee. (2013). Productivity of small farmers, large field model in Thoi Lai District, Can Tho City in 2013.
De, N. N. (2005). *Farmers, agriculture and rural development in the Mekong Delta of Vietnam.* Vietnam: Education Publishing House.
Dean, M. (1999). *Governmentality: Power and rule in modern society*. Thousand Oaks: Sage.
Dicken, P. (2011). *Global shift: Mapping the changing contours of the world economy* (6th ed.). New York: The Guilford Press.
Food and Agriculture Organization. (2013). Vietnam country programming framework.
Foucault, M. (2007). In M. Senellart, F. Ewald & A. Fontana (Eds.), *Security, territory, population*. Palgrave Macmillan.
Garschagen, M., Revilla-Diez, J., Nhan, D., & Kraas, F. (2012). Socio-economic development in the Mekong Delta: Between the prospects for progress and the realms of reality. In F. Renaud & C. Künzer (Eds.), *The Mekong Delta system: Interdisciplinary analyses of a river delta* (pp. 83–132). Dordrecht: Springer Environmental Science and Engineering.
Gorman, T. (2014). Moral economy and the upper peasant: The dynamics of land privatization in the Mekong Delta. *Journal of Agrarian Change, 14*(4), 501–521.
GlobalGAP. (2014). GLOBALGAP. Accessed March 23, 2015 from http://www.globalgap.org
Graffham, A. (2006). EU legal requirements for imports of fruits and vegetables (a suppliers guide). Fresh Insights no. 1, DFID/IIED/NRI. Available via Agrifoodstandards. Accessed January 15, 2015 from http://www.agrifoodstandards.net
Hansen, H., & Trifković, N. (2014). Food standards are good—For middle-class farmers. *World Development, 56*, 226–242.
Hess, M. (2004). "Spatial" relationships? Towards a reconceptualization of embeddedness. *Progress in Human Geography, 28*(2), 165–186.
Hess, M., & Coe, N. M. (2006). Making connections: Global production networks, standards, and embeddedness in the mobile telecommunications industry. *Environment and Planning A, 38* (7), 1205–1227. doi:10.1068/a38168
Lee, Y. (2009). Balanced development in globalizing regional development? Unpacking the new regional policy of South Korea. *Regional Studies, 43*(3), 353–367.
Lee, Y., Heo, I., & Kim, H. (2014). The role of the state as an inter-scalar mediator in globalizing liquid crystal display industry development in South Korea. *Review of International Political Economy, 21*(1), 102–129.
Neilson, J., Pritchard, B., & Yeung, H. W. (2014). Global value chains and global production networks in the changing international political economy: An introduction. *Review of International Political Economy, 21*(1), 1–8.
Nguyen, D. C., Le T. D., Nguyen V. S., & Miller, F. (2007). Livelihoods and resource use strategies of farmers in the Mekong Delta. In T. B. Tran, T. S. Bach & L. Miller (Eds.), *Challenges to sustainable development in the Mekong Delta* (pp. 69–98). Bangkok: The Sustainable Mekong Research Network.
Raymond, C. (2008). "No responsibility and no rice": The rise and fall of agricultural collectivization in Vietnam. *Agricultural History, 82*(1), 43–61.
Scott, S. (2009). Agrarian transformations in Vietnam: Land reform, markets, and poverty. In M. Spoor (Ed.), *The political economy of rural livelihoods in transition economies: Land, peasants and rural poverty in transition* (pp. 175–200). London: Routledge.
Tallontire, A. (2007). CSR and regulation: Towards a framework for understanding private standards initiatives in the agri-food chain. *Third World Quarterly, 28*(4), 775–791.
Ut, T., & Kajisa, K. (2006). The impact of green revolution on rice production in Vietnam. *The Developing Economies, 44*(2), 167–189.
Yang, C. (2009). Strategic coupling of regional development in global production networks: Redistribution of Taiwanese personal computer investment from the Pearl River delta to the Yangtze River delta, China. *Regional Studies, 43*(3), 385–407.

Yang, C. (2014). Market rebalancing of global production networks in the post-Washington consensus globalizing era: Transformation of export-oriented development in China. *Review of International Political Economy, 21*(1), 130–156.

Yang, D. Y., Hsu, J., & Ching, C. (2009). Revisiting the Silicon Island? The geographically varied "strategic coupling" in the development of high-technology parks in Taiwan. *Regional Studies, 43*(3), 369–384.

Yeung, H. W. (2009). Regional development and the competitive dynamics of global production networks: An East Asian perspective. *Regional Studies, 43*(3), 325–351.

Yeung, H. W. (2014). Governing the market in a globalizing era: Developmental states, global production networks and inter-firm dynamics in East Asia. *Review of International Political Economy, 21*(1), 70–101. doi:10.1080/09692290.2012.756415

Chapter 6
The Role of Maps in Capturing Distal Drivers of Deforestation and Degradation: A Case Study in Central Mozambique

Yaqing Gou

Abstract Remote sensing provides unprecedented abundance of earth observation images from plot to global scale. Modeling land use and land cover change, however, remains challenging due to complex drivers and processes involved. This challenge is aggravated by the merging of distal drivers, which play an increasingly important and complex role in altering local land cover and land use. Distal drivers, in the form of global markets, NGOs, international governments, and institutions, significantly contribute to shaping the landscape of central African states—the focus of this chapter—by controlling capital, information, knowledge flow, and international development initiatives. In this chapter, I will explore the applications and constraints of Remote Sensing (RS) and Geographic Information System (GIS) in capturing distal drivers of deforestation and forest degradation in the Beira corridor, Manica, Mozambique. I investigate how this analysis fits into the conceptual and methodological framework of distal drivers. In the Beira corridor, it is primarily the scarcity and uneven distribution of forest resources shaping land use competition in woodland areas. This competition is intensified by dramatic population increases, demands from remote markets, and conflicts of interest between local government and international organizations to preserve or develop certain forests and not others. Remote sensing images can capture the patterns of land use change as 'balanced' results of different land uses within a geographic area with time, which reflects the respective outcomes of competition between these different land use activity types. Examples explain drivers can be connected to their spatial patterns by combining regional biomass change maps derived from optical and radar remote sensing imagery, and knowledge of local deforestation and degradation processes. Moreover, we discuss the biases and uncertainties that result from processing RS images, as well as interpolating land use maps from RS images. I conclude with a brief discussion of the biases and uncertainties that may affect our perception of land use change, and how maps and their attributes themselves may feedback into land use competition.

Y. Gou (✉)
School of Geosciences, University of Edinburgh, Edinburgh, Scotland, UK
e-mail: Y.Gou@ed.ac.uk

J. Niewöhner et al. (eds.), *Land Use Competition*,
Human-Environment Interactions, DOI 10.1007/978-3-319-33628-2_6

Keywords Remote sensing · Uncertainty · Land use practices · Governance · Radar

6.1 Introduction

6.1.1 Why Do We Need Maps of Deforestation and Forest Degradation?

Although humans have been modifying land to obtain food and other essentials for thousands of years, the concept of land use competition in its current understanding emerged in land use and land cover change (LUCC) research only relatively recently with the perceived looming scarcity of land due to booming populations and increasing demand for resources. In the 1960s, land use competition centered on urban frontiers to explore the competition of land resources between agriculture use and urban/industrial expansion (Wibberley 1959; Best 1968; Snyder 1966). Since 1970s, with increasing concerns on global climate change and the realization of terrestrial ecosystems as important sources and sinks of carbon (Sagan et al. 1979; Woodwell et al. 1983; Stone et al. 1983; Houghton et al. 1985; Woodwell et al. 1986), LUCC studies began to focus on changes of forest and woodland—the major formation of terrestrial ecosystems. Since early 1990s, theoretical and methodological guidelines of calculating carbon emission from change of forest sinks are provided by international conventions to monitor the impact of human activities on global climate change. Land Use, Land-Use Change and Forestry (LULUCF) sectors within the United Nations Framework Convention on Climate Change (UNFCCC), and further stipulated under the Kyoto Protocol in 1997 (UNFCCC 2014b), regulate the definition and accounting for carbon from certain activities (UNFCCC 2001).

With global forest lost at an alarming rate of 13 million hectares per year from 1990 to 2005 (FAO 2005), deforestation and degradation have become an important topic. In 2005, a special agenda focus on reducing carbon emissions from deforestation and degradation in developing countries was introduced into the Conference of the Parties (COP) agenda, entitled Reducing Emissions from Deforestation and Forest Degradation (REDD) (UNFCCC 2006), which was then adopted in the Warsaw Framework in 2013, as reducing emissions from deforestation and forest degradation and the role of conservation, sustainable management of forests, and enhancement of forest carbon stocks in developing countries (REDD+) (UNFCCC 2014a). Meanwhile, world deforestation still continues at an alarmingly high rate according to Global Forest Resources Assessment (FRA) in 2010 (FAO 2010b). Thus, there is urgent need for accurate forest change maps to improve carbon accounting from deforestation and degradation for international regulations and applications in ecological, social, political, and scientific fields.

6.1.2 Mapping Forests: A Remotely Sensed Approach

It is not surprising that the thriving of LUCC studies is closely linked with the development of remote sensing (RS) and Geographic Information Systems (GIS) technologies. RS and GIS can provide crucial information for LUCC studies, namely data about spatial patterns and temporal dynamics of land use and land cover change, mainly in terms of maps. RS is a general term describing the process of obtaining information on a certain target or area from a distance, typically from an airborne or satellite platform (Campbell 2002). The information that can be gleaned from remotely sensed data is constrained by the type of both target object and the instrument or sensor employed. Optical, synthetic aperture RADAR (SAR) and Light Detection and Ranging (LIDAR) are the most commonly used sensors in detecting forest change.

Applications of optical remote sensing imagery in classifying land cover types date back to the early 1970s (Anderson 1976), with Landsat imagery widely used in forest cover change detection in Global Forest Resources Assessment since 1990. The Landsat satellite series provides the scientific community with the world's longest continuously acquired imagery at a high resolution (30 m), 16-day repeat cycle, and global coverage since 1992 (USGU 2014). Other frequently used optical imagery for earth observation includes The Moderate Resolution Imaging Spectroradiometer (MODIS), Satellite Pour l'Observation de la Terre (SPOT), RapidEye, etc. This 'family' is expanding steadily with newly launched Landsat 8 and scheduled Sentinel 2. Based on the Landsat dataset, the first global forest change map from 2000 to 2012 at a 30 m resolution has been published last year (Hansen et al. 2013) (Fig. 6.1a).

Radar sensors, on the other hand, observe forests with a totally different algorithm. Rather than 'seeing' forest canopy, microwave can penetrate through clouds and forest canopy and interact with trunks, brunches, and undergrowth, depending on the wavelength in use. This enables radar backscatter images that closely represent woody biomass and forest structure (Woodhouse et al. 2012). Long wave length, such as L-band (Olander et al. 2008; Mitchard et al. 2011; Stedham 2012) and P-band (ESA 2012), has proved to be most effective in mapping forest extent, especially in tropical regions where constant cloud cover may constrain the use of optical imagery. Available L-band SAR imagery includes Japanese Earth Resources Satellite 1 (JERS-1), Phased Array type L-band Synthetic Aperture Radar on Advanced Land Observing Satellite (ALOS PALSAR), and the newly launched ALOS-2. In 2014, Japan Aerospace Exploration (JAXA) published the first 25 m/50 m global forest/non-forest map (2007–2010) based on the L-band ALOS PALSAR dataset (JAXA 2014) (Fig. 6.1b). With L-band Satellites for Observation and Communications (SAOCOM) and BIOMASS mission scheduled to launch in the coming 5 years, abundance of data specified to detect different aspects of forests will be increased greatly.

LiDAR was introduced to forest ecosystem studies in early 2000 (Lefsky et al. 2002). Because of its ability to measure the three-dimensional distribution of plant

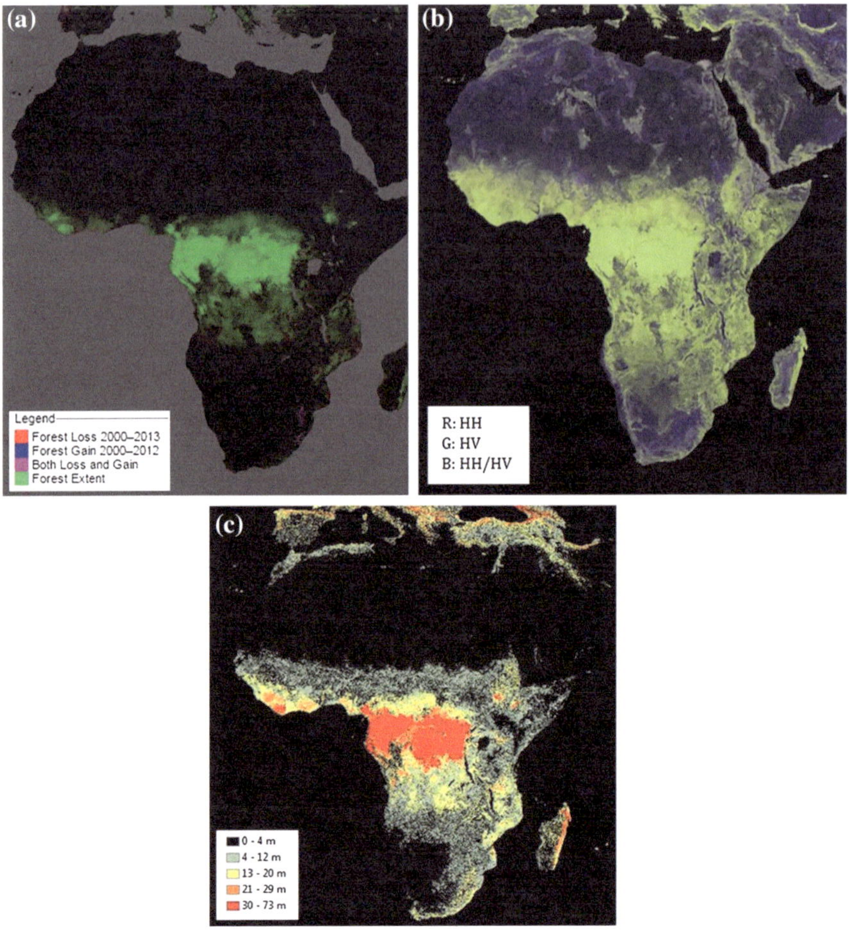

Fig. 6.1 Map of forest cover in Africa, cropped from the global forest map generated from different sensors. **a** Forest cover change map from 2000 to 2013 based on optical imagery, *data source* Hansen et al. (2013); **b** forest cover map generated from L-band radar, *data source and explanation of HH/HV nomenclature* JAXA (2014); and **c** forest height map generated from Lidar and optical, *data source* Simard et al. (2011)

canopies, LiDAR images are widely used to accurately estimate vegetation structural attributes, such as tree height, tree density, leaf area index (LAI), and above-ground biomass (AGB) (Sun et al. 2011; Huang et al. 2013; Mitchard et al. 2012) (Fig. 6.1c). Constrained by a smaller footprint and relatively higher cost, airborne LiDAR is usually applied at a regional scale ranging from one to several thousand hectares. However, multisource remote sensing data fusion technology made the first Global Forest Height map possible in 2010. This dataset is produced by combining LiDAR (ICESat) with optical (MODIS) imagery with 1 km resolution (Simard et al. 2011). Also, the first spaceborne LiDAR, Global Ecosystem

Dynamics Investigation (GEDI), will be launched by the National Aeronautics and Space Administration (NASA) in 2018 monitoring tree height and canopy density at a global scale (NASA 2014).

6.1.3 Linking Deforestation and Degradation to Its Drivers: The Role of GIS

While RS provides essential information of forest cover and forest biomass at various scales, GIS serves as an important tool for exploring their spatial patterns and change dynamics. 'Spatial pattern' refers to the geographic variation across space, which describes the perceptual structure, placement, or distribution of forest or forest change (Burrough and McDonnell 1998), such as the number, shape, clustering, or line arrangement of deforestation and degradation patches. These patterns, which capture combined effects of complex LUCC drivers, can be further manipulated with descriptive spatial analysis models, such as hotspot analysis or land use change matrices (which describes the proportion of forest taken up by other land cover and land use types).

Another primary function of GIS is to determine the spatial relationships between features that occupy the same location (Burrough and McDonnell 1998). This can be achieved by overlaying ancillary maps (census data, soil maps, vegetation map, terrain maps, etc.) with land use or land cover maps derived from RS, such as to explore the relationships between forest change and environmental, ecological, economic, and social variables. Based on these spatial relationships, various predictive spatially explicit models have been built to explain past patterns of LUCC or predict future trends. These models can be location-based statistical and cellular models, which emphasize neighborhood effects (Committee on Needs and Research Requirements for Land Change Modeling 2013), or agent-based with a focus mainly on processes.

6.1.4 Who Else Has Participated in Changing the Forests?

Spatial patterns of deforestation and forest degradation, which are captured by GIS and RS, are the direct and indirect consequences of ecological, social, and economic drivers. These drivers have shifted local land use for centuries through different processes. Over the past decades, competition for forest resources has been intensified predominantly by changing population dynamics, agricultural expansion, changing lifestyles, and new distributional dynamics of globalization. Geist and Lambin (2002) proposed to categorize these drivers into proximate causes and underlying drivers. Recently, a process-based conceptual framework has been proposed to bring the focus of LUCC studies in urban areas away from geographical colocated phenomena and toward a better understanding of distal drivers, which are linked to local events through a complex set of dynamic processes (Seto et al. 2012, and Chaps. 1 and 2 for more detail).

Distal drivers, also defined to address the increasing importance of distal connections and flows of driving current land use change, increasingly shape local land use disputes as global markets (Rueda and Lambin 2013) and flows of goods (Garrett et al. 2013), information, and knowledge interact with local constellations. Distal drivers need to receive more attention, because neither population increases nor resource scarcity alone constitute the sole and major underlying causes of land cover change (Lambin et al. 2001). In this chapter, I explore if looking through a lens of remote sensing and GIS will allow me to zoom in on a specific local situation of land use competition, but extend the scope of the analysis to discuss distal drivers of that competition.

6.2 Capture Distal Drivers in the Beira Corridor, Manica, Mozambique

6.2.1 Land Cover and Land Use History in the Beira Corridor

The Beira corridor is historically an important road and railway network, which links Beira—the second largest city in Mozambique and one of the most important ports of southern Africa—with Zambia, Zimbabwe, and Malawi (Fig. 6.2). This area, which covers the districts of Manica and Sofala, suffered greatly from the civil war between 1977 and 1992 and is still largely underdeveloped. Since 1992, this area is undergoing rapid land use change, accompanied by large population movement mainly as resettlement in previously abandoned rural areas (Stedham 2012). For decades, small-scale agriculture, notably for maize production, has been the dominant form of income for rural people and the main driver of deforestation and degradation of Miombo woodlands (Bradley and Dewees 1993; Cavendish 2000). Recently, charcoal making has become an additional main driver, particularly in Mozambique (FAO 2010c). With economic and population growth, Mimobo woodland is disappearing at a rate of 219,000 ha (3.1 %) per year (FAO 2010c). This trend is also captured in the Beira corridor by the Global Forest Change map from 2000 to 2012.

Although the deforestation and degradation rate of Mozambique is not as astonishing as Indonesia or Brazil, Manica has been identified as a pilot area of REDD+. This is because the original biomass stock of Miombo woodland is relatively low (characterized by an open tree canopy and a continuous grass layer) (Campbell 1996; Mayaux et al. 2004), and small biomass loss will make a large difference for the ecosystem services that Miombo woodland can provide, such as carbon sequestration and biodiversity preservation (Shackleton et al. 2002; Bourguignon 2006). Moreover, local people's livelihood heavily depends on forest

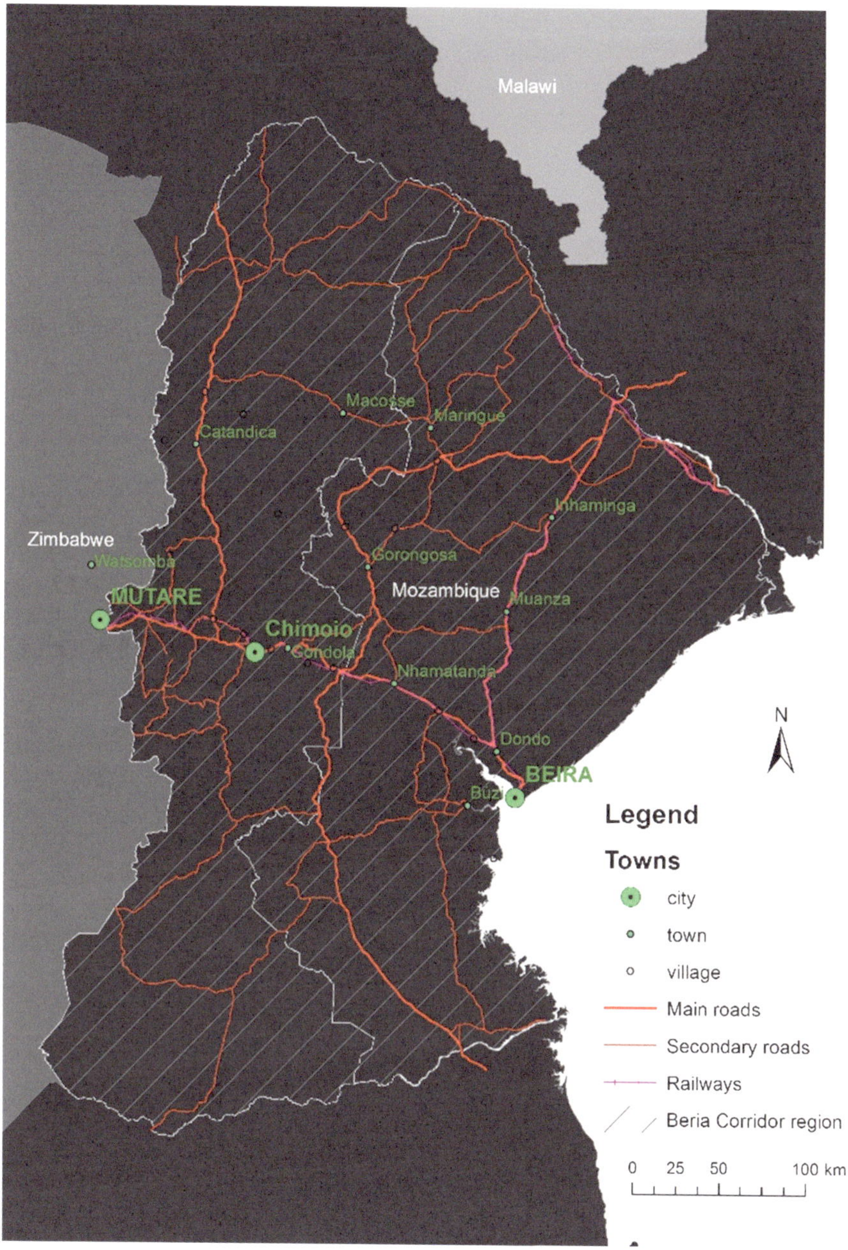

Fig. 6.2 Location of the Beira corridor

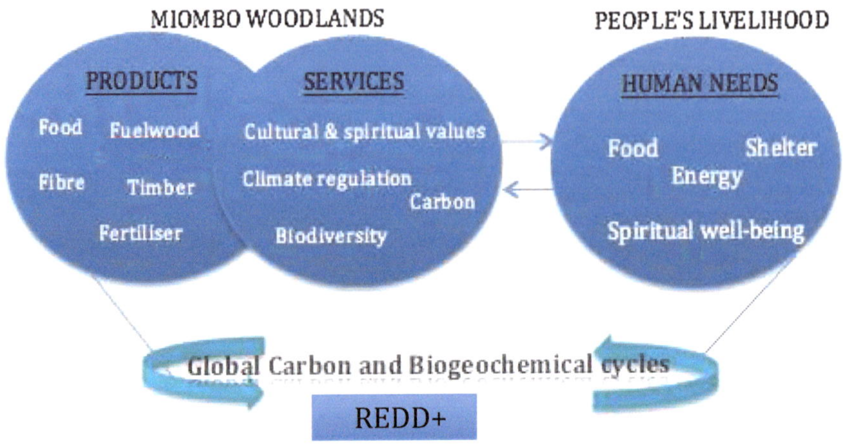

Fig. 6.3 Actors related to deforestation and degradation of Miombo woodland in the Beira corridor

produce, such as food and fuelwood (FAO 2010a). 75 million people inhabit miombo woodland regions, with an additional 25 million urban dwellers relying on miombo wood or charcoal as a source of energy (Dewees et al. 2010).

To alleviate poverty, the government of Mozambique launched the Beira Agricultural Growth Corridor initiatives in 2010 (BAGC 2014), aiming at structural change among the smallholder farmers by boosting agricultural productivity and promoting commercial agriculture, which currently consists of less than 0.3 % of arable land in the Beira corridor region. These initiatives stimulated cooperation between government, the private sector, local farmers, and the international community in land use decision-making processes, increasing the competition primarily between woodland conservation interests and other land uses. Actors connected with Miombo woodland are shown in Fig. 6.3.

6.2.2 Can Spatial Patterns Reveal Hidden Distal Drivers?

6.2.2.1 Patterns of Deforestation and Degradation in Beira Corridor Region

Human activities may alter the biophysical land surface with certain characteristics in spatial patterns in terms of shape, size, intensity, and proximity to other features such as roads or houses. For example, in central Mozambique, clearance for small-scale agriculture (Fig. 6.4b) in Miombo woodland (Fig. 6.4a) usually results in high-intensity above-ground woody biomass loss with plot sizes between ~1 and 2 ha. Clearance for commercial farming, with similar high intensity of biomass loss, usually results in plots over 4 ha (Ryan et al. 2014).

Fig. 6.4 Typical types of human disturbance to Miombo woodland. Photo source: Iain McNicol

Patterns of charcoal production, however, are more complex as our interviews in the area indicate. It can involve selective removal of certain tree species (such as *Brachystegia spiciformis*) or medium size stems (Fig. 6.4c) from an area of ~0.2 ha surrounding kilns (Ryan et al. 2014). Yet, it can also involve the clearance of a patch of woodland. In this case, patterns of charcoal making are featured with ~2 ha intensive biomass loss in the center, with low biomass loss in the surrounding areas. Interviews also suggested a tight relationship between charcoal making and clearance for farming. The majority of the kilns are built temporarily at the first stage of setting up a new farm, when woodland is cleared and trees are burnt and sold as charcoal. In this case, charcoal making is more like a by-product of farming, such that the income can be used for buying necessities for agriculture. At the same time, however, such charcoal-making activities may degrade the surrounding woodland, which may stimulate the transition into framing even further.

Both optical and radar remote sensing imagery are capable of detecting change in Miombo woodland, with advantages and limitations to both data types. Hansen et al. (2013) forest change maps in the study area from 2007 to 2010 (based on optical Landsat imagery, 30 m resolution, Fig. 6.5a) and radar ALOS PALSAR biomass change maps at 25 m resolution (Fig. 6.5b) show that both of these two

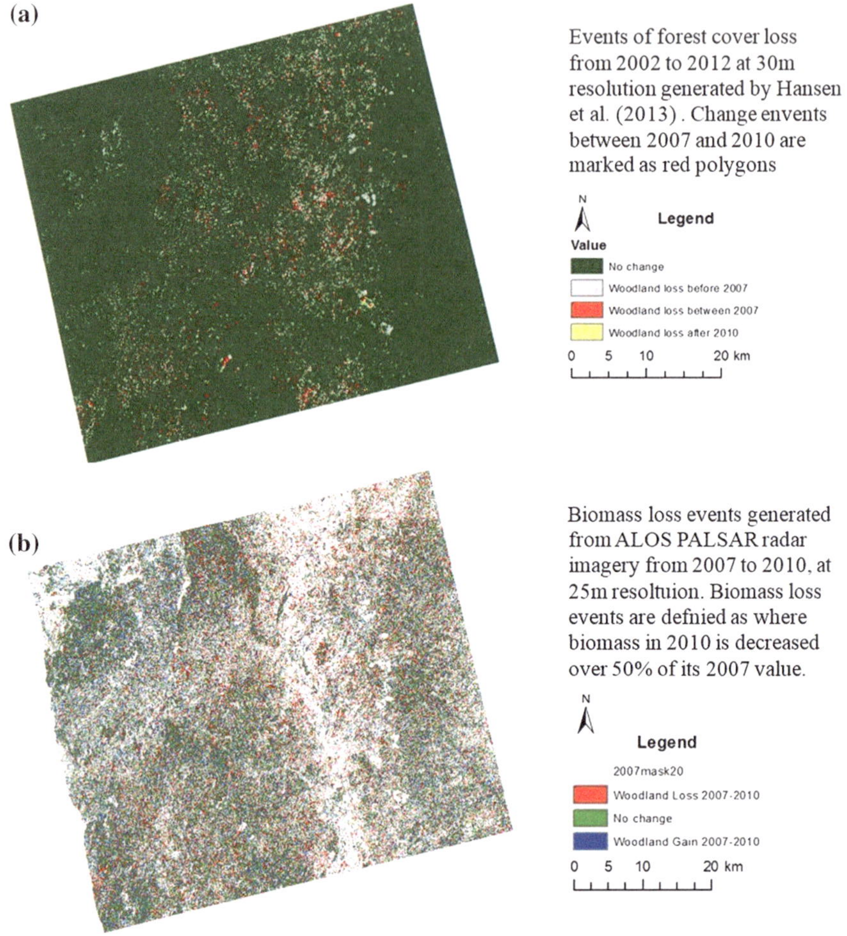

Fig. 6.5 Woodland change maps from optical and radar sensor for a subsection of the Beira corridor in Central Mozambique

sensors capture the high-intensity biomass loss (deforestation) from clearance for agriculture or charcoal. These woodland disturbance activities are mostly carried out at a small scale with plot size for all detected forest loss events less than 1 ha (Table 6.1). From the data, obvious signs of commercial farmland cannot be discerned with less than 1 % detected forest loss events exceeding 4 ha. This result is supported by the Beira agricultural initiative report 2013 (BAGC 2014), as well as field interviews in 2013. Large-scale commercial agriculture is still rare in this area.

Statistcs (Table 6.1) show that ALOS PALSAR radar imagery captures six times as many change events as Hansen optical imagery, especially for woodland loss at a very small plot size (less than 0.04 ha). This result is likely due to the fact that optical sensors provide information at the top-of-canopy level, while radar waves

Table 6.1 Statistic and spatial patterns of woodland loss events captured by optical and radar sensor

	Number of events	Size of forest loss events (ha)				Distance to roads	Distance to households
		Max	Medium	Min	Average		
Hansen map	81,901	9.98	0.09	0.05	0.39	2037 m	259 m
ALOS PALSAR	586,877	9.17	0.04	0.04	0.09	2741 m	315 m

can penetrate woodland canopy and detect subtle changes. Yet, the discrepancy may also result from differences in method, e.g., different definitions of forest and deforestation classifications, use of different minimum mapping units, or different levels of uncertainties. To exclude the effects from different forest definitions, we masked out change events from ALOS PALSAR imagery that are located in areas with a biomass value of less than 20 tC/ha to match the forest definition used by the Hansen map. The obvious difference persists.

6.2.2.2 Effects of Maize and Charcoal Markets

Woody biomass loss is related to small-scale agriculture and charcoal making, which are identified as main reasons for woodland deforestation and degradation in Mozambique in various international reports and studies. To explore this relationship, annual biomass loss from 2007 to 2010 has been plotted against the change of charcoal price at the local market and the change of maize price at both local and global markets. Constrained by the availability of reliable data on charcoal price, an estimation of 37 % annual increase is used (EUEI 2012; Herd 2007). Results are inconclusive. Changes in price of charcoal and maize do not readily correlate with the trend of biomass loss captured by radar.[1] Limited by the frequency, period, and accuracy of data available for the Beira Corridor region, it is difficult to conclude about dynamics from these data. This demonstrates, first and foremost, that small-scale agricultural activity and local markets operate according to dynamics that are not easily captured with data that are spatially detailed but suffer from too coarse a temporal and social or institutional resolution (see also Chap. 7). With more frequent RS imagery covering a longer time span, it might be possible to capture the connection between markets and woodland deforestation and degradation, or the interconnections between charcoal making and opening a new farm (farmland arising as a collateral land cover from charcoal making, or charcoal making as a by-product of clearance for farmland) (Fig. 6.6).

[1]Optical imagery showed a different trend than radar imagery, which might due to the characteristics of the two sensors, differences in forest and forest loss events, time of year when the image is captured, and various other factors.

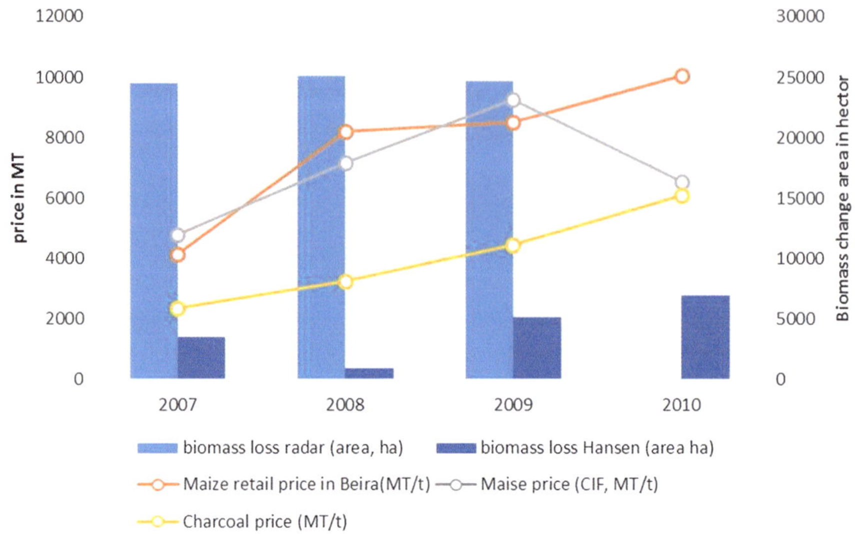

Fig. 6.6 Woodland loss detected from RS imagery plotted with change in price of maize and charcoal. Retail price of maize in Beira is taken from FAO (2013); price in charcoal is estimated from Herd (2007) and EUEI (2012). The trend of biomass loss, rather than area itself is comparable between change maps detrived from optical and radar imagery, due to the characteristics of the sensor and different detecting algorithm

6.2.3 RS and GIS: An Unbiased Microscope or Distorting Mirror?

6.2.3.1 Linking Patterns to Processes

Recently, developments in spatial modeling, remote sensing, and computing performance have provided unprecedented amounts of data to capture, monitor, and model the spatial and temporal patterns of LUCC. This makes it more feasible to relate discernible patterns of land cover to the processes of land use and their drivers. For example, deep time series of optical imagery has been proved to be effective to reveal ecological processes (Kennedy et al. 2014). Woodland deforestation and degradation-related processes, meanwhile, are vastly complicated by the influences of global markets, exchange of knowledge and information, and international cooperation between actors in structures of governance.

The framework of distal drivers expands the focus of LUCC studies from a piece of land and its geographically neighboring areas to include new dimensions, both spatially and temporally. The process of telecoupling (Friis et al. 2015) alerts to land use drivers that cross spatial scales and urge analysts to take into account land use competition at various levels of institutional hierarchy and diverse spatial and temporal scales (see also Chap. 2). The action of distal drives on local contexts is commonly processed by selected local mediators (see Chap. 4). Distal drivers may create unique LUCC patterns as they amplify or attenuate local institutional factors, such as the distinctive spatial patterns shown for distally driven commercial farming and significantly more local, traditional small-scale agriculture for the Beira corridor. In other cases, distal driver may introduce a process altogether different from existing local factors, such as REDD+. In a less developed region as the Beira corridor region, the primary concern is to foster economic growth and market integration, alleviate poverty, and improve people's livelihood, which usually heavily relies on resources that woodland provides (including land). These needs conflict with the aims of REDD+ with its focus on decelerating the deforestation and degradation rate in developing countries. Understanding and managing REDD+'s 'distal' needs and mediate cooperation with local actors by means of compensation, conserving, and sustainable management is still a major challenge. Depending on its implementation, REDD+, as a distal driver, may affect land use decision-making process at government level, improve agricultural efficiency, and exert a subtle awareness of forest preservation for local people.

6.2.3.2 Uncertainties of Deforestation and Degradation Estimation

Information captured in a remotely sensed image is not only constrained by the characteristic of the sensor itself (type and resolution), but also largely affected by the bias and uncertainties imposed during image processing and modeling processes. Concern has been raised over the accuracy of baseline maps used in REDD+ content for years (IPCC 2006), and recently research from Mitchard et al. (2013) showed that large uncertainties resulted from the comparison of two global biomass maps, especially when integrated only with few field data. Figure 6.7 summaries how backscatter signal from ALOS PALSAR imagery is interpreted to woodland deforestation and degradation maps. Because radar backscatter is a direct measurement neither of woodland nor of woody biomass, ground plots are needed to build empirical (or semiempirical) relationships between woody biomass and radar backscatter. Three types of uncertainties may be introduced in the woodland deforestation and degradation maps produced in this process: plot-level uncertainty, landscape-level uncertainty, and interpretation uncertainty. Plot-level uncertainty includes errors from ground measurement and allometric models. Plot-level uncertainty together with errors from remote sensing images, and the errors from unrepresentativeness of ground plots compared to the heterogeneity of actual landscape comprise landscape-level uncertainty. Landscape-level uncertainty is perhaps most closely associated with biomass maps produced from radar backscatter. To further interpret biomass to woodland, woodland deforestation, and degradation, or other thematic maps, classification thresholds for biomass are needed to define these terms. Here, interpretation uncertainty is introduced through the potential discrepancies between general thresholds applicable in all contexts and local thresholds defined for a specific local system. Integrating profound knowledge of the local system will help to reduce interpretation uncertainty.

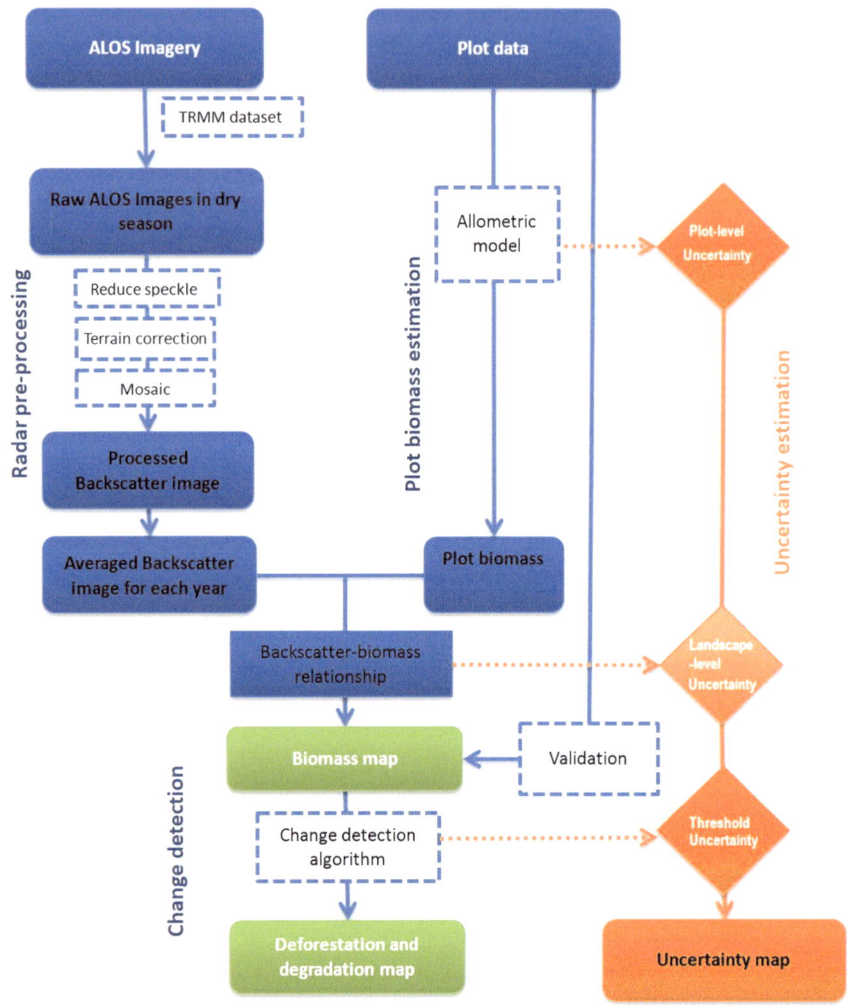

Fig. 6.7 Processing chain of building deforestation and degradation maps from ALOS PALSAR radar imagery, with uncertainty estimation

6.2.3.3 Can a Map Itself Become a Distal Driver?

As remote sensing-based global forest maps are largely available at low cost, forest maps have been widely used in ecologic, social, and political fields, where the map users are generally not involved in the process of map building. The process of producing remote sensing-based maps is not a neutral intermediary but a mediator that shapes the representation of land surfaces in particular, systematically biased ways (cf. Latour 2005). Problems may arise when such mediators are treated as if they were intermediaries, i.e., free of biases. It is therefore important for map users to be (made) aware of the enclosed potential biases so that decisions are made in full awareness of the choices, which needed to be made to produce the map and which may not be visible or legible in the final map.

Firstly, despite the limitation of sensors and effects of noise, remote sensing images provide a mediated but nevertheless 'direct' visual representation of the world, compared to land use data from, e.g., interviews, where much longer chains of translation, perception, and interpretation are involved in connecting land with map. RS images do not contain biases toward any particular subset of land use processes. Instead, they deliver a 'balanced' result in as far as they show the actual outcome of negotiations between various LUCC processes. Interpreting these RS images to a thematic or land use classification map, however, is always a process of selection such that only some of the available information is focused and extracted. Secondly, uncertainty of LULC classification should be considered in any application of the maps. Because maps, especially global scale maps, are normally based on understanding from a limited number of areas. A universal definition of categories across the entire study region may not be suitable for one particular area. A good illustration of this effect is the previous discussion on how different forest definitions used in Hansen's deforestation map and a radar map result in very different areas being represented in the map as deforestation and degradation patches (Fig. 6.6). It is worth mentioning in this context that over 1000 forest definitions are currently in use in various regions and study fields (Lund 2012). Thirdly, it appears that the vulnerability of analyses to bias increases with the various kinds of 'distalness' (Chap. 2) and thus the complexity of LUCC drivers. A simple example: In the Beira corridor, the tight and complex relationship between charcoal making and clearing forest for a new farm makes classifying land use into any of these two categories questionable; particularly when these attributions are treated as mutually exclusive from then on. Fourthly, the focus and thus the labeling of a map depend to a significant degree on the interests and perceptions of the mapmaker. Any particular land use at a certain location (pixel) always has physical, economic, and political dimensions that can be made more or less visible not least through the map labels. In the case presented here, the forest degradation map could also be labeled as a charcoal burning events map or smallholder farmers' livelihood map, depending on the perspective of actors and agencies. What appears as environmental problems in one map shows as local economic performance in another based on the same set of data. Labeling processes embed one particular land use interest in land use maps, which may obstruct other understandings of land use competition in that particular location. This is not to say that maps are simply social constructions in any trivial sense of the term. It is merely pointing out the looping effects between maps, the world they try to represent and the map users developing interventions into the world based on an understanding of land use competition dynamics that is influenced by maps among other information. In this sense, the maps available for any particular area will have an impact on the understanding of the dynamics of land use competition in that area. In many cases, they are not a major driver such as markets for land-based products and they certainly should not be understood in any deterministic sense. Yet, they introduce expertise into political decision-making processes and as such mediate land use competition.

6.3 Conclusion

Remote sensing provides us with the unprecedented ability to sense various aspects of land surface change at global scale and high temporal resolution. While the advantages of remote sensing have been widely recognized in land use competition studies, revealing the underlying processes that shape patterns of change is still challenging. This is exacerbated as geographically, socially, and institutionally distal factors play an increasingly important role in shaping local landscapes. These factors, such as global markets, international organizations, and NGOs, alter local land use through mediating and shaping flows of capital, material, and information, often across spatial and temporal scales. This creates complex LUCC patterns on the land surface. To capture them, 'sensing' needs to operate on different spatial and temporal scales; i.e., a profound understanding of local LUCC processes needs to be integrated with various ancillary datasets including GIS, statistics, and time series information.

What can we 'see' from RS and GIS?	What can we not 'see'?	Aperture of what we can observe
(1) Spatial patterns of LUCC from local to global scale, showing a 'balanced' outcome from various LUCC processes. (2) Time series analysis may reveal distal driver that creates certain temporal and spatial patterns. (3) Data fusing from various RS sources and spatial analysis combing ancillary data may help reveal patterns of LUCC	(1) Processes of LUCC, such as information flow, decision-making process, capital flow, etc (2) Agencies and drivers participating LUCC processes. (3) Pathways that distal driver mediated with local factors (4) RS provides land cover information rather than land use. Detected land cover information is also constrained by the characteristics of sensor in use.	(1) Land use classification categories can be selective, which only reflects part of the information captured by RS in that pixel, or the most important information that interests the map builder (2) Map with the same label might present land cover information with varies degree of bias, due to inconsistent definition used among different map makers to simplify complex processes for a certain case. (3) The focus and thus the labeling of a map depend to a significant degree on the interests and perceptions of the mapmaker.

References

Anderson, J. R. (1976). *A land use and land cover classification system for use with remote sensor data*, Vol. 964. US Government Printing Office.

BAGC. (2014). *Beira agricultural growth corridor.*

Best, R. H. (1968). Competition for land between rural and urban uses. *Land use and resources: Studies in applied geography. A memorial study to Sir Dudley Stamp*, 89–100.

Bourguignon, H. (2006). Enhancing the role of forests in the socio-economic development of forested African countries. *International Journal of Forestry Review, 8*(1), 126–129. doi:10.1505/ifor.8.1.126.

Bradley, P., & Dewees, P. (1993). Indigenous woodlands, agricultural production and household economy in the communal areas. *World Bank Technical Paper*, 63–63.

Burrough, P. A., & McDonnell, R. A. (1998). *Principles of GIS*. London: Oxford University Press.

Campbell, B. M. (1996). *The Miombo in transition: Woodlands and welfare in Africa*. Cifor.

Campbell, J. B. (2002). *Introduction to remote sensing*. Boca Raton: CRC Press.

Cavendish, W. (2000). Empirical regularities in the poverty-environment relationship of rural households: Evidence from Zimbabwe. *World Development, 28*(11), 1979–2003.

Committee on Needs and Research Requirements for Land Change Modeling GSC. (2013). Board on Earth Sciences and Resources, Division on Earth and Life Studies. *Advancing land change modeling: Opportunities and research requirements*.

Dewees, P. A., Campbell, B. M., Katerere, Y., Sitoe, A., Cunningham, A. B., Angelsen, A., & Wunder, S. (2010). Managing the Miombo woodlands of southern Africa: policies, incentives and options for the rural poor. *Journal of Natural Resources Policy Research, 2*(1), 57–73.

ESA. (2012). *BIOMASS, an Earth Explorer to observe forest biomass*. Report for Mission Selection: Biomass.

EUEI. (2012). *Mozambique Biomass Energy Strategy*.

FAO. (2005). *Global forest resources assessment 2005*.

FAO. (2010a). *Global forest land-use change 1990–2005*. Global Forest Resources Assessment.

FAO. (2010b). *Global forest resources assessment 2010*.

FAO. (2010c). *Global forest resources assessment 2010*. Country Report Mozambique.

FAO. (2013). *Analysis of incentives and disincentives for maize in Mozambique*.

Friis, C., Nielsen, J. Ø., Otero, I., Haberl, H., Niewöhner, J., & Hostert, P. (2015). From teleconnection to telecoupling: Taking stock of an emerging framework in land system science. *Journal of Land Use Science*, 1–23. doi:10.1080/1747423X.2015.1096423

Garrett, R. D., Lambin, E. F., & Naylor, R. L. (2013). Land institutions and supply chain configurations as determinants of soybean planted area and yields in Brazil. *Land Use Policy, 31*, 385–396. doi:10.1016/j.landusepol.2012.08.002.

Geist, H. J., & Lambin, E. F. (2002). Proximate causes and underlying driving forces of tropical deforestation. *BioScience, 52*(2), 143–150. doi:10.1641/0006-3568(2002)052[0143:Pcaudf]2.0.Co;2.

Hansen, M. C., Potapov, P. V., Moore, R., Hancher, M., Turubanova, S. A., Tyukavina, A., et al. (2013). High-resolution global maps of 21st-century forest cover change. *Science, 342*(6160), 850–853. doi:10.1126/science.1244693.

Herd, A. R. C. (2007). *Exploring the socio-economic role of charcoal and the potential for sustainable production in the Chicale Regulado*. Mozambique: University of Edinburgh.

Houghton, R. A., Boone, R. D., Melillo, J. M., Palm, C. A., Woodwell, G. M., Myers, N., et al. (1985). Net flux of carbon-dioxide from tropical forests in 1980. *Nature, 316*(6029), 617–620. doi:10.1038/316617a0.

Huang, W. L., Sun, G. Q., Dubayah, R., Cook, B., Montesano, P., Ni, W. J., & Zhang, Z. Y. (2013). Mapping biomass change after forest disturbance: Applying LiDAR footprint-derived models at key map scales. *Remote Sensing of Environment, 134*, 319–332. doi:10.1016/j.rse.2013.03.017.

IPCC. (2006). *2006 IPCC guidelines for national greenhouse gas inventories*.

JAXA. (2014). *New global 25m-resolution PALSAR mosaic and forest/non-forest map* (2007–2010)—version 1. http://www.eorc.jaxa.jp/ALOS/en/palsar_fnf/fnf_index.html

Kennedy, R. E., Andrefouet, S., Cohen, W. B., Gomez, C., Griffiths, P., Hais, M., et al. (2014). Bringing an ecological view of change to landsat-based remote sensing. *Frontiers in Ecology and the Environment, 12*(6), 339–346. doi:10.1890/130066.

Lambin, E. F., Turner, B. L., Geist, H. J., Agbola, S. B., Angelsen, A., Bruce, J. W., et al. (2001). The causes of land-use and land-cover change: moving beyond the myths. *Global environmental change* 11(4): 261–269.

Latour, B. (2005). *Reassembling the social: An introduction to actor-network-theory*. Oxford: Oxford University Press.

Lefsky, M. A., Cohen, W. B., Parker, G. G., & Harding, D. J. (2002). Lidar remote sensing for ecosystem studies. *BioScience, 52*(1), 19–30. doi:10.1641/0006-3568(2002)052[0019:Lrsfes]2.0.Co;2.

Lund, H. G. (2012). *Definitions of forest, deforestation, afforestation, and reforestation.*

Mayaux, P., Bartholomé, E., Fritz, S., and Belward, A. (2004). A new land-cover map of Africa for the year 2000. *Journal of Biogeography,* 31(6): 861–877.

Mitchard, E., Saatchi, S., Lewis, S., Feldpausch, T., Woodhouse, I., Sonké, B., et al. (2011). Measuring biomass changes due to woody encroachment and deforestation/degradation in a forest–savanna boundary region of central Africa using multi-temporal L-band radar backscatter. *Remote Sensing of Environment, 115*(11), 2861–2873.

Mitchard, E. T., Saatchi, S. S., Baccini, A., Asner, G. P., Goetz, S. J., Harris, N., & Brown, S. (2013). Uncertainty in the spatial distribution of tropical forest biomass: A comparison of pan-tropical maps. *Carbon Balance and Management.*

Mitchard, E. T. A., Saatchi, S. S., White, L. J. T., Abernethy, K. A., Jeffery, K. J., Lewis, S. L., et al. (2012). Mapping tropical forest biomass with radar and spaceborne LiDAR in Lope National Park, Gabon: Overcoming problems of high biomass and persistent cloud. *Biogeosciences, 9*(1), 179–191. doi:10.5194/bg-9-179-2012.

NASA. (2014). *Global ecosystem dynamics investigation lidar* (GEDI). http://science.nasa.gov/missions/gedi/. Accessed January 3, 2015.

Olander, L. P., Gibbs, H. K., Steininger, M., Swenson, J. J., & Murray, B. C. (2008). Reference scenarios for deforestation and forest degradation in support of REDD: A review of data and methods. *Environmental Research Letters, 3*(2), 025011.

Rueda, X., & Lambin, E. F. (2013). Linking globalization to local land uses: How eco-consumers and gourmands are changing the Colombian coffee landscapes. *World Development, 41*, 286–301. doi:10.1016/j.worlddev.2012.05.018.

Ryan, C. M., Berry, N. J., & Joshi, N. (2014). Quantifying the causes of deforestation and degradation and creating transparent REDD+ baselines: A method and case study from central Mozambique. *Applied Geography, 53*, 45–54.

Sagan, C., Toon, O. B., & Pollack, J. B. (1979). Anthropogenic albedo changes and the earths climate. *Science, 206*(4425), 1363–1368. doi:10.1126/science.206.4425.1363.

Seto, K. C., Reenberg, A., Boone, C. G., Fragkias, M., Haase, D., Langanke, T., et al. (2012). Urban land teleconnections and sustainability. *Proceedings of the National Academy of Sciences, 109*(20), 7687–7692.

Shackleton, S., Shackleton, C., Netshiluvhi, T., Geach, B., Balance, A., & Fairbanks, D. (2002). Use patterns and value of savanna resources in three rural villages in South Africa. *Economic Botany, 56*(2), 130–146.

Simard, M., Pinto, N., Fisher, J. B., & Baccini, A. (2011). Mapping forest canopy height globally with spaceborne lidar. *Journal of Geophysical Research: Biogeosciences, 116*. doi:Artn G04021. doi:10.1029/2011jg001708

Snyder, J. H. (1966). A new program for agricultural land use stabilization: The California Land Conservation Act of 1965. *Land Economics*, 29–41.

Stedham, B. (2012). Forest carbon stocks in Malawi 2007–10. *An interim report to LTS International*, Vol. 1. University of Edinburgh.

Stone, T. A., Houghton, R. A., Melillo, J. M., & Woodwell, G. M. (1983). Deforestation in the Amazon Basin measured by satellite—A release of CO_2 to the atmosphere. *Biological Bulletin, 165*(2), 511.

Sun, G. Q., Ranson, K. J., Guo, Z., Zhang, Z., Montesano, P., & Kimes, D. (2011). Forest biomass mapping from lidar and radar synergies. *Remote Sensing of Environment, 115*(11), 2906–2916. doi:10.1016/J.Rse.2011.03.021

UNFCCC. (2001). *Land use, land-use change and forestry decision* 11/CP.7.

UNFCCC. (2006). *Reducing emissions from deforestation in developing countries.*

UNFCCC. (2014a). *Key decisions relevant for reducing emissions from deforestation and forest degradation in developing countries (REDD+).*

UNFCCC. (2014b). *Reporting and accounting of LULUCF activities under the Kyoto Protocol.* http://unfccc.int/methods/lulucf/items/4129.php

USGU. (2014). *Landsat project description.* http://landsat.usgs.gov/about_project_descriptions.php

Wibberley, G. P. (1959). *Agriculture and urban growth, a study of the competition for rural land.*

Woodhouse, I. H., Mitchard, E. T. A., Brolly, M., Maniatis, D., & Ryan, C. M. (2012). CORRESPONDENCE: Radar backscatter is not a 'direct measure' of forest biomass. *Nature Climate Change, 2*(8), 556–557.

Woodwell, G. M., Hobbie, J. E., Houghton, R. A., Melillo, J. M., Moore, B., Peterson, B. J., & Shaver, G. R. (1983). Global deforestation—Contribution to atmospheric carbon-dioxide. *Science, 222*(4628), 1081–1086. doi:10.1126/science.222.4628.1081.

Woodwell, G. M., Houghton, R. A., Stone, T. A., & Park, A. B. (1986). Changes in the area of forests in Rondônia, Amazon Basin, measured by satellite imagery. *The changing carbon cycle* (pp. 242–257). Berlin: Springer.

Chapter 7
Nuts About Gold: Competition for Land in Madre de Dios, Peru

Neha Joshi

Abstract Natural forests are being lost at unsustainable rates as frontiers of development advance and global demand for natural resources increases. This chapter examines two challenges to tropical forest conservation—the demand for use of land for multiple and continuous spatially and temporally linked purposes, and the gaps in our knowledge to fully understand and characterize resulting land cover change dynamics. This chapter uses the case of the biodiversity hot spot of Madre de Dios in Peru as an example. Here, land-use competition is largely resultant from mutually exclusive activities of Brazil nut harvesting, which depends on intact primary forest, development through construction of the Interoceanic Highway and legal and illegal mining of gold. The analysis compares areas with authorized land-use allocations against remotely sensed satellite data to identify past land cover change events and areas of potential land-use competition and conflict. Based on the case study, this chapter illustrates the increasing importance of distal and transnational commodity demands as a driver for land-use change. Further, it discusses the advantages and caveats of spatially and temporally explicit quantification of change using remote sensing techniques. For the latter, it focusses on issues of simplicity and system boundary definitions that may overlook relevant connections to distal drivers and forest/non-forest functions. This chapter concludes that teleconnected systems that drive local land-use and land cover change (LULCC) must be accounted for when mapping land disturbance dynamics, and discusses the potential paths that remote sensing analysis may take to understand and address these challenges.

Keywords Natural resources · Biodiversity · Conflict · Remote sensing · Radar · Telecoupling

---

N. Joshi (✉)
Department of GeoSciences and Natural Resource Management,
University of Copenhagen, Copenhagen, Denmark
e-mail: npjo@ign.ku.dk

J. Niewöhner et al. (eds.), *Land Use Competition*,
Human-Environment Interactions, DOI 10.1007/978-3-319-33628-2_7

7.1 Introduction: Land Uses in a Biodiversity Hotspot

The region of Madre de Dios, a "Capital of Biodiversity" (Peruvian law No. 26311) in south-eastern Peru, is nestled in the Amazon basin and hosts remarkably diverse flora and fauna and a large uninterrupted expanse of primary humid tropical rainforests (Myers et al. 2000). Although it is the least densely populated region in the country, it has experienced the largest proportional increase in population in Peru, approximately 63.5 %, since 1993 (INEI 2008). Its strategic location offers a roadway, the Interoceanic Highway (Carretera Interoceánica Sur, IOH) paved since 2006 as means of economic integration and development, connecting Brazil and Bolivia to the Pacific (Southworth et al. 2011). As one of the best-preserved natural forests in Peru and Amazonia, monitoring human-induced deforestation and forest degradation (i.e. the diffuse reduction of forest canopy cover and woody biomass) following development initiatives is hence a well-recognized priority in the region. Madre de Dios is a relevant area for globally negotiated mechanisms aimed at reducing emissions from deforestation and forest degradation, with conservation, sustainable management and enhancement of forest carbon stocks in developing countries (REDD+) (UNFCCC 2007, 2011; Nunes et al. 2012).

To date, land-use and land cover change (LULCC) in Madre de Dios has largely been influenced by a number of government policies incentivising agricultural expansion, and local and global economic boom-and-bust cycles of commodities such as rubber, Brazil nuts, gold and timber (Chavez and Perz 2012). Between 1985 and 1990, regional policy incentives for cattle and pasture expansion were launched (Alvarez and Naughton-Treves 2013). However, the policies were withdrawn after 1990 since associated credit programs were not sustainable following a fiscal crisis in the country and then reintroduced in the 2000s with a focus on production expansion and improved agricultural product quality (Chávez Michaelsen et al. 2013). In parallel, over 20,000 local inhabitants of Madre de Dios are also traditionally employed in the industry of, and dependent on, the sustainable extraction of high-economic value Brazil nuts (Bertholletia excelsa, Lecythidaceae) from the region's intact tropical forests. The harvest of Brazil nut trees for timber is prohibited; however, legalized selective timber harvesting of other species within the forests is commonly used as a complementary income supplement (Cossío-Solano et al. 2011). Further, the rapid expansion of gold mining in the 2000s, linked to the increase in global gold prices and influx of migrant artisanal miners to the region (Swenson et al. 2011), was an important driver of regional deforestation and land cover change (Asner et al. 2013). The importance that gold extraction has played in influencing land cover change in the last decade marks a transition from regional or national-level drivers of deforestation to global and telecoupled systems, influenced by fluctuations and reversibility, driving deforestation in Madre de Dios. Investments in infrastructure, such as Interoceanic Highway, and the rapid expansion of gold mining pose a threat to the sustainability of Brazil nut extraction and selective timber harvesting in the region (Fraser 2013).

The resulting physical land-use dynamics observed in Madre de Dios is inherently complex, involving (i) clearance for sustainable agriculture and pastoral farms, commonly followed by land abandonment and rapid regrowth of secondary forests; (ii) selective removal of timber species from intact forests, which may be followed by canopy cover recovery in 1–2 years (Asner et al. 2006); and (iii) legal and illegal mining for gold, including very small-scaled clandestine operations (Asner et al. 2013). A common method of detecting and monitoring these dynamics involves the use of satellite data for mapping and the use of ground data for training and validating maps. Such techniques are well established and have been used to map deforestation across the tropics (Hansen et al. 2013). In Peru and specifically in Madre de Dios, deforestation data sets include the Asner et al. (2010), Hansen et al. (2013) and Potapov et al. (2014) map products, which identify and spatially demarcate the clearance of land at a specific time period (MINAM 2014). Their representation of land cover change from one type to another (forest to non-forest or non-forest to forest), however, broadly conceives change as a simple and irreversible conversion between two land cover or land-use types (Mertens and Lambin 2000) and may be inadequate to describe the complexity of the processes that underlie land cover changes. For example, successional forest dynamics (Christensen 2014) post-clearance, such as secondary forest regrowth that may be important to biomass accumulation, may be ignored by such representations (Joshi et al. 2015). Particularly relevant to the land-use processes is also the slow and diffused loss of woody biomass from forests arising from forest degradation. A universal definition of forest degradation is absent, and degraded lands may be anything in between "forests" and "bare ground" based on numerous understandings of degradation-related changes to land (Schoene et al. 2007; Sasaki and Putz 2009; Guariguata et al. 2009). Mapping degradation is a difficult task, since it does not involve full forest stand-replacement, as is typical of deforestation (Joshi et al. 2015). Two broad challenges are hence identified in assessing LULCC in Madre de Dios: the complexities and evolution of the drivers of change, which in turn lead to dynamic physical changes in land-use and cover patterns. The objective of this chapter is twofold: (i) to examine LULCC and its local and distal drivers in Madre de Dios and (ii) to present the complexities of land-use dynamics associated with different deforestation purposes, and assess the ability of remote sensing data to detect and characterize these dynamics. We discuss the benefits of accounting for land cover change trajectories (Mertens and Lambin 2000), which are better able to represent LULCC as a spatially and temporally dynamic process.

7.2 Observing the Conflict of Nuts, Gold and Other Land Resources

Forest cover changes are an accumulation of continuous events of local deforestation over time, which vary in intensity and are influenced by a variety of factors including geophysical land attributes, access to infrastructure, population growth, and local and global socioeconomic organization and relations (Rosa et al. 2013).

Globally, space-borne remote sensing and geographic information systems (GISs) have helped represent LULCC over the last four decades, providing a valuable database of historic and current anthropogenic impacts on land surfaces. Using the Hansen et al. (2013) estimated annual satellite-based deforestation product across Madre de Dios, it was found that approximately 0.1 % of the total region was disturbed per year (i.e. 8979 ha per year) from 2001 to 2013.

A land-use allocation system has played a key role in both regulating and influencing deforestation and forest degradation for specific land-use activities in Madre de Dios. These land allocations include conservation-prioritized indigenous lands, protected natural areas, Brazil nut and conservation concessions, as well as non-conservation lands including forest, reforestation and mining concessions, covering over 85 % of the region's area (Fig. 7.1). It was found that the rates of recent deforestation (Hansen et al. 2013) differed largely in areas with different land allocations. Rates were higher in land allocated specifically for mining activities, a trend similar to that observed in other studies in the region (Nunes et al. 2012; Vuohelainen et al. 2012), and slightly lower in areas allocated to Brazil nut extraction and other sustainable management allocations (Fig. 7.2). Further, deforestation rates rise sharply after the 2006–2007 period, as also observed in the previous studies (Asner et al. 2013, 2010; Swenson et al. 2011).

The significantly differing rates of deforestation in different areas of land-use allocation are of concern, since several agencies have assigned overlapping and contradicting designations on the same land area, and these have been shown to reduce the efficacy of planned conservation outcomes (Scullion et al. 2014). In 2006, Peru was the largest gold produced in Latin America, with gold mining in

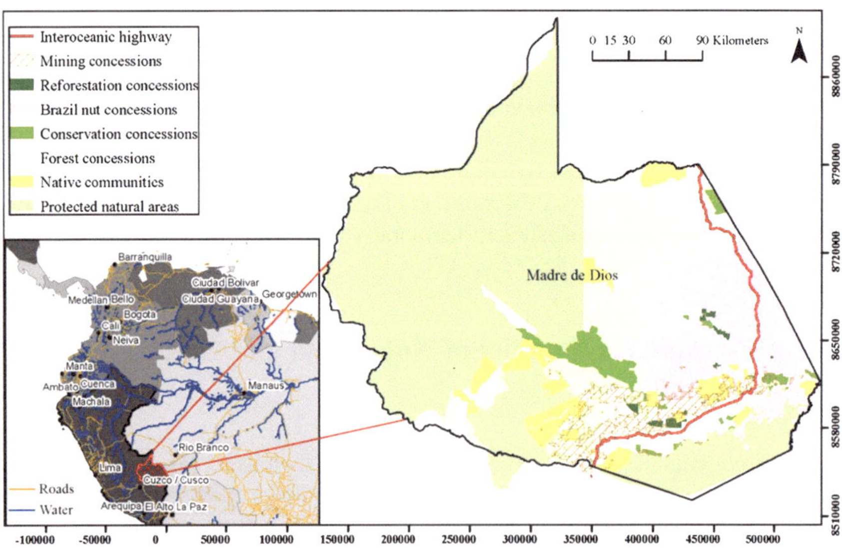

Fig. 7.1 Region of Madre de Dios and government designated landuse concessions in the area

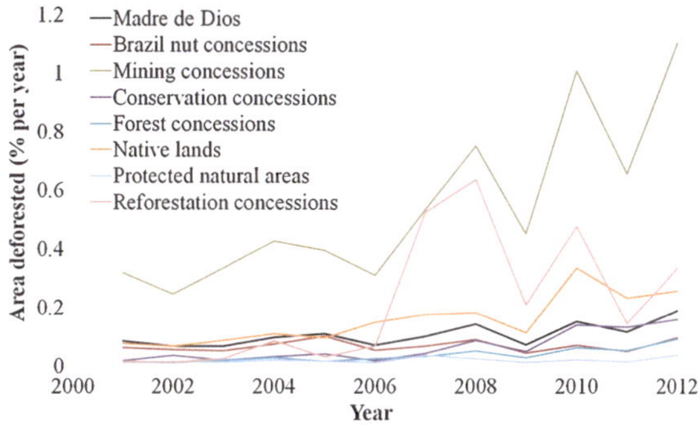

Fig. 7.2 Estimated annual rate of deforestation in Madre de Dios as percentage of total area as provided in the Hansen et al. (2013) study

Madre de Dios accounting for majority of the country's artisanal gold production (Brooks et al. 2007). Thousands of locals are employed in the mining industry, generating revenues that surpass those from other local industries and making the activity a dominant socio-political force in the region. New gold mining expansion inside authorized mining concessions also, however, fell inside conservation areas, Brazil nut concessions and lands for native communities. GIS and remote sensing information from various studies was obtained to visualize areas of potential conflict, as exemplified in Fig. 7.3. In the region shown, activities leading to land cover change due to deforestation are caused primarily from gold mining. The dominant proportion of land cover change events are concentrated within designated mining

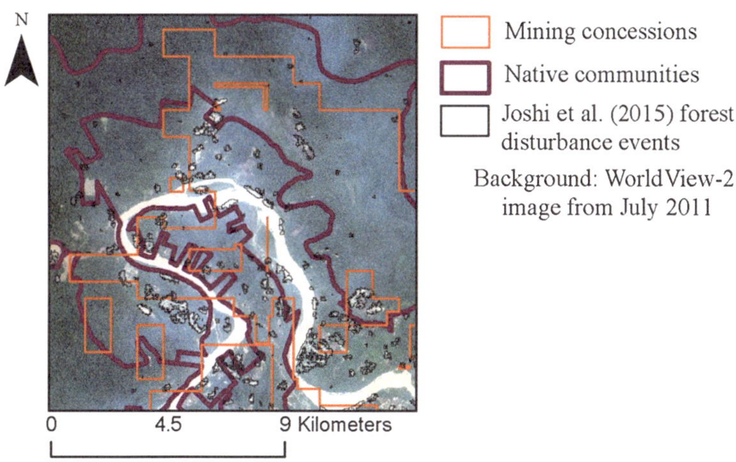

Fig. 7.3 Anthropogenic forest disturbances [obtained from the radar satellite-based disturbance maps of Joshi et al. (2015)] between the years 2007 and 2010 in a region along the Madre de Dios River. These disturbances were primarily from gold mining and encroach upon land designated for native communities, partially due to overlapping and contradicting landuse allocations

areas; however, some overlap lands allocated for native communities. There is also an evident encroachment of activities across the borders of land designations.

Satellite-based observations of the resulting land cover changes, as exemplified here, raise questions as to how land allocations influence land-use dynamics. The poor enforcement of their legal boundaries may lead to land invasion and result in unsustainable resource extraction (Chávez Michaelsen et al. 2013). It has previously been noted that overlapping land-use allocations have fuelled social and political conflict through a dispute between Brazil nut harvesters, who depend on large standing forests, and gold miners whose activities typically involve clearing forest cover along the alluvial banks of the Madre de Dios River (Scullion et al. 2014; Fraser 2013). Broadly, although well-established, satellite-based observations go only so far so as to physically demarcate potential areas of conflict, and often detailed ground knowledge of lands affected is required to back data interpretation. Instead, linking remote sensing data to the activities that are causing deforestation and degradation and their underlying drivers is what may be required to construct a complete picture of regional LULCC and their associated dynamics and evolution. In addition, predicting future scenarios to determine the scope for effectively reducing or regulating LULCC requires such links to be examined and extrapolated (Ryan et al. 2014)—a component of mapping LULCC that is largely understudied to date and requires urgent consideration. A newer and more representative approach to conceptualizing change mapped by satellite data is hence required.

7.3 Local to Distal Drivers of Change and Their Impacts

The drivers of land-use change in Madre de Dios are a combination of local and non-local factors, as well as the domestication of distal drivers that challenge the spatial boundaries of land allocations. For example, according to the Vuohelainen et al. (2012) study, agricultural expansion and illegal logging were identified as the main local drivers of deforestation in protected natural areas, while the proximity of these areas to population centres was the main indirect driver of deforestation. A key finding was that the illegal occupation of land by migrants, who were largely engaged in agriculture- and mining-related activities, was an important driver of land-use change (Vuohelainen et al. 2012). Economic incentives drove high rates of immigration, actuated by the gold rush following the global financial crisis of 2008 (Swenson et al. 2011; Fraser 2009) and the paving of the tri-national Interoceanic Highway since 2006 (Southworth et al. 2011), and subsequently led increasing natural resource extraction in Madre de Dios in the years following. A diagnostic study of the commercial flow of extracted resources, such as gold, indicates that the product is destined for international markets (Mosquera et al. 2009; Swenson et al. 2011). This indicates that the contemporary drivers of change in Madre de Dios largely transcend political and geographic boundaries. They are influenced at both short- and long-time scales by decisions taken within a network of actors and processes that are distributed (unevenly) across the globe and not necessarily fully regulated by global institutions, highlighting the importance of accounting for distal and teleconnected systems.

Further, it is important to account for coupled impacts of some land-use changes on other local natural resource systems. For example, the influx of migrants does not solely impact land resources due to their main purpose of migration (e.g. gold mining), since economically driven migrants also demand land resources for basic needs such as agricultural and pastoral farming. Not accounting for such uses of land would risk underestimating the indirect and local impacts of distal drivers of change.

Another relevant example is the extensive lethal exposure among untrained artisanal miners to mercury, which is commonly used to form an amalgamation with gold during extraction from sediments. Gold mining activities directly release mercury into rivers and sediments, and subsequent bioaccumulation and magnification pose large human health and environmental risks (Fraser 2009; Telmer and Veiga 2009).

The two examples above indicate that both the interests in land resource exploitation and its feedbacks to the system will dictate the dynamics that link the drivers of LULCC to local observations. The disconnect of the sources of demand, production spaces, and the subsequent distal impacts of LULCC manifests as a more complex system than the traditional understanding of the system of land cover changes; that is, what has gone so far will continue at a similar rate and pattern, driven by population, land access, resource demands, etc. A recent study on predictive deforestation modelling in the neighbouring Brazilian Amazon suggests that the emergence of deforestation "can change through time in ways that are not readily anticipated when viewing the phenomenon at the regional scale" (Rosa et al. 2013). As long as the drivers of change are proximate, projecting pathways of change in the future can be reasonably attempted. However, with increasing distalness and disconnects, drivers' volatility increases and local control over their sources and impacts gradually decreases. Local decisions no longer have the potential to reduce volatility; hence, increasing the vulnerability of land resources to shocks or tipping points exemplified well by the rapid gold price changes and subsequently rapid forest lost in Madre de Dios around the years 2007 to 2009 (Asner et al. 2010; Joshi et al. 2015).

7.4 Land Disturbances Conceptualized as Dynamic Processes

The complexity in the drivers of LULCC also requires a change in our way of conceptualizing and mapping the processes of changes to land, based on how the drivers manifest themselves. Traditionally, a land surface can be represented as a set of spatial units that are influenced by anthropogenic activities and hence changed from one unit to another. The classified representation of land in this way allows for concision and clarity (Lambin 1999), and the potential to quantify and report the amount of land cover change through binary classifications of land into "changed" or "unchanged," as may be useful for mechanisms such as REDD+. For example,

REDD + monitoring based on a calculation of activity data × emission factor would benefit from maps with definite binary classifications. However, although practical and simple, landscapes do not necessarily undergo irreversible conversions from one unit to another in reality, especially in the light of the influence of numerous and rapidly evolving drivers of change. Anthropogenic influences on land may vary widely on spatial and temporal scales, giving rise to processes or trajectories of land cover change that are influenced by fluctuations and reversibility over time (Mertens and Lambin 2000). Some studies have used distinguishable categories of land conversions, i.e. "the complete replacement of one cover type by another" or modifications, "i.e. more subtle changes that affect the character of the land cover without changing its overall classification" (Lambin 1999).

An example of binary classifications of land into "changed" or "unchanged" due to anthropogenic disturbances is provided by the global Hansen et al. (2013) study and demonstrated in Fig. 7.4 for the purpose of understanding the impact of using land classifications for forest monitoring. The deforestation product of Hansen et al. (2013) is based on optical remote sensing data from the Landsat satellites (NASA 2014), and it assigns changed land units to a specific year of change occurrence. A comparison of the deforestation units to a higher resolution WorldView-2 satellite image indicates few important features: (i) Assigning discrete spatial boundaries to deforestation seems to disconnect it from what appears to be a more spatially

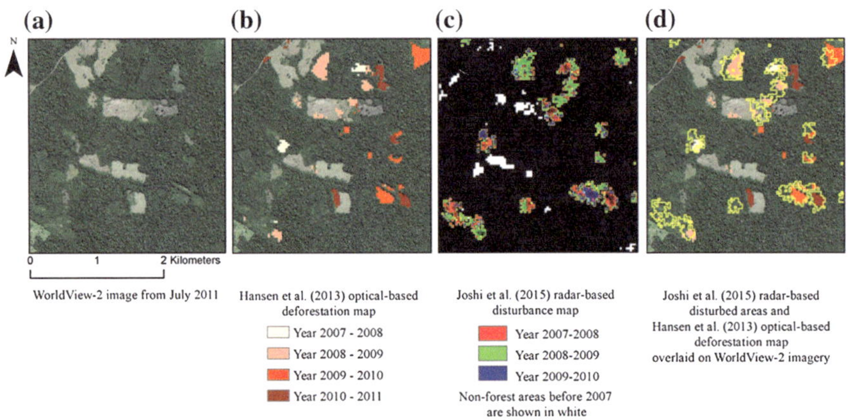

Fig. 7.4 **a** A visualization of the methods of mapping and delineating deforestation in Madre de Dios. A WorldView-2 satellite image is provided as a backdrop to the maps. Deforestation maps from the Hansen et al. (2013) study are provided in **b**, deforestation maps from the Joshi et al. (2015) study are provided in **c**, and both maps are overlaid in **d** for comparison. The values of the image pixels in **b** are the magnitude of radar signal change following human-induced forest disturbance. A colour combination of *red–green–blue* is used for pixels values, and pixels with mixed colour combination hence indicate where repeated disturbances have taken place. Note, only areas that have changed are detected in both products, and hence, the area of non-forest land (undisturbed between 2007 and 2011) in the WorldView-2 satellite images appears greater than the area detected as disturbed (between 2007 and 2011) in the Joshi et al. (2015) and Hansen et al. (2013) study. For detailed comparative statistics between the two products, see Joshi et al. (2015)

continuous gradient of forest disturbance; (ii) the severity of change (e.g. whether it resulted in deforestation or degradation) is not distinguishable through binary classifications (some changed areas appear "greener" on the July 2011 WorldView-2 image than others); (iii) changed areas in each annual time-slice do not overlap, indicating that the classification assumes that change is a temporally explicit occurrence; and (iv) the recovery or successional dynamics (Christensen 2014) post-clearance, which rapidly begin after land is abandoned and might fluctuate through time, is not always captured by binary classifications. It must be noted here that the Hansen et al. (2013) study is based on the fraction of forest cover loss in the study area and hence does provide an indication of the magnitude of loss, rather than only the exemplified binary classification provided in Fig. 7.4. The examples are simply a demonstration of classification approaches. Further, it must also be noted that the Hansen et al. (2013) study provides a separate classification map for forest gain; however, rapid clearance and growth dynamics are not captured (see http://www.globalforestwatch.org/map). The representation of land in terms of continuous variable, such as forest cover, is an alternative to discrete spatial units. It views land as a continuum of biophysical variables (such as vegetation cover and biomass) that are impacted by anthropogenic activities. This permits some clarification in the processes involving land cover change, since, for example, it can potentially distinguish deforestation from small-scale and diffuse forest disturbances or degradation. This distinguishment relies heavily on a high spatial and temporal resolution of remote sensing data, such that rapidly changing processes of forest growth, mortality and anthropogenic disturbance may be repeatedly and accurately tracked.

From the example above, it is evident that assigning a spatially or temporally explicit tag to a land unit broadly risks losing information that is inherent to continuous land cover transformations. This is particularly relevant to processes of forest degradation. A summary review of studies that address forest degradation have referred to the process as "gradients or progression (of degradation) in space and time" leading to "persistent declines of carbon stocks over time" (Guariguata et al. 2009). Further, others have stressed the relevance of contagion of deforestation or degradation such that locations "surrounded by recently deforested land are likely to be more likely to suffer deforestation themselves", showing a deforestation "spread through space" (Rosa et al. 2013). There is a strong indication, therefore, that a shift in the conceptualization of deforestation from a process that can be represented as a binary classification, to one that has a continuum of spatio-temporal progression, is a relevant requirement to land-use change mapping.

An approach to visualizing this conceptualization of land cover change in Madre de Dios is taken in the Joshi et al. (2015) study using radar satellite data from the Advanced Land Observing Satellite (ALOS PALSAR). Radar waves differ from optical-based satellite signals in wavelength. Long-wavelength radar signals (e.g. the 23.6-cm L-band from ALOS PALSAR) interact with the larger elements of vegetation, such as branches and stems, and may hence be sensitive to small-magnitude changes within and below forest canopies (Woodhouse 2005). The advantages of the use of radar include being able to specifically pick up low-magnitude forest degradation (Mitchard et al. 2011, 2013; Ryan et al. 2012,

2014), in addition to deforestation that is commonly captured by optical-based satellite data. Further, measuring the magnitude of radar signal change, such as increases and decreases in radar backscatter (Woodhouse 2005), enables capturing both deforestation and following successional dynamics post-clearance. Radar also has the additional advantage of acquiring cloud- and smoke-independent images, since long wavelengths are unaffected by atmospheric effects.

Using the methodology provided in the Joshi et al. (2015) study and exemplified in Fig. 7.4c, it is observed that (i) there is a continuous progression of deforestation through time (here years 2007–2010), with some areas seeing repeated and overlapping disturbance in more than one time period; (ii) there is a more diffuse spatial gradient of disturbance that disseminates from the centres of deforested areas; and (iii) the spatial boundaries of disturbance detected by radar exceed those detected by traditional optical-based deforestation detection algorithms (Fig. 7.4d). The study reports that the area of forest disturbances detected by radar is over two times that detected in the Hansen et al. (2013) study [for detailed comparative statistics between the two products, see Joshi et al. (2015)]. However, some caution must be taken in this interpretation, since radar picks up small-scale and small-magnitude disturbances that may not be complete forest loss as the exemplified map from the Hansen et al. (2013) study. Further, it must be noted that the study is highly adapted to specific disturbances seen in Madre de Dios, as opposed to the global product provided by the Hansen et al. (2013) study, and hence, comparisons between the two must be made with caution. The lack of a dense series of satellite-based long-wavelength radar images, which cover large spatial scales, currently makes it impossible to produce global deforestation and degradation maps using radar which would be comparable to the results of Hansen et al. (2013). Indeed, both radar and optical products have strengths and deficits, and future research must be focussed on combining the two in efficient ways to minimize data loss when detecting forest disturbances.

The Joshi et al. (2015) results however highlight the need for accounting for a progressive and dynamic form of land cover change. Although the idea of accounting for changes as continuous processes is not new (e.g. Mertens and Lambin 2000), the methods adopted in the Joshi et al. (2015) study, including the use of radar data, provide a more realistic representative conceptualization of changes than numerous previous studies. The representations help us rethink the social–ecological constellation of common changes to land surfaces, and the interventions are required to reduce such changes. For example, direct intervention plans that specifically target the reduction of forest degradation, as is a part of a large number of countries' REDD + readiness actions (Salvini et al. 2014), may now be focussed on lands at different grades of forest loss. Compliance monitoring in lands that are prone to more subtle changes in forest cover, as well as those that experience rapid successive dynamics (secondary forest regrowth) post-disturbance, may be aided by such representations of change. Short-term LULCC events that are lost in coarse temporal resolution and spatial designations may be picked up. More importantly, as the volatility in trajectories of change increase, there is an accompanying increasing need to understand regional socio-ecological responses, such as the short-term impacts of the changes in international gold prices on local-scale land dynamics in Madre de Dios.Depicting LULCC as trajectories of change necessarily demand more detail in LULCC analyses, such as the improved conceptualization of forest disturbances using radar satellite data that Joshi et al. (2015) has attempted to provide

7.5 Concluding Remarks

While the need for a simple understanding of quantified land cover change in the past and present requires little evaluation of the drivers of the changes, the need for more representative maps of LULCC increases as these drivers evolve to complex systems of disconnected and temporally dynamic sources of demand and production spaces. The land dynamics observed in Madre de Dios serve as an example of how distal drivers of change challenge simplistic binary change assumptions (i.e. forest to non-forest or vice versa), as well as the systems' boundary definitions. As satellite-based remote sensing of LULCC become increasingly reliable and established, there is an urgent requirement to focus research towards accounting for land disturbance dynamics, including rapid repeatability and reversibility of changes.

Acknowledgments The author would like to thank DigitalGlobe for providing the high-resolution WorldView-2 image; Natalia Woo and Jorge Torres from Bosques Amazonicos (http://www.bosques-amazonicos.com/) for providing valuable insights into the land-use and land cover changes in Madre de Dios; and Dr. Edward Mitchard (University of Edinburgh) and Prof. Dr. Patrick Hostert (IRI THESys, Humboldt-Universität zu Berlin) for reviewing this chapter.

References

Alvarez, N. L., & Naughton-Treves, L. (2013). Linking national agrarian policy to deforestation in the Peruvian Amazon: A case study of Tambopata, 1986–1997. *Ambio, 32*(4), 269–274.

Asner, G. P., Broadbent, E. N., Oliveira, P. J. C., Keller, M., Knapp, D. E., & Silva, J. N. M. (2006). Condition and fate of logged forests in the Brazilian amazon. *Proceedings of the National Academy of Sciences, 103*(34), 12947–12950.

Asner, G. P., Llactayo, W., Tupayachi, R., & Luna, E. R. (2013). Elevated rates of gold mining in the amazon revealed through high-resolution monitoring. *Proceedings of the National Academy of Sciences, 110*(46), 18454–18459.

Asner, G. P., Powell, G. V. N., Mascaro, J., Knapp, D. E., Clark, J. K., Jacobson, J., et al. (2010). High-resolution forest carbon stocks and emissions in the Amazon. *Proceedings of the National Academy of Sciences, 107*(38), 16738–16742.

Brooks, W. E., Sandoval, E., Yepez, M. A. & Howard, H. (2007). USGS Peru mercury inventory 2006. U.S. *Geological Survey Open File Report* 20071252. URL: http://pubs.usgs.gov/of/2007/1252/ofr2007-1252.pdf

Chavez, A., & Perz, S. (2012). Adoption of policy incentives and land use: Lessons from frontier agriculture in south-eastern Peru. *Human Ecology, 40*(4), 525–539.

Chávez Michaelsen, A., Huamani Briceño, L., Fernandez Menis, R., Bejar Chura, N., Valera Tito, F., Perz, S., et al. (2013). Regional deforestation trends within local realities: Land cover change in south-eastern Peru 1996–2011. *Land, 2*(2), 131–157.

Christensen, N. L, Jr. (2014). An historical perspective on forest succession and its relevance to ecosystem restoration and conservation practice in north America. *Forest Ecology and Management, 330*, 312–322.

Cossío-Solano, R., Guariguata, M., Menton, M., Capella, J., Ríos, L., & Peña, P. (2011). *El aprovechamiento de madera en las concesiones castañeras (bertholletia excelsa) en Madre de Dios, Peru: un análisis de situación normativa*. Technical report.

Fraser, B. (2009). Peruvian gold rush threatens health and the environment. *Environmental Science and Technology, 43*(19), 7162–7164.

Fraser, B. (2013). Conflicting land-use rights threaten Brazil nut harvesters. CIFOR Forests News Blog.

Guariguata, M. R., Nasi, R., & Kanninen, M. (2009). Forest degradation: It is not a matter of new definitions. *Conservation Letters, 2*(6), 286–287.

Hansen, M. C., Potapov, P. V., Moore, R., Hancher, M., Turubanova, S. A., Tyukavina, A., et al. (2013). High-resolution global maps of 21st-century forest cover change. *Science, 342*(6160), 850–853.

INEI. (2008). *Censos Nacionales 2007, XI de Poblacin y VI de Vivienda: Per_l Sociode mográfico del Peru*. Lima: Instituto Nacional de Estadística e Informática.

Joshi, N., Mitchard, E. T., Woo, N., Torres, J., Moll-Rocek, J., Ehammer, A., et al. (2015). Mapping dynamics of deforestation and forest degradation in tropical forests using radar satellite data. *Environmental Research Letters, 10*(3), 034014.

Lambin, E. F. (1999). Monitoring forest degradation in tropical regions by remote sensing: Some methodological issues. *Global Ecology and Biogeography, 8*(3/4), 191–198.

Mertens, B., & Lambin, E. F. (2000). Land-cover-change trajectories in southern Cameroon. *Annals of the Association of American Geographers, 90*(3), 467–494.

MINAM. (2014). *Geoservidor: Convenio De Colaboracion Interinstitucional Entre La Institucion Carnegie Para La Ciencia*. Departamento De Ecologia Global Y El Ministerio Del Ambiente.

Mitchard, E., Saatchi, S., Lewis, S., Feldpausch, T., Woodhouse, I., Sonk, B., Rowland, C., & Meir, P. (2011). Measuring biomass changes due to woody encroachment and deforestation/degradation in a forest-savanna boundary region of central Africa using multi-temporal l-band radar backscatter. *Remote Sensing of Environment, 115*(11), 2861–2873 (DESDynI VEG-3D Special Issue).

Mitchard, E. T., Meir, P., Ryan, C. M., Woollen, E. S., Williams, M., Goodman, L. E., et al. (2013). A novel application of satellite radar data: measuring carbon sequestration and detecting degradation in a community forestry project in Mozambique. *Plant Ecology and Diversity, 6*(1), 159–170.

Mosquera, C., Chavez, M. L., Pachas, V. H., & Moschella, P. (2009). *Estudio Diagnostico de la Actividad Minera Artesanal en Madre de Dios*. [English: Diagnostic Study of Artisanal Mining Activities in Madre de Dios] Lima, Peru: Cooperación, Caritas, Conservation International. ISBN: 978-9972-865-07-7. url http://www.minam.gob.pe/mn-ilegal/images/files/estudio_diagnostico_mineria_artesanal_madredios.pdf

Myers, N., Mittermeier, R., Mittermeier, C., da Fonseca, G., & Kent, J. (2000). Biodiversity hotpots for conservation priorities. *Nature, 403*, 853–858.

NASA. (2014). *Landsat Missions*. url http://landsat.usgs.gov/

Nunes, F., Soares-Filho, B., Giudice, R., Rodrigues, H., Bowman, M., Silvestrini, R., & Mendoza, E. (2012). Economic benefits of forest conservation: Assessing the potential rents from Brazil Nut concessions in Madre de Dios, Peru, to channel REDD+ investments. *Environmental Conservation, 39*, 132–143.

Potapov, P. V., Dempewolf, J., Talero, Y., Hansen, M. C., Stehman, S. V., Vargas, C., et al. (2014). National satellite-based humid tropical forest change assessment in Peru in support of REDD+ implementation. *Environmental Research Letters, 9*(12), 124012.

Rosa, I. M. D., Purves, D., Souza, C, Jr, & Ewers, R. M. (2013). Predictive modelling of contagious deforestation in the Brazilian amazon. *PLoS ONE, 8*(10), e77231.

Ryan, C. M., Berry, N. J., & Joshi, N. (2014). Quantifying the causes of deforestation and degradation and creating transparent REDD+ baselines: A method and case study from central Mozambique. *Applied Geography, 53*, 45–54.
Ryan, C. M., Hill, T., Woollen, E., Ghee, C., Mitchard, E., Cassells, G., et al. (2012). Quantifying small-scale deforestation and forest degradation in African woodlands using radar imagery. *Global Change Biology, 18*(1), 243–257.
Salvini, G., Herold, M., Sy, V. D., Kissinger, G., Brockhaus, M., & Skutsch, M. (2014). How countries link REDD+ interventions to drivers in their readiness plans: Implications for monitoring systems. *Environmental Research Letters, 9*(7), 074004.
Sasaki, N., & Putz, F. E. (2009). Critical need for new definitions of "forest" and "forest degradation" in global climate change agreements. *Conservation Letters, 2*(5), 226–232.
Schoene, D., Killmann, W., von Lüpke, H., & LoycheWilkie, M. (2007). *Definitional issues related to reducing emission from deforestation in developing countries forests and climate change*. Working paper no. 5. Rome: UNFAO.
Scullion, J. J., Vogt, K. A., Sienkiewicz, A., Gmur, S. J., & Trujillo, C. (2014). Assessing the influence of land-cover change and conflicting land-use authorizations on ecosystem conversion on the forest frontier of Madre de Dios, Peru. *Biological Conservation, 171*, 247–258.
Southworth, J., Marsik, M., Qiu, Y., Perz, S., Cumming, G., Stevens, F., et al. (2011). Roads as drivers of change: Trajectories across the tri-national frontier in map, the south-western Amazon. *Remote Sensing, 3*(5), 1047–1066.
Swenson, J., Carter, C., Domec, J., & Delgado, C. (2011). Gold mining in the Peruvian Amazon: Global prices, deforestation, and mercury imports. *PLoS ONE, 6*(4), e18875.
Telmer, K. H., & Veiga, M. M. (2009). World emissions of mercury from small scale and artisanal gold mining. In N. Pirrone & R. Mason (Eds.), *Mercury fate and transport in the global atmosphere: Emissions, measurements and models*. Berlin: Springer.
UNFCCC. (2007). *United Nations framework convention on climate change. Report of the conference of the parties on its thirteenth session*. The United Nations Climate Change Conference Bali.
UNFCCC. (2011). *United Nations framework convention on climate change*. Fact sheet: Reducing emissions from deforestation in developing countries: Approaches to stimulate action.
Vuohelainen, A. J., Coad, L., Marthews, T. R., Malhi, Y., & Killeen, T. J. (2012). The effectiveness of contrasting protected areas in preventing deforestation in Madre de Dios, Peru. *Environmental Management, 50*(4), 645–663.
Woodhouse, I. (2005). *Introduction to microwave remote sensing*. Boca Raton, London: CRC Press, Taylor & Francis Group.

Part II
Land Use Competition and Ecosystem Services

Chapter 8
Competition for Land-Based Ecosystem Services: Trade-Offs and Synergies

Daniel Müller, Helmut Haberl, Lara Esther Bartels, Matthias Baumann, Marvin Beckert, Christian Levers, Florian Schierhorn, Jana Zscheischler, Petr Havlik, Patrick Hostert, Ole Mertz and Pete Smith

Abstract In this chapter, we define, conceptualize, and exemplify competition for ecosystem services derived from land. Competition for land-based ecosystem services arises when utilization of an ecosystem service by one actor reduces the possibility of other actors to use the same or other ecosystem services. Therefore, we focus on trade-offs and synergies between ecosystem services and argue that the functional relationships between the trade-offs are crucial in shaping how land-use competition plays out. We use this conceptualization of competition for land-based ecosystem services as well as the closely related concepts of trade-off and synergies as a framework to interpret the five case studies in this section, which provide a

---

Daniel Müller and Helmut Haberl contributed equally to this chapter.

D. Müller (✉) · F. Schierhorn
Halle (Saale) and IRI THESys, Leibniz Institute of Agricultural Development in Transition Economies, Berlin, Germany
e-mail: mueller@iamo.de

H. Haberl
Institute of Social Ecology, Alpen-Adria University, Vienna, Austria
e-mail: helmut.haberl@aau.at

P. Hostert
Department of Geography and IRI THESys, Humboldt-Universität Zu Berlin, Berlin, Germany
e-mail: patrick.hostert@geo.hu-berlin.de

L.E. Bartels
Governance & Sustainability Lab, Trier University, Trier, Germany

M. Baumann · C. Levers
Department of Geography, Humboldt-Universität zu Berlin, Berlin, Germany

M. Beckert · P. Smith
Institute of Biological and Environmental Sciences, University of Aberdeen, Aberdeen, Scotland

J. Niewöhner et al. (eds.), *Land Use Competition*,
Human-Environment Interactions, DOI 10.1007/978-3-319-33628-2_8

unique and rich overview of land-use competition in terms of actors involved, geographic coverage, spatial scale, and ecosystem services. These studies convey important insights into opportunities and challenges of intervening into competition for land-based resources and can inform efforts to improve land governance.

Keywords Governance · Complex patterns · Interdependencies · Scale · Land use

8.1 Introduction

In the last decade, prices of agricultural commodities and food have shown impressive price spikes, thereby deviating from the trend of falling real prices in preceding decades (Swinnen and Squicciarini 2012; GEA 2012). It has become evident that global agricultural production must rise substantially in the next decades to meet the demands of a growing and increasingly affluent world population. Additional pressure on land resources originates from the rising demand for areas devoted to biodiversity conservation, carbon sequestration, and bioenergy production (Ray et al. 2013; Tilman et al. 2011; Alexandratos and Bruinsma 2012). The growing demands for land-based services have contributed to discussions about a "looming land scarcity" (Lambin and Meyfroidt 2011) and resulted in a rising research interest in better understanding land-use competition (e.g., Haberl et al. 2014; Smith et al. 2010).

Already today, competition for scarce resources is a key trigger for civil strife and international conflicts (Homer-Dixon 1994), drawing attention to questions such as how to identify and anticipate hot spots of competition for land-based resources and services and how to attain equitable, efficient, and sustainable resource use under increasing competition? Despite the growing pressure on land resources, land competition is often addressed with sectorial approaches in policy-making and land-use planning. For example, the complex nature of land use results in a situation where services provided by land are often treated separately, usually by different decision-makers or administrative bodies. Likewise, separate ministries typically exist for agriculture, forestry, finance, energy, and economic cooperation. Moreover, land use is governed at the level of nations, provinces, or municipalities, or in some cases at the regional level (e.g., the European Union). Many globally concerted

J. Zscheischler
Institute of Socio-Economics, Leibniz Centre for Agricultural Landscape Research (ZALF), Müncheberg, Germany

P. Havlik
International Institute for Applied Systems Analysis (IIASA), Laxenburg, Austria

O. Mertz
Department of Geosciences and Natural Resource Management, University of Copenhagen, Copenhagen, Denmark

attempts to steer land use toward socially desirable pathways have had limited effects, e.g., the Kyoto protocol or the United Nations' program to reduce emissions from deforestation and forest degradation (REDD+, cf. Schlamadinger et al. 2007; Corbera and Schroeder 2011). The increasing globalization of land competition with drivers originating in distant locations will remain a key challenge for the future provision of land-based services (Lambin and Meyfroidt 2011; Eakin et al. 2014; Sect. 8.1). Efforts to make global land governance more effective are required if the aim is to mitigate the negative outcomes of globalizing land use and steer land use toward sustainable pathways (Sikor et al. 2013).

This chapter aims to provide an analytical framework for better understanding competition for land-based ecosystem services. We draw on a collection of five case studies that examine competition for land-based ecosystem services from different angles, in different geographic contexts, at different spatial and temporal scales, and originating from different actors (see Table 8.1 for a summary of the case studies). The case studies complement each other in their empirical contribution and, in that way, allow for a rich conceptual synthesis. We intensively discussed all case studies during a two-week workshop in Berlin in September 2014 and jointly derived the concepts presented in this chapter on the basis of the empirical material of the case studies.

8.1.1 *What Is Competition?*

8.1.1.1 Competition Occurs Between Actors

Competition arises when two or more actors strive for resources that are not accessible in the desired or demanded quantity and quality. If the availability of resources is limited, utilization of a resource by one actor may reduce or even preclude the use of that resource by other actors. Resources may also be limited to specific actors because they are too costly or because power relations prevent certain

Table 8.1 Summary of case studies in this section of the book

Case	Description	Chapter
Tanzania	Contested land in Loliondo: the eastern border of the Serengeti National Park between conservation, hunting tourism, and pastoralism	Bartels (2016)
Russia–Brazil	How the collapse of the beef sector in post-Soviet Russia displaced competition for ecosystem services to the Brazilian Amazon	Schierhorn et al. (2016)
Scotland	Of trees and sheep: trade-offs and synergies in farmland afforestation in the Scottish uplands	Beckert et al. (2016)
Germany	Land-use competition related to woody biomass production on arable land in Germany	Zscheischler et al. (2016)
South America	Land competition in the South American Chaco	Baumann et al. (2016)

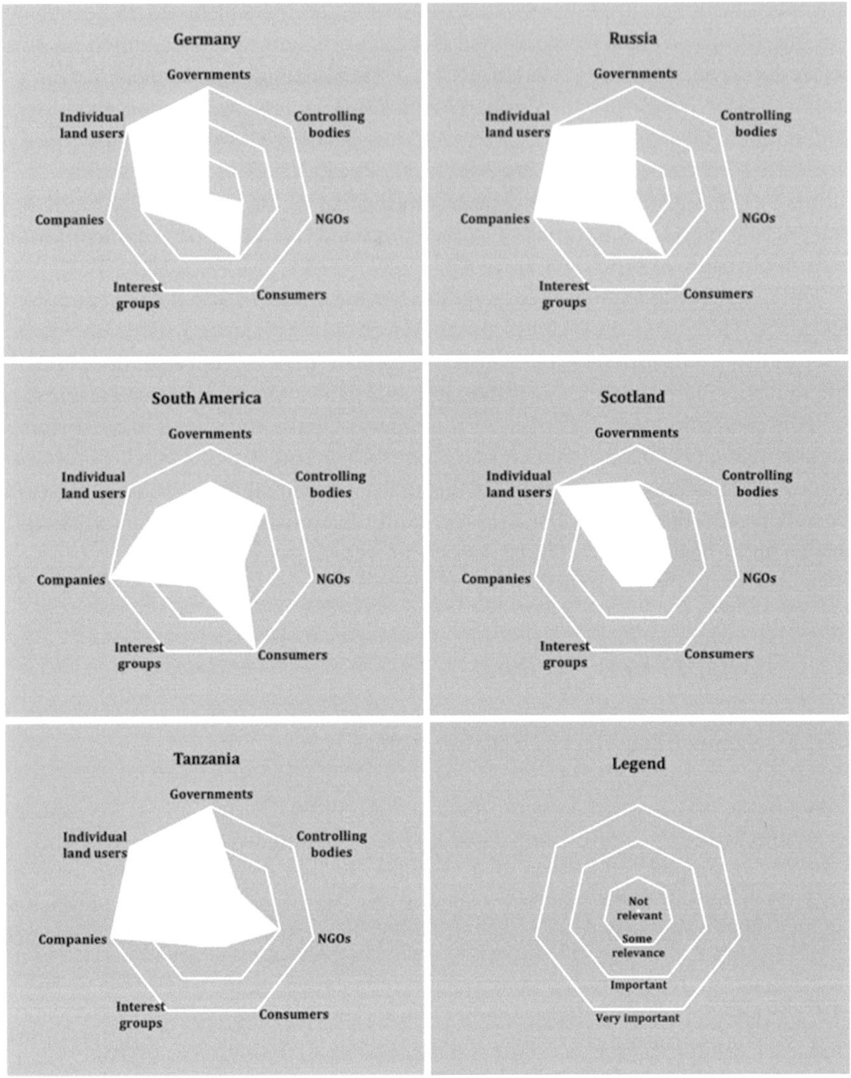

Fig. 8.1 Types of actors and the ranking of their importance for land-use competition. Rankings are based on and discussed in detail in the case studies of this section (examples for controlling bodies are UNFCCC, the convention on biodiversity, or international mechanisms for REDD+ projects)

actors from accessing the resources. In this chapter, we focus on land as a resource, in particular on the products or services delivered by terrestrial ecosystems (i.e., "land-based ecosystem services"), as discussed below. Several groups of actors may be involved in the competition for land-based ecosystem services. Figure 8.1 exemplifies the involved actors from the case studies of this book section.

8.1.1.2 Disciplinary Approaches to Competition

The notion of competition is common in several scientific disciplines, including ecology, sociology, and economics. In each discipline, competition has a venerable tradition as a concept guiding scientific inquiry. In ecology, competition is thought to play a major role within species ("intraspecific competition"), where it is important in regulating population density, and between species ("interspecific competition") as a major force structuring biotic communities (Birch 1957; Gause 1934). While it is clear that competition negatively affects the inferior competitor due to reduced resource access, ecologists also recognize positive aspects of competition such as its role in population regulation and evolution (Cain et al. 2008). Ecologists also distinguish factors from resources: While factors such as temperature may affect species and thereby regulate their distribution in space and time, they are not "used" or "depleted" and hence are not considered as resources (Cain et al. 2008). However, factors can turn into resources when they are becoming scarce, such as oxygen, which is an (de facto) unlimited factor in free air conditions but can be depleted in the soil and is consequently competed for by heterotrophic organisms living there.

In sociology, the basic concepts of "competition" and "conflict" are used to describe dissociative processes of social interaction, in contrast to "cooperation." In this context, competition is defined as an impersonal, unconscious, continuous struggle between individuals or groups (Park and Burgess 1921) for goods and services, which are scarce or limited in quantity. The notion of competition differs from that of conflicts, which are defined as social interactions that are personal, conscious, and not continuous. Whereas competition is subject to norms and rules, conflicts are characterized by a breakdown of social rules (Sharma 2008; Sutherland et al. 1956).

Economic theory suggests that prices of resources reflect their scarcity. In the (exceptional) case of perfect markets (many buyers and sellers, homogenous goods, perfect information, full market transparency), competition would lead to the allocation of resources to their most productive use (Smith 1778). However, market conditions are often far from being perfect. For example, imperfect competition may arise when monopolistic structures prevail (when economies of scale prevent the existence of several competing suppliers, a situation called "natural monopoly"; Sharkey 1982), when prices fail to reflect the full costs of a resource, in particular when they do not reflect "external costs" (i.e., costs borne by actors not accounted for in the market transaction in question; Coase 1960), when the quality and value of a product are difficult to measure or compare (Shapiro 1983), or when competitors have asymmetric access to information (Pattanayak et al. 2010). In the event of imperfect competition, prices fail to indicate the degree of competition and other mechanisms beyond interactions on markets are needed to effectively and efficiently manage resource use and resource allocation (Rohlf 2010).

The Organisation for Economic Co-operation and Development OECD (2001) borrowed the concept of joint production from agricultural production economics (Shumway et al. 1984) to construct an analytical framework that captures the economics of simultaneous provision of commodities and non-commodities by agriculture. This concept explains trade-offs or synergies between different ecosystem services through the prevalent relationships with respect to production inputs. Non-allocable inputs, such as tractors, which, once acquired, can be used both for landscape management (e.g., mowing) and for crop production, entail complementarities and synergies in production (Sakai 1974). Competition in production then arises because of the prevalence of allocable limited inputs such as land (Moschini 1989). Havlík et al. (2005) implemented these concepts to analyze the provisioning of biodiversity by extensive suckler cow production. They argued that for low intensities of production, an increasing cattle herd, as a non-allocable input, may increase both beef production and biodiversity. At higher levels of stocking densities, however, biodiversity on the grazing land declines and, hence, the effect of the land as allocable fixed factor prevails, which in turn leads to competition between beef and biodiversity production.

Competition for land and land-based ecosystem services rarely plays out in perfect market conditions, as postulated in the basic assumptions of neoclassical economic theory, for a variety of reasons:

1. Perfect markets require homogenous goods, and land is all but homogenous. Land quality may be determined differently in many respects, ranging from the suitability for agriculture, proficiency for forestry to its scenic value (Delbecq et al. 2014; Colwell and Sirmans 1978). The valuation of these qualities is strongly dependent on cultural and contextual factors.
2. Even for proprietors, the use of land can be legally constrained by spatial planning rules. For example, in many countries land markets are divided into submarkets with different rights of use, such as for construction, farmland, or forestry. Ownership of a parcel of land often does not entitle proprietors to alter land use as owners may wish. In the EU, for example, if the land is dedicated to agriculture or forestry, the proprietor may not be allowed to build a house. Or if land is legally designated as forest, the proprietor is not allowed to convert this forest to agriculture without permission. Changes in rights of use by regulators can therefore affect land values (Jaeger et al. 2012).
3. Land is an immobile resource but the products or services it yields or provides can typically be transported. The value of land therefore depends on its connectedness to other locations, e.g., related to infrastructures such as roads or urban centers (Guiling et al. 2009). Public investments into infrastructure may therefore change the value of land and alter conditions under which competition for land plays out. Land-use competition is hence also dependent on transport costs, technology, and infrastructure.
4. Land resources are quasi non-renewable and cannot be enlarged or restocked easily (restoration measures after land degradation may improve land suitability in the medium to long term, making land a renewable resource in some cases). Moreover, unsustainable use of resources can result in resource degradation to the point of irreversibility and effectively prevent certain uses, e.g., cropping (Sabbi and Salvati 2014).

5. The maximization of some land-based ecosystem services may negatively or positively affect the quantity or quality of other services from the same land (Power 2010), which may be seen as an "externality" of land-use practices. Such externalities are rarely monetized and compensated for, resulting in insufficient provision of land-based services with positive externalities (e.g., landscape beauty) and excessive provision of land-based services that are associated with negative externalities (e.g., CO_2 emissions).
6. Land demand originates from a variety of drivers, and these drivers act from different and increasingly distant locations. This makes the attribution of cause and effect difficult and complicates the understanding and the analyses of land competition (Seto et al. 2012; Eakin et al. 2014 and Sect. 8.1). For example, Russia imports substantial amounts of meat from South America and is thus one key actor of the increasing land competition between forest and agricultural lands in distant places, such as the Amazon (Chap. 10).
7. Property rights of land resources, and the services they provide, are contested, overlapping, unclear, undocumented, or non-existent in many locations. Multiple claims from different actors often overlap on the same physical plot of land, resulting all too often in conflicts over land or land-based ecosystem services (Sikor and Lund 2009).
8. Land distribution is strongly affected by power relations that often favor large land owners who have better access to production factors and output markets and who are better connected to authorities of land governance (Sikor and Müller 2009).
9. Land is an attractive source of taxation for governments. This drives efforts to survey and title land, which then makes land resources eligible to use as collateral for loans or for market exchanges. While titling can secure land rights of poor people (Williamson 2010), it may also lead to landlessness and more competition for land because resource-rich individuals or corporations strive to invest in land (Bromley 2009).

In sum, neoclassic economic theory may help to understand competition for land, but it is certainly not sufficient as a framework to completely untangle this issue. Understanding competition for land requires approaches that account for the peculiarities and the multifunctional characteristics of land-based production. Land competition, driven by varying and volatile demands of different actors for land-based resources, may also influence rates of innovation. On the one hand, it may increase innovation since it provides economic incentives to outperform competitors. On the other hand, competition may obstruct innovation by reducing the rents of a successful innovator (Aghion et al. 2001). Moreover, competition may not result in socially beneficial outcomes in the provision of products and services that are characterized by non-excludability, i.e., when those who use but do not pay for a service cannot be excluded from the use (Ruttan and Hayami 1984).

8.1.1.3 Conceptualization of Competition for Land-Based Ecosystem Services

Figure 8.2 summarizes our conceptualization of competition for land-based ecosystem services. Land-based ecosystem services depend on climate and other environmental conditions but also on land-use activities (Spangenberg et al. 2014). The contributions of land use to mitigate climate change are also conceived as ecosystem services (e.g., carbon sequestration and climate regulation) demanded by actors such as governments or international bodies, but also increasingly by land users. Therefore, the framework in Fig. 8.2 distinguishes the "land users" who can make decisions on how to use the land, based on their rights of use and often in interaction with other land users, from "other actors" who affect those decisions through various mechanisms, e.g., by influencing the demand for products or services or by changing laws or regulations. The "other actors" may be consumers demanding certain products, regulators imposing zoning laws or designating certain areas as nature conservation sites, or actors who implement programs such as to reduce emissions from deforestation and forest degradation or to enhance forest carbon stocks (REDD+) or design energy policies to introduce biofuels.

Figure 8.2 therefore distinguishes different layers of interaction processes that contribute to competition for land in different ways and that show how synergies and trade-offs between ecosystem services influence processes of land-use competition.

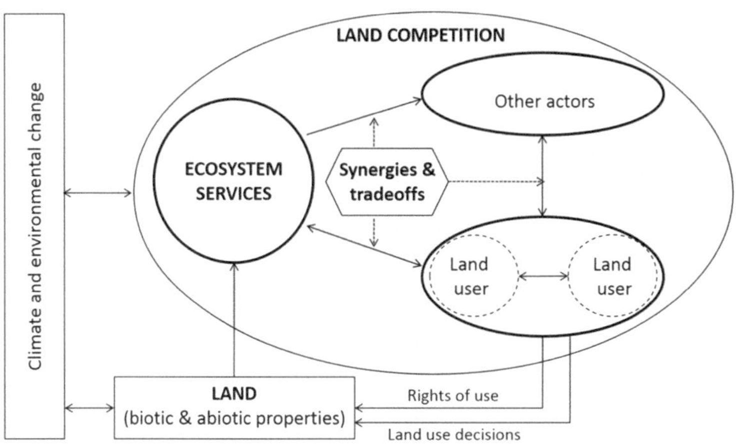

Fig. 8.2 Conceptualization of the competition for land-based ecosystem services used in this chapter and the associated case studies. See *text* for explanation

8.1.1.4 Land-Use Competition and Trade

Land resources may be scarce in one place, but abundant elsewhere. In theory, an optimal global allocation of land-based production activities could substantially reduce the area required for agricultural production and leave room for other land-based ecosystem services such as carbon sequestration, biodiversity, or bioenergy (Müller et al. 2011; Johnson et al. 2014). This suggests that reallocation of land-based production from less to more suitable areas would diminish pressure on land and could reduce trade-offs. However, the notion of an optimal global allocation of production factors is a theoretical concept that would face many institutional hurdles as well as obstacles related to power relations, as evidenced by the current land grab and environmental justice debates (e.g., Haberl 2015; Messerli et al. 2013; Hornborg and Jorgensen 2010). Moreover, an empirical analysis of globally traded cropland products suggests that reducing yield gaps holds greater potential to decrease the future demand for cropland area than a spatial reallocation of cropping could achieve (Kastner et al. 2014).

Despite its potential to create distributional conflicts, trade is an obvious mechanism for overcoming resource limitations in any specific locale. Indeed, during the agrarian–industrial transition, the maintenance of supply for growing urban populations requires more efficient means of transport to industrialized centers (e.g., railroads instead of horse-drawn carriages) as well as improvements of agricultural labor efficiency (Fischer-Kowalski and Haberl 2007; Sieferle et al. 2006; Sieferle and Osmann 2010). In other words, trade allows to raise consumption and to deliver services above the limits imposed by the capacity of locally available resources (Erb 2012), which complicates the analysis of ecological boundaries at local or regional scales (Dearing et al. 2014). Furthermore, trade does not eliminate land-use competition but rather results in much more complex processes of land-use competition that are mediated, among others, by global markets. Land-use competition may today proceed between actors that are unlikely to ever meet in person, but who are linked through product flows (trade), markets, communication, knowledge systems, and other processes that are currently being investigated under the heading of "teleconnections" or "telecouplings" (Eakin et al. 2014; Seto et al. 2012; see also Sect. 8.1 for a more detailed discussion).

8.1.1.5 Trade-Offs and Synergies in Competition for Land-Based Ecosystem Services

The use of one land-based ecosystem service can affect the availability of other land-based ecosystem services in a multitude of ways. This reflects that ecosystem services are less a spontaneous, autonomous outcome of ecosystem processes (Spangenberg et al. 2014), but are at least as much, if not more, the outcome of human activities to manage the land (or of human colonization of ecosystems; see Fischer-Kowalski and Haberl 1993). One may also say that ecosystem services are socially constructed, both in terms of their dependence on active social interventions and management and in terms of how they are communicated and valued in society (Ernstson 2013). Such a notion of ecosystem services reflects that land use can result in different services that may influence each other in various ways (Fig. 8.3). This implies that the interdependencies between ecosystem services strongly affect the way in which competition for land plays out between various actors and how benefits are distributed between different social groups (Ernstson 2013).

In Fig. 8.3, we show possible types of relationships between two quantifiable ecosystem services, which is the simplified view that we adopt in order to facilitate discussion. We acknowledge that in real-world situations, more than two ecosystem services need to be considered. Moreover, many land-based ecosystem services are difficult to quantify and compare. Nevertheless, we exemplify real-world examples for each stylized trade-off relationship in Table 8.2.

Trade-offs (Fig. 8.3a) are a common interdependency between ecosystem services: In the simplified two-dimensional system, the use of one ecosystem service negatively affects the availability of the other service (Fig. 8.3a; see Polasky et al. 2008; Moilanen et al. 2011; Macchi et al. 2013). Competition ensues from such trade-off situations when the demand for services originates from different actors; e.g., an expansion of agricultural production on forest land will reduce the global service of carbon sequestration. Accordingly, we argue that a trade-off situation can only lead to competition if two or more actors are involved. Single actors may also face trade-off situations when deciding on how to use a particular piece of land. We understand this as an internal decision-making process rather than competition because one actor, by definition, cannot compete with itself. However, actors such as industry, governmental bodies, and interest groups can influence the decision of

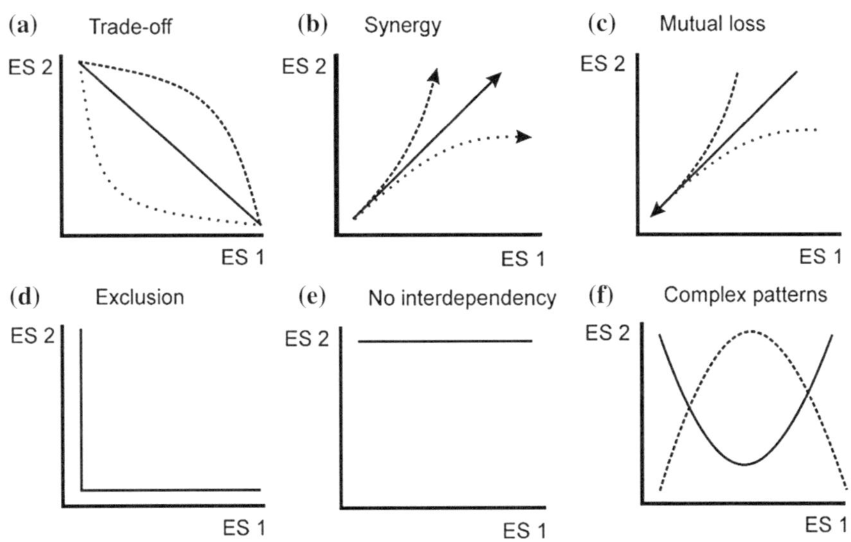

Fig. 8.3 Stylized examples of static relations between two ecosystem services (ES) in a given location. See *text* for explanation

Table 8.2 Examples for the stylized trade-off curves described in Fig. 8.3

Term	Explanation	Example	Degree of competition
Trade-off	Increasing the supply of one service negatively affects the other	Concave: utilization of a small fraction of land for wind power plants strongly reduces landscape beauty and habitat for bats (Cryan and Barclay 2009)	High
		Linear: utilization of maize for biofuel or food (Wu et al. 2012)	Medium
		Convex: agroforestry; increase in tree density reduces the ability of grazing land for sheep in a nonlinear fashion (Chap. 11)	Low
Synergy	Increasing the supply of one ecosystem service increases the supply of the other	Restoration of degraded dry lands in Australia helps to maintain food output and at the same time results in considerable bioenergy output and carbon sequestration (Harper et al. 2012)	None
		Extensive livestock grazing can suppress fire susceptibility (Dubinin et al. 2010) and result in higher grassland biodiversity (Isselstein et al. 2005)	None
Mutual loss (lose–lose)	Simultaneous reduction of two ecosystem services	Overgrazing reduces food production, causes some sensitive species to disappear, and favors the spread of more robust species, thus decreasing biodiversity (Lykke et al. 2004; Petz et al. 2014)	Negative synergistic
Exclusion (win–lose)	Provisioning of one ecosystem service excludes the other	Soil sealing excludes all ecosystem services based on plant production (Wratten et al. 2013)	Extreme
No interdependency	Provision of one ecosystem service is independent of the other	Raising one service (e.g., agricultural output) does not affect another service (e.g., pollination) (Bennett et al. 2009)	None
Complex pattern	Various nonlinear relationships between ecosystem services	Patches of managed lands introduced in a uniform natural landscape increase biodiversity and agricultural production up to an inflection point at which further intensification results in uniform landscapes with little biodiversity but high production (Wrbka et al. 2004)	Context dependent

a single actor. This may be the case when environmental policy pressures agricultural landowners to improve the environmental value of their land, while the landowner's objective is to maximize profits from agricultural production.

Synergies (Fig. 8.3b) are situations where two or more ecosystem services increase simultaneously, i.e., one ecosystem service raises the provision of the other on the same unit of land. Following our earlier definition of competition, a synergistic provision of ecosystem services does not result in competition (which does not preclude the possibility that land-use conflicts could emerge). Trade-offs and synergies can be linear but also take other functional forms such as concave or convex (Table 8.2). Bennett et al. (2009) define a synergy as positive bidirectional interactions where two ecosystem services are enhanced simultaneously. As an example for a bidirectional interaction, they use the preservation of uncultivated land that increases pollinator abundance and, therefore, enhances agricultural production in a confined area. In our conceptualization, this is an example for a synergistic relation between ecosystem services. Another example is extensive livestock grazing that allows meat production while reducing the risk of large-scale grassland fires (Dubinin et al. 2010). Extensive livestock grazing has also been shown to result in higher grassland biodiversity (Isselstein et al. 2005).

A mutual loss relationship (Fig. 8.3c) is "negatively synergistic" when less of one service implies less of the other service. Overuse resulting in extreme ecosystem degradation is a case in point (Table 8.2). An extreme trade-off situation occurs when the use of one service results in the total loss of the other service. In this case, the use of the resources is mutually exclusive (Fig. 8.3d). For example, building a house depletes the carbon sequestration capacity of the ecosystem.

Figure 8.3e sketches the case when the provision of one ecosystem service is independent of the other ecosystem service. Bennett et al. (2009) refer to this as a unidirectional, positive interaction where one service increases, while the other remains the same. Therefore, this represents a perfect example for "no interdependency" (Fig. 8.3e) that is labeled as "no change" by Haase et al. (2012). Finally, Fig. 8.3f visualizes the possible relationships for complex, in this case monotonous and nonlinear, connections between two ecosystem services that can result in combinations of the aforementioned patterns (e.g., first synergy, then trade-off). Some examples include cultural landscapes and related biodiversity provision in extensively managed grasslands, where the level of biodiversity will increase with moderate intensity of agricultural production, but will start to decrease once a certain level of intensity is achieved (Balent et al. 1998).

Synergies and trade-offs can occur within a single land use. For instance, a forest can provide regulating services (e.g., carbon sequestration), but a trade-off situation occurs when this forest is harvested. On the other hand, carbon sequestration within a forest does not compete with recreation experiences (a cultural service), which results in congruence with a single land use and a synergistic outcome. Finding synergies between different ecosystem services can therefore be seen as a way to overcome competition. For example, agroforestry systems have been put forward as land-use activities providing synergies between agricultural and forestry land uses (Ramachandran Nair et al. 2010; Udawatta and Jose 2012; Smith et al. 2013).

Competition can occur not only between different land uses but also between different ecosystem services and between different categories of ecosystem services (e.g., between provisioning vs. non-provisioning services). For example, afforestation for carbon may lead to more habitats for certain species, resulting in the synergistic provision of two services. Likewise, a direct provisioning service such as bioenergy may also lead to an enhancement in biodiversity and carbon (Lemus and Lal 2005).

8.2 Spatial Scale

The representation of spatial scales is a central feature characterizing the dynamics of land competition. Trade-offs, synergies, and other relations between ecosystem services can extend from the surfaces of single units of land to regions and from on-site relationships to distant areas. It is therefore crucial for our understanding of competition for land-based ecosystem services to consider the spatial scale at which trade-offs and synergies can take place. Scales are concepts that are socially constructed and crucially influenced by perspectives, knowledge, agendas, and living conditions. They are mental constructs and support the understanding of systems. In this respect, spatial scales are simplified representations of the "real" world that help analyze and understand observed dynamics (Massey 2005 and introduction to Sect. 8.1).

Spatial scale in the context of competition for land-based ecosystem services can refer to ecological units (i.e., scales of ecosystem) or to institutional entities (i.e., scales of socioeconomic systems, regulatory frameworks, or the scale of land users; see Hein et al. 2006). These different analytical scales are important to consider when discussing land-use competition as the emergent outcome of interaction of actors at different spatial scales (Smith et al. 2010). As multiple actors and drivers operate on different spatial scales, competition also plays out differently at different scales, which complicates the understanding of land competition.

8.2.1 Scaling in the Case Study Chapters of This Section

An example for local competition is provided by the case study in Tanzania (Chap. 9). This case exemplifies how land competition is shaped by different and competing local interests for land-based ecosystem services. However, land-use competition and land-use change at the local scale is frequently affected by regional and global drivers and interdependencies. These and other top-down pressures, such as the demand from affluent outsiders for hunting in undisturbed ecosystems (Chap. 9), the demand for soy-based food and fodder in Western societies (Chap. 13), the national and international political support for bioenergy generation (Chap. 12) or for carbon sequestration (Chap. 11), may all trigger competition for land-based ecosystem services and land-use change at the local level. Conversely, the sum of local land-use decisions may trigger or relax competition at distant places and different scales. For example, the sum of local-level land-use changes in post-Soviet Russia has triggered land-use competition in the Brazilian Amazon (Chap. 10).

8.3 Intervention and Governance

As outlined above, competition only partially follows classic economic theory and may result in socially inferior outcomes regarding the provision of products and services that are characterized by non-excludability. In such cases, intervention may allow steering the use of limited land resources and their functions, thereby achieving more socially beneficial outcomes while avoiding social tensions, for example, by preventing degradation and functional loss of land. Moreover, competition for land-based ecosystem services often involves trade-offs that cause undesirable outcomes for groups of actors, who may, in response, demand policy or management intervention to safeguard their stakes or to attain higher rents from land use. This can particularly occur when trade-offs between services are not synergistic and are demanded by at least two different actors (cf. Fig. 8.3a).

Competition can be steered through a range of mechanisms including regulation (e.g., laws, zoning), incentives (e.g., incentive payments for service provision, labeling, or geographic indication), price policies (e.g., subsidies or taxes for input or output prices), or negotiations (e.g., agreements, protocols). A typical goal of such interventions is to manage the trade-offs in order to achieve desirable outcomes for specific interest groups or for society at large. Other interventionist goals can be to avoid conflicts over contested land-based services. However, managing the trade-offs requires that intervening bodies or institutions attach value judgments to the services that are competed for and claimed by different actors. In other words, the intervening body has to attach concrete beliefs or quantitative values to the ecosystem service axes in Fig. 8.3. Sometimes, this can be achieved by monetary valuation of the service (Azqueta and Sotelsek 2007; Costanza et al. 1997), e.g., by attaching the price of carbon to carbon fluxes, which in turn facilitates a quantitative comparison between different ecosystem services (Nelson et al. 2009). For example, commercial farmers desire to maximize their profits from crop production, while the global community, represented by multilateral institutions, aims to reduce emissions from land use through REDD+. REDD+ is a contemporary approach designed to intervene in such competition by compensating land users for maintaining or increasing carbon in vegetation and soil. In principle, the aim of REDD+ is to make carbon-saving land-use activities more profitable than competing carbon-depleting land-use activities (Angelsen et al. 2009).

However, attaching a monetary value to a specific service is problematic in many cases. This is particularly evident in the valuation of priceless societal benefits such as the cultural value of a landscape. Even if a value can be attached for a particular ecosystem service, this value may not reflect the value attached by society, as is the case for the price of carbon on currently active carbon markets. For example, the carbon prices in the EU Emission Trading System (ETS) are far below the prices that would be required to provide the necessary incentive for substantially reducing current levels of carbon emissions.

Another common mechanism to steer competition is zoning of land to conserve nature by excluding destructive land uses. The establishment of protected areas is typically imposed by governments in response to the demand for safeguarding particular tracts of land, e.g., from development for agriculture or mining or for nature conservation. These demands may originate from local, regional, national, or international agencies. Frequently, protected areas are concentrated on less fertile land in order to reduce competition with agricultural land uses (Joppa et al. 2008). In other instances, however, zoning collided with customary uses of land that have not been formally acknowledged and titled or that have been ignored by zoning authorities, such as in the case of pastoralism (Chap. 9). Zoning initiatives that stir competition for land have led to the impoverishment of people (Cernea and Schmidt-Soltau 2006), may increase land-use competition by reducing areas available for cultivation (Homewood 2004), and can even result in land-use conflicts in adjacent areas (West et al. 2006).

As many actors may compete for the same land-based ecosystem services, calls for intervention may originate from different angles and sides. For example, well-organized interest groups such as commercial farmers in Europe or the USA can rely on the support of professionally organized, well-connected farmer associations with considerable political power. Such organizations can be very successful in lobbying for their clients, often with global implications for land use. The US biofuel lobby, for instance, managed to obtain and uphold substantial political support for growing corn and processing it to ethanol (Tyner 2008). Conversely, the large group of small-scale farmers that dominate land use in developing countries lacks the political power and the powerful associations that support their needs. The strong variation in the ability of actor groups to lobby for interventions in their own interest may lead to strong biases in favor of politically powerful actors in governing the competition for land-based ecosystem services.

Intervention into competition may also have unexpected and unintended consequences that occur later in time or in other locations and which are difficult and often impossible to foresee (Müller et al. 2014). For example, subsidies for increasing agricultural production may be well intended and initially uncontroversial but later turn out to be detrimental to society. One example are the subsidy schemes of the EU that aimed at attaining self-sufficiency in food production, but that remained in place despite the massive production increases that turned the EU from a net importer to a net exporter of agricultural produce. The maintenance of EU agricultural subsidies caused increasing competition on world markets and disadvantaged food-exporting developing countries (Josling 2008). Similarly, national policy interventions that aim at promoting carbon sequestration in vegetation and soil at the expense of increasing food production through agricultural expansion can have long-lasting, unexpected consequences. The foregone production may have a price-increasing effect, which provides incentives to expand cultivation elsewhere with negative overall effects on the carbon balance (Agrawal et al. 2011). Such uncertainty in decision-making may be very pronounced in the land sector, where volatile market conditions, technological change, and sectorial policies complicate the formulation of strategies that steer future land use onto sustainable pathways.

Finally, interventions into competition for land have allocative consequences, because the aim of interventions is typically to steer resource use to more desirable outcomes from the perspective of the authority that intervenes. The common justification for such interventions is to correct for market failures that result from land use by means of internalizing the negative environmental costs imposed on third parties. However, these corrective interventions and deliberate changes in governance can have manifold consequences for the efficiency and equity of resource use, unforeseen off-site effects, result in rent-seeking behavior of interest groups, and may lead to path dependency of policy directions (Byerlee and Deininger 2013; Veldkamp 2009; Sikor et al. 2013). These challenges warrant a careful a priori analysis of the type of competition, the actors involved, the relationships between the services that are competed for, and the spatial scales at which competition occurs.

8.4 Synthesis

Land competition, defined in this chapter as the competition between different actors for scarce land-based ecosystem services, is increasingly making the headlines because of the growing demands for limited natural resources and expectations of escalating future resource scarcities. However, the often complex processes of land-use competition remain poorly understood. We have conceptualized the competition for land-based ecosystem services using literature drawn from ecology, sociology, economics, and geography, as well as with scholarly work from various fields related to sustainability science. The intellectual contribution of this conceptual chapter is the categorization of different types of competition into functional relationships that guide trade-offs between ecosystem services. We used this conceptual framework to interpret the five case studies in this book section that all examine varying types of competition with different tools and at different spatial and temporal scales. The conceptual ideas can inform interdisciplinary research in other settings by helping to frame the analysis of land-use competition. We hope that our ideas contribute to generating more nuanced insights into opportunities and challenges of intervening into competition for land-based resources and can ultimately encourage better land governance.

Acknowledgments HH gratefully acknowledges funding by Humboldt University's KOSMOS program, which has allowed him to spend four months in Berlin in 2014. The authors are also grateful to contributions by the EU-FP7 projects VOLANTE (FP7-ENV-2010-265104),

HERCULES (FP7-ENV-2013-627), I-REDD+ (FP7-ENV-265286), German Research Foundation (DFG), the Leibniz Association's "Pakt für Forschung", and the German Federal Ministry of Education and Research (BMBF) (01 LN 1316 A). The research contributes to the Global Land Project (GLP).

References

Aghion, P., Harris, C., Howitt, P., & Vickers, J. (2001). Competition, imitation and growth with step-by-step innovation. *Review of Economic Studies, 68*(3), 467–492. doi:10.1111/1467-937x.00177.

Agrawal, A., Nepstad, D., & Chhatre, A. (2011). Reducing emissions from deforestation and forest degradation. *Annual Review of Environment and Resources, 36*(1), 373–396. doi:10.1146/annurev-environ-042009-094508.

Alexandratos, N., & Bruinsma, J. (2012). World agriculture towards 2030/2050: The 2012 revision. *ESA working paper,* No. 12-03. FAO, Rome.

Angelsen, A., Brockhaus, M., Kanninen, M., Sills, E., Sunderlin, W. D., Wertz-Kanounnikoff, S., & Abdel Nour, H. O. (2009). *Realising REDD+: National strategy and policy options*. Center for International Forestry Research (CIFOR), Bogor, Indonesia.

Azqueta, D., & Sotelsek, D. (2007). Valuing nature: From environmental impacts to natural capital. *Ecological Economics, 63*(1), 22–30.

Balent, G., Alard, D., Blanfort, V., & Gibon, A. (1998). Activités de pâturage, paysages et biodiversité. *Annales de Zootechnie, 47*(5–6), 419–429.

Bartels, L. E. (2016). Contested land in loliondo: The eastern border of the serengeti national park between conservation, hunting tourism, and pastoralism. In J. Niewöhner et al. (Eds.), *Land use competition*. doi:10.1007/978-3-319-33628-2_9.

Baumann, M., Piquer-Rodríguez, M., Fehlenberg, V., Pizarro, G. G., & Kuemmerle, T. (2016). Land-use competition in the South American chaco. In J. Niewöhner et al. (Eds.), *Land use competition*. doi:10.1007/978-3-319-33628-2_13.

Beckert, M., Smith, P., Chapman, S. (2016). Of Trees and Sheep: Trade-offs and synergies in farmland afforestation in the scottish uplands. In J. Niewöhner et al. (Eds.), *Land use competition*. doi:10.1007/978-3-319-33628-2_11.

Bennett, E. M., Peterson, G. D., & Gordon, L. J. (2009). Understanding relationships among multiple ecosystem services. *Ecology Letters, 12*(12), 1394–1404. doi:10.1111/j.1461-0248.2009.01387.x.

Birch, L. C. (1957). The meanings of competition. *The American Naturalist, 91*(856), 5–18. doi:10.2307/2458507.

Bromley, D. W. (2009). Formalising property relations in the developing world: The wrong prescription for the wrong malady. *Land Use Policy, 26*(1), 20–27. doi:10.1016/j.landusepol.2008.02.003.

Byerlee, D., & Deininger, K. (2013). Growing resource scarcity and global farmland investment. *Annual Review of Resource Economics, 5*(1), 13–34. doi:10.1146/annurev-resource-091912-151849.

Cain, M. L., Bowman, W. D., & Hacker, S. D. (2008). *Ecology*. Incorporated: Sinauer Associates.

Cernea, M. M., & Schmidt-Soltau, K. (2006). Poverty risks and national parks: policy issues in conservation and resettlement. *World Develepment, 34*(10), 1808–1830. doi:10.1016/j.worlddev.2006.02.008.

Coase, R. H. (1960). The problem of social cost. *Journal of Law and Economics, 3*, 1–44.

Colwell, P. F., & Sirmans, C. F. (1978). Area, time, centrality and the value of urban land. *Land Economics, 54*(4), 514–519.

Corbera, E., & Schroeder, H. (2011). Governing and implementing REDD+. *Environmental Science & Policy, 14*(2), 89–99. doi:10.1016/j.envsci.2010.11.002.

Costanza, R., d'Arge, R., de Groot, R., Farber, S., Grasso, M., Hannon, B., et al. (1997). The value of the world's ecosystem services and natural capital. *Nature, 387*(6630), 253–260.

Cryan, P. M., & Barclay, R. M. R. (2009). Causes of bat fatalities at wind turbines: Hypotheses and predictions. *Journal of Mammalogy, 90*(6), 1330–1340. doi:10.1644/09-MAMM-S-076R1.1.

Dearing, J. A., Wang, R., Zhang, K., Dyke, J. G., Haberl, H., Hossain, M. S., et al. (2014). Safe and just operating spaces for regional social-ecological systems. *Global Environmental Change, 28*, 227–238. doi:10.1016/j.gloenvcha.2014.06.012.

Delbecq, B. A., Kuethe, T. H., & Borchers, A. M. (2014). Identifying the extent of the urban fringe and its impact on agricultural land values. *Land Economics, 90*(4), 587–600.

Dubinin, M., Potapov, P., Lushchekina, A., & Radeloff, V. C. (2010). Reconstructing long time series of burned areas in arid grasslands of southern Russia by satellite remote sensing. *Remote Sensing of Environment, 114*(8), 1638–1648. doi:10.1016/j.rse.2010.02.010.

Eakin, H., DeFries, R., Kerr, S., Lambin, E. F., Liu, J., Peter J., Marcotullio, Messerli, P., Reenberg, A., Ximena, R., Swaffield, S. R., Wicke, B., & Zimmerer, K. (2014). Significance of telecoupling for exploration of land-use change. In K. Seto & A. Reenberg (Eds.), Rethinking global land use in an urban era. Cambridge: MIT Press.

Erb, K.-H. (2012). How a socio-ecological metabolism approach can help to advance our understanding of changes in land-use intensity. *Ecological Economics, 76*, 8–14. doi:10.1016/j.ecolecon.2012.02.005.

Ernstson, H. (2013). The social production of ecosystem services: A framework for studying environmental justice and ecological complexity in urbanized landscapes. *Landscape and Urban Planning, 109*(1), 7–17. doi:10.1016/j.landurbplan.2012.10.005.

Fischer-Kowalski, M., & Haberl, H. (1993). Metabolism and colonization. Modes of production and the physical exchange between societies and nature. *Innovation: The European Journal of Social Science Research, 6*(4), 415–442. doi:10.1080/13511610.1993.9968370.

Fischer-Kowalski, M., & Haberl, H. (2007). *Socioecological transitions and global change: Trajectories of social metabolism and land use*. London: Edward Elgar Publishing, Incorporated.

Gause, G. F. (1934). *The struggle for existence*. Baltimore: The Williams & Wilkins Company.

GEA. (2012). *Global energy assessment—Toward a sustainable future*. New York, NY, USA and Laxenburg, Austria: Cambridge University Press and International Institute for Applied Systems Analysis.

Guiling, P., Brorsen, B. W., & Doye, D. (2009). Effect of urban proximity on agricultural land values. *Land Economics, 85*(2), 252–264. doi:10.3368/le.85.2.252.

Haase, D., Schwarz, N., Strohbach, M., Kroll, F. & Seppelt, R. (2012). Synergies, trade-offs, and losses of ecosystem services in urban regions: An integrated multiscale framework applied to the Leipzig-Halle region, Germany. *Ecology and Society, 17* (3). doi:10.5751/ES-04853-170322

Haberl, H. (2015). Competition for land: A sociometabolic perspective. *Ecological Economics, 119*, 424–431. doi:10.1016/j.ecolecon.2014.10.002.

Haberl, H., Mbow, C., Deng, X., Irwin, E. G., Kerr, S., Kuemmerle, T., et al. (2014). Finite Land Resources and Competition. In K. Seto & A. Reenberg (Eds.), *Rethinking global land use in an urban era*. Cambridge: MIT Press.

Harper, R. J., Okom, A. E. A., Stilwell, A. T., Tibbett, M., Dean, C., George, S. J., et al. (2012). Reforesting degraded agricultural landscapes with Eucalypts: Effects on carbon storage and soil fertility after 26 years. *Agriculture, Ecosystems & Environment, 163*, 3–13. doi:10.1016/j.agee.2012.03.013.

Havlík, P., Veysset, P., Boisson, J.-M., Lherm, M., & Jacquet, F. (2005). Joint production under uncertainty and multifunctionality of agriculture: Policy considerations and applied analysis. *European Review of Agricultural Economics, 32*(4), 489–515. doi:10.1093/erae/jbi027.

Hein, L., van Koppen, K., de Groot, R. S., & van Ierland, E. C. (2006). Spatial scales, stakeholders and the valuation of ecosystem services. *Ecological Economics, 57*(2), 209–228. doi:10.1016/j.ecolecon.2005.04.005.

Homer-Dixon, T. F. (1994). Environmental scarcities and violent conflict: Evidence from cases. *International Security, 19*(1), 5–40. doi:10.2307/2539147.

Homewood, K. M. (2004). Policy, environment and development in African rangelands. *Environmental Science & Policy, 7*(3), 125–143.

Hornborg, A., & Jorgensen, A. K. (2010). *International trade and environmental justice: Toward a global political ecology*. New York: Nova Science Publishers.

Isselstein, J., Jeangros, B., & Pavlu, V. (2005). Agronomic aspects of biodiversity targeted management of temperate grasslands in Europe–A review. *Agronomy Research, 3*(2), 139–151.

Jaeger, W. K., Plantinga, A. J., & Grout, C. (2012). How has Oregon's land use planning system affected property values? *Land Use Policy, 29*(1), 62–72. doi:10.1016/j.landusepol.2011.05.005.

Johnson, J. A., Runge, C. F., Senauer, B., Foley, J., & Polasky, S. (2014). Global agriculture and carbon trade-offs. *Proceedings of the National Academy of Sciences of the United States of America, 111*(34), 12342–12347. doi:10.1073/pnas.1412835111.

Joppa, L. N., Loarie, S. R., & Pimm, S. L. (2008). On the protection of "protected areas". *Proceedings of the National Academy of Sciences of the United States of America, 105*(18), 6673–6678.

Josling, T. (2008). External influences on CAP reforms: An historical perspective. In J. F. M. Swinnen (Ed.), *The perfect storm: The political economy of the fischler reforms of the common agricultural policy*. Brussels: Centre for European Policy Studies.

Kastner, T., Erb, K.-H., & Haberl, H. (2014). Rapid growth in agricultural trade: Effects on global area efficiency and the role of management. *Environmental Research Letters, 9*(3), 034015. doi:10.1088/1748-9326/9/3/034015.

Lambin, E. F., & Meyfroidt, P. (2011). Global land use change, economic globalization, and the looming land scarcity. *Proceedings of the National Academy of Sciences of the United States of America, 108*(9), 3465–3472. doi:10.1073/pnas.1100480108.

Lemus, R., & Lal, R. (2005). Bioenergy crops and carbon sequestration. *Critical Reviews in Plant Sciences, 24*(1), 1–21. doi:10.1080/07352680590910393.

Lykke, A. M., Kristensen, M. K., & Ganaba, S. (2004). Valuation of local use and dynamics of 56 woody species in the Sahel. *Biodiversity and Conservation, 13*(10), 1961–1990. doi:10.1023/B:BIOC.0000035876.39587.1a.

Macchi, L., Grau, H. R., Zelaya, P. V., & Marinaro, S. (2013). Trade-offs between land use intensity and avian biodiversity in the dry Chaco of Argentina: A tale of two gradients. *Agriculture, Ecosystems & Environment, 174*, 11–20. doi:10.1016/j.agee.2013.04.011.

Massey, D. (2005). *For Space*. London: SAGE Publications.

Messerli, P., Heinimann, A., Giger, M., Breu, T., & Schönweger, O. (2013). From 'land grabbing' to sustainable investments in land: Potential contributions by land change science. *Current Opinion in Environmental Sustainability, 5*(5), 528–534. doi: 10.1016/j.cosust.2013.03.004.

Moilanen, A., Anderson, B. J., Eigenbrod, F., Heinemeyer, A., Roy, D. B., Gilling, S., et al. (2011). Balancing alternative land uses in conservation prioritization. *Ecological Applications, 21*(5), 1419–1426. doi:10.1890/10-1865.1.

Moschini, G. (1989). Normal inputs and joint production with allocatable fixed factors. *American Journal of Agricultural Economics, 71*(4), 1021–1024. doi:10.2307/1242678.

Müller, C., Lotze-Campen, H., Huber, V., Popp, A., Svirejeva-Hopkins, A., Krause, M., & Schellnhuber, H. J. (2011). Towards a great land-use transformation? In Ú. O. Spring, C. Mesjasz, J. Grin, et al. (Eds.), *Coping with global environmental change, disasters and security: Threats, challenges, vulnerabilities and risks* (pp. 23–28). Berlin: Springer.

Müller, D., Sun, Z., Vongvisouk, T., Pflugmacher, D., Xu, J., & Mertz, O. (2014). Regime shifts limit the predictability of land-system change. *Global Environmental Change, 28*, 75–83. doi:10.1016/j.gloenvcha.2014.06.003.

Nelson, E., Mendoza, G., Regetz, J., Polasky, S., Tallis, H., Cameron, D., et al. (2009). Modeling multiple ecosystem services, biodiversity conservation, commodity production, and tradeoffs at landscape scales. *Frontiers in Ecology and the Environment, 7*(1), 4–11. doi:10.1890/080023.

OECD. (2001). *Multifunctionality: Towards an analytical framework*. Organisation for Economic Co-operation and Development (OECD), Paris.
Park, R. E., & Burgess, E. W. (1921). *Introduction to the science of sociology*. Chicago, IL: The University of Chicago Press.
Pattanayak, S. K., Wunder, S., & Ferraro, P. J. (2010). Show me the money: Do payments supply environmental services in developing countries? *Journal of Environmental Economics and Policy*. doi:10.1093/reep/req006.
Petz, K., Alkemade, R., Bakkenes, M., Schulp, C. J. E., van der Velde, M., & Leemans, R. (2014). Mapping and modelling trade-offs and synergies between grazing intensity and ecosystem services in rangelands using global-scale datasets and models. *Global Environmental Change, 29*, 223–234. doi:10.1016/j.gloenvcha.2014.08.007.
Polasky, S., Nelson, E., Camm, J., Csuti, B., Fackler, P., Lonsdorf, E., et al. (2008). Where to put things? Spatial land management to sustain biodiversity and economic returns. *Biological Conservation, 141*(6), 1505–1524. doi:10.1016/j.biocon.2008.03.022.
Power, A. G. (2010). Ecosystem services and agriculture: Tradeoffs and synergies. *Philosophical Transactions of the Royal Society of London B: Biological Sciences, 365*(1554), 2959–2971. doi:10.1098/rstb.2010.0143.
Ramachandran Nair, P. K., Nair, V. D., Mohan Kumar, B., & Showalter, J. M. (2010). Carbon sequestration in agroforestry systems (Chap. 5). In L. S. Donald (ed) *Advances in agronomy* (Vol 108, pp. 237–307). New York: Academic Press. doi:10.1016/S0065-2113(10)08005-3.
Ray, D. K., Mueller, N. D., West, P. C., & Foley, J. A. (2013). Yield trends are insufficient to double global crop production by 2050. *PLoS ONE, 8*(6), e66428. doi:10.1371/journal.pone.0066428.
Rohlf, W. (2010). *Introduction to economic reasoning* (8th ed.). Boston, MA: Addison-Wesley.
Ruttan, V. W., & Hayami, Y. (1984). Toward a theory of induced institutional innovation. *Journal of Development Studies, 20*(4), 203–223.
Sabbi, A., & Salvati, L. (2014). Searching for a downward spiral? Soil erosion risk, agro-forest landscape and socioeconomic conditions in Italian local communities. *Land Use Policy, 41*, 388–396. doi:10.1016/j.landusepol.2014.06.023.
Sakai, Y. (1974). Substitution and expansion effects in production theory: The case of joint production. *Journal of Economic Theory, 9*(3), 255–274. doi:10.1016/0022-0531(74)90051-9.
Schlamadinger, B., Bird, N., Johns, T., Brown, S., Canadell, J., Ciccarese, L., et al. (2007). A synopsis of land use, land-use change and forestry (LULUCF) under the Kyoto Protocol and Marrakech Accords. *Environmental Science & Policy, 10*(4), 271–282.
Schierhorn, F., Gittelson, A. K., Müller, D. (2016) How the collapse of the beef sector in post-soviet russia displaced competition for ecosystem services to the brazilian amazon. In J. Niewöhner et al. (Eds.), *Land use competition*. doi:10.1007/978-3-319-33628-2_10.
Seto, K. C., Reenberg, A., Boone, C. G., Fragkias, M., Haase, D., Langanke, T., et al. (2012). Urban land teleconnections and sustainability. *Proceedings of the National Academy of Sciences of the United States of America, 109*(20), 7687–7692. doi:10.1073/pnas.1117622109.
Shapiro, C. (1983). Premiums for high quality products as returns to reputations. *The Quarterly Journal of Economics, 98*(4), 659–679. doi:10.2307/1881782.
Sharkey, W.W. (1982). *The theory of natural monopoly*. Cambridge: Cambridge University Press.
Sharma, R. K. (2008). *Fundamentals of sociology*. New Dehli: Atlantic Publishers.
Shumway, C. R., Pope, R. D., & Nash, E. K. (1984). Allocatable fixed inputs and jointness in agricultural production: Implications for economic modeling. *American Journal of Agricultural Economics, 66*(1), 72–78. doi:10.2307/1240617.
Sieferle, R. P., Krausmann, F., Schandl, H., & Winiwarter, V. (2006). *Das Ende der Fläche: Zum gesellschaftlichen Stoffwechsel der Industrialisierung*. Köln: Böhlau.
Sieferle, R. P., & Osmann, M. (2010). *The subterranean forest: Energy systems and the industrial revolution*. White Horse Press.
Sikor, T., Auld, G., Bebbington, A. J., Benjaminsen, T. A., Gentry, B. S., Hunsberger, C., Izac, A.-M., Margulis, M. E., Plieninger, T., Schroeder, H., & Upton, C. (2013). Global land

governance: From territory to flow? *Current Opinion in Environmental Sustainability, 5*(5), 522–527. doi:10.1016/j.cosust.2013.06.006.
Sikor, T., & Lund, C. (2009). Access and property: A question of power and authority. *Development and Change, 40*(1), 1–22. doi:10.1111/j.1467-7660.2009.01503.x.
Sikor, T., & Müller, D. (2009). The limits of state-led land reform: An introduction. *World Development, 37*(8), 1307–1316. doi:10.1016/j.worlddev.2008.08.010.
Smith, A. (1778). *An inquiry into the nature and causes of the wealth of nations*. London: Strahan and Cadell.
Smith, P., Gregory, P. J., van Vuuren, D., Obersteiner, M., Havlík, P., Rounsevell, M., et al. (2010). Competition for land. *Philosophical Transactions of Royal Society of London B: Biological Sciences, 365*(1554), 2941–2957. doi:10.1098/rstb.2010.0127.
Smith, J., Pearce, B. D., & Wolfe, M. S. (2013). Reconciling productivity with protection of the environment: Is temperate agroforestry the answer? *Renewable Agriculture Food Systems, 28* (01), 80–92. doi:10.1017/S1742170511000585.
Spangenberg, J. H., Görg, C., Truong, D. T., Tekken, V., Bustamante, J. V. & Settele, J. (2014). Provision of ecosystem services is determined by human agency, not ecosystem functions. Four case studies. *International Journal of Biodiversity Science, Ecosystem Services & Management, 10*(1), 40–53. doi:10.1080/21513732.2014.884166
Sutherland, R. L., Woodward, J. L., & Maxwell, M. A. (1956). *Introductory sociology*. J.B. Lippincott Company: Chicago, IL.
Swinnen, J., & Squicciarini, P. (2012). Mixed messages on prices and food security. *Science, 335* (6067), 405–406. doi:10.1126/science.1210806.
Tilman, D., Balzer, C., Hill, J., & Befort, B. L. (2011). Global food demand and the sustainable intensification of agriculture. *Proceedings of the National Academy of Sciences of the United States of America, 108*(50), 20260–20264. doi:10.1073/pnas.1116437108.
Tyner, W. E. (2008). The US ethanol and biofuels boom: Its origins, current status, and future prospects. *BioScience, 58*(7), 646–653. doi:10.1641/b580718.
Udawatta, R., & Jose, S. (2012). Agroforestry strategies to sequester carbon in temperate North America. *Agroforestry Systems, 86*(2), 225–242. doi:10.1007/s10457-012-9561-1.
Veldkamp, A. (2009). Investigating land dynamics: Future research perspectives. *Journal of Land Use Science, 4*(1), 5–14.
West, P., Igoe, J., & Brockington, D. (2006). Parks and peoples: The social impact of protected areas. *Annual Review of Anthropology, 35*(1), 251–277. doi:10.1146/annurev.-anthro.35.081705.123308.
Williamson, C. R. (2010). The two sides of de Soto: Property rights, land titling, and development. In E. Chamlee-Wright (Ed.), *The annual proceedings of the wealth and well-being of nations, 2009–2010* (pp. 95–108). Beloit, WI: Beloit College Press.
Wratten, S., Sandhu, H., Cullen, R., & Costanza, R. (2013). *Ecosystem services in agricultural and urban landscapes*. New york: Wiley.
Wrbka, T., Erb, K.-H., Schulz, N. B., Peterseil, J., Hahn, C., & Haberl, H. (2004). Linking pattern and process in cultural landscapes. An empirical study based on spatially explicit indicators. *Land Use Policy, 21*(3), 289–306. doi:10.1016/j.landusepol.2003.10.012.
Wu, F., Zhang, D., & Zhang, J. (2012). Will the development of bioenergy in China create a food security problem? Modeling with fuel ethanol as an example. *Renewable Energy, 47*, 127–134. doi:10.1016/j.renene.2012.03.039.
Zscheischler, J., Gaasch, N., Manning D. B., & Weith, T. (2016). Land use competition related to woody biomass production on arable land in Germany. In J. Niewöhner et al. (Eds.), *Land use competition*. doi:10.1007/978-3-319-33628-2_12.

Chapter 9
Contested Land in Loliondo: The Eastern Border of the Serengeti National Park between Conservation, Hunting Tourism, and Pastoralism

Lara Esther Bartels

Abstract This case study analyzes a land conflict in Loliondo, an area adjacent to the eastern border of the Serengeti National Park in northern Tanzania. The pastoralists of the area and a Dubai-based hunting company are competing over a strip of land of ~135,000 ha alongside the Serengeti. At the same time, the Tanzanian government claims the land in order to conserve the Greater Serengeti Ecosystem. The government intends to allocate the strip of land solely for nature conservation and hunting while excluding pastoralism. This chapter argues that the competition between the pastoralists and the hunting company is one over land-based ecosystem services. The intention of the government to allocate the contested strip of land to nature conservation and hunting only has fostered the competition, which has evolved into a conflict. It is shown that pastoralists are able to derive specific provisioning ecosystem services from the contested land also during the dry season. This makes the area most valuable for the food security of the pastoralists of the area, which will be destabilized by the current policies of the government.

Keywords Tanzania · Competition · Ecosystem services · Pastoralism

9.1 Introduction

About 44 % of the total land area in Tanzania is under some form of protection or conservation (United Republic of Tanzania (URT) 2009a). The East African country is particularly famous for the Serengeti National Park, but has also the

L.E. Bartels (✉)
Governance and Sustainability Lab, University of Trier, Trier, Germany
e-mail: bartelsl@uni-trier.de

L.E. Bartels
Integrative Research Institute on Transformations of Human-Environment Systems (IRI THESys), Humboldt-Universität zu Berlin, Berlin, Germany

J. Niewöhner et al. (eds.), *Land Use Competition*,
Human-Environment Interactions, DOI 10.1007/978-3-319-33628-2_9

largest protected area in Africa, the Selous Game Reserve (Baldus et al. 2003). Indeed, Tanzania is the country with the largest area under protection in Africa both in absolute and relative terms (Brockington 2008: 577).

Nature conservation in Tanzania is closely coupled to the tourism industry as tourism is mainly based on the wildlife that can be found in the country (Wade et al. 2001). Thereby, not only photographic safaris, but also hunting tourism is an increasing sector within the tourism industry (Lindsey et al. 2007). In 2004, tourism contributed 16 % to the gross domestic product (GDP) and is thus the second most important sector after agriculture (Skof 2008). In addition, the tourism industry generated almost 25 % of the total export earnings of the country in 2004 (ibid.).

Nature conservation in Tanzania is not only linked to tourism but also has a history of dispossession of local people (Brockington 1999; Neumann 1998). Pastoralists are especially affected by the establishment of nature conservation areas (Exner et al. 2015). The great majority of the livestock kept in Tanzania, around 43.8 million animals in 2008, is herded by smallholders, mostly following pastoralist practices (URT 2012; FAO 2005). About 90 % of the fodder for these livestock comes from natural grasslands where pastoralists is herding their animals (Sarwatt and Mollel 2000). A growing portion of these grasslands are allocated to Tanzania's nature conservation areas putting pastoralists under threat. An increasing body of literature draws attention to the eviction of pastoralists and other local communities in order to establish conservation areas such as the Mkomazi, Serengeti, Lake Manyara, and Tarangire National Park (Geisler and de Sousa 2001; Goldman 2006; Brockington 2002; Neumann 1998). Recently, the conservation practice in Tanzania has been described as a form of *landgrabbing* (Benjaminsen et al. 2011; Ngoitiko et al. 2010; Nelson et al. 2012; Benjaminsen and Bryceson 2012; Gardner 2012). Indeed, nature conservation is one of the main drivers of increasing competition over land and land conflicts in Tanzania (Benjaminsen et al. 2011; Exner et al. 2015). At the same time, Maasai pastoralists are used as "tourism exhibits" in the growing cultural tourism sector (Azarya 2004: 963).

Nature conservation, tourism, and pastoralism constitute a critical, closely related triangle in Tanzania. This triangle is at the center of the case study illustrated in this chapter. The case study draws attention to a land conflict in the Loliondo division in northern Tanzania (division is a administrative unit in Tanzania). The pastoralists of the area and a Dubai-based hunting company are competing over a strip of land of ~135,000 ha alongside the eastern border of the Serengeti. This strip of land is already part of a nature conservation area, that covers the whole Loliondo division. Yet, the government intends to reduce the nature conservation area to the contested strip of land. At the same time the government aims to permit hunting while excluding pastoralism within the reduced nature conservation area.

In the first section of this chapter, the case study is presented by reconstructing different periods of the conflict. It is argued that the case study exemplifies competition between the pastoralists and the trophy hunting company over land-based ecosystem services that are aimed to be obtained from the strip of land which is located within a nature conservation area. The government's latest intention of permitting hunting and prohibiting pastoralism within the contested strip of land has increased the competition to the point of conflict. The second section of this chapter focuses on the perspectives of the pastoralists. The Maasai pastoralists claim the contested land because they derive specific provisioning services from the strip of land alongside the Serengeti during the dry season. The section explains why the plans of the government will destabilize the local grazing system and thus the livelihood of the pastoralists of the area.

This chapter is based on a four-month long fieldwork conducted during the second half of 2012 and in the beginning of 2014 in Tanzania (see Bartels 2013 for more details). Together with available literature and media reports on the land conflict in Loliondo, the chapter draws on expert interviews conducted with representatives of nongovernmental organizations (NGOs) and government officials in Arusha and Dar es Salaam and semi-structured interviews and focus group discussions conducted with pastoralists in the village Ololosokwan located in Loliondo. A fieldwork assistant translated interviews held in local languages (i.e. in Maa, the language of the Maasai, and Kiswahili) and contributed to the understanding of local practices.

9.2 Contested Land at the Eastern Border of the Serengeti National Park

The case study is situated in the Loliondo division, an area in northern Tanzania that is surrounded by mostly conservation areas (see Fig. 9.1), the Serengeti National Park to the west, to the south the Ngorongoro Conservation Area and to the east the Lake Natron Basin. The Lake Natron Basin is designated as Ramsar site, i.e., is protected under the Ramsar Convention on Wetlands with international importance (Gibson 2011). The north of the division touches the Kenyan border, close to the Maasai Mara National Reserve of Kenya. Loliondo is at the same time a division and a Game Controlled Area (GCA), i.e., a wildlife-protected area. The Loliondo division/GCA is 400,000 ha large and consists of several villages which are mainly inhabited by Maasai pastoralists (URT 2010).

Loliondo has not only become one of the most important areas for tourism development in the country, but has been shaped through conservation ideologies and practices since colonial times (Gardner 2012). This case study concentrates on an ongoing conflict concerning a strip of the Loliondo division/GCA which stretches across the eastern border of the Serengeti. This strip of land has been contested for more than 20 years. The conflict has peaked following the decision of the Tanzanian government to exclude grazing and thus effectively the Maasai from the contested area. The following sections reconstruct periods of the conflict focusing on different policies of the government that have contribute to shape the conflict.

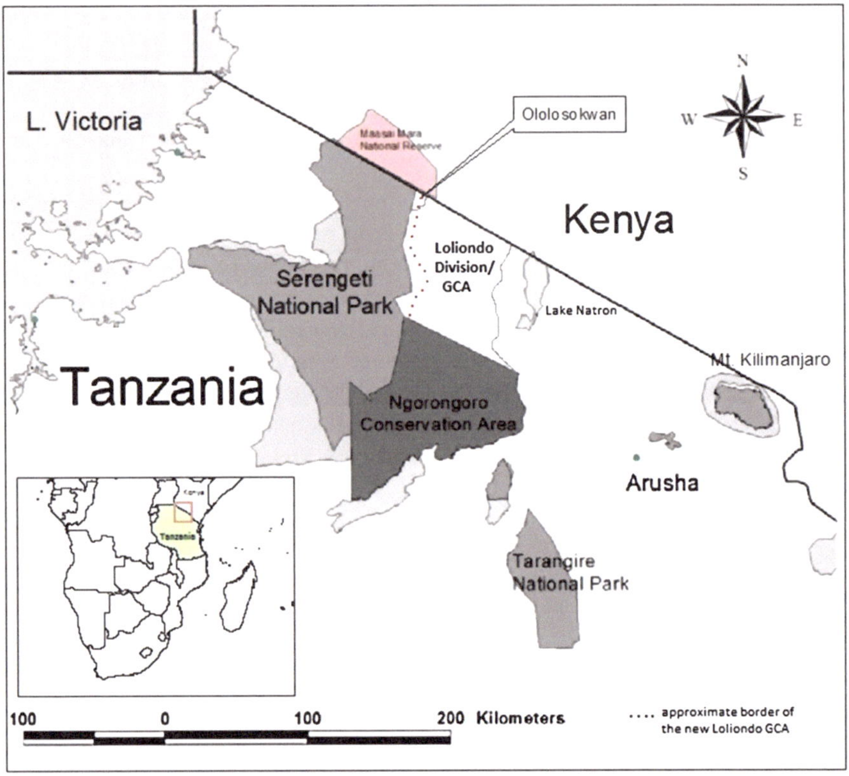

Fig. 9.1 Northern Tanzania and Loliondo Division/Game Controlled Area (GCA) with the approximate border of the new Loliondo GCA and the village Ololosokwan, Map adopted from Tanzania Natural Resource Forum (TNRF) and Maliasili Initiatives (2011: 5) with kind permission for reproduction by TNRF and Maliasili Initiatives

9.2.1 The Loliondo Division/GCA becomes a Hunting Ground for the Ortello Business Corporation

In 1959, the Loliondo division was declared a Game Reserve by the British Colonial government under the Fauna Conservation Ordinance. Later, in 1974, the area was categorized as a Game Controlled Area (GCA) to allow commercial trophy hunting (URT 1974; Ojalammi 2006). The Wildlife Conservation Act of 1974 defines a GCA as a “protected area” in which hunting is allowed provided that a license is granted (URT 1974). In the early 1990s, the Tanzanian government granted a hunting license to the Dubai-based hunting company Ortello Business Corporation (OBC). Although the license is valid for the whole Loliondo GCA/division, OBC has used only the strip of land alongside the eastern border of the Serengeti National Park as their intensive hunting ground in the past (Ojalammi 2006; Nelson and Ole Makko 2003). This might be due to the fact that this strip is particularly rich in wildlife because the border of the Serengeti is not fenced and thus game roams into the area. The granting of the hunting license to OBC can be seen as the starting point of the long-lasting land conflict in Loliondo. Besides some irregularities in the granting of the license and the establishment of permanent structures such as gates, permanent buildings, and an airstrip by OBC without any clear authorization, the presence of OBC has induced protest by local communities. In fact, the presence of OBC has restricted their access to the grazing areas (Nelson et al. 2012). This is especially true for villages situated adjacent to the Serengeti and thus located in the intensive hunting area of OBC (Nelson et al. 2012; Nelson and Ole Makko 2003; Alexander 1993). The land conflict has reached its first peak in 2009 when pastoralists were evicted from the intensive hunting ground of OBC by the Tanzanian government's Field Force Unit and OBC's security forces in order to turn it into a "livestock-free zone" (Nelson et al. 2012: 14).

9.2.2 Changing Legislation: The New Wildlife Conservation Act of 2009

In 2009 a new Wildlife Conservation Act has been released changing the legal basis of the conflict. While the former Wildlife Conservation Act of 1974 allowed not only hunting within GCAs, but also other land uses such as grazing and crop cultivation (URT 1974); the 2009 act explicitly prohibits grazing and crop cultivation and only permits hunting within GCAs (URT 2009b). This act has created a new legal basis for the land conflict in Loliondo, since pastoralists are not allowed any longer to graze their livestock within the intensive hunting ground of OBC. Moreover, the new Wildlife Conservation Act has added a new dimension to the conflict since the inhabitants of Loliondo are not allowed to cultivate or practice pastoralism within the entire Loliondo GCA/division.

The tightening of the nature conservation legislation for GCAs due to the prohibition of land uses such as cultivation and grazing contradicts the title status of some villages, which are located within the Loliondo GCA. For instance, Ololosokwan village, located in Loliondo at the corner of the Serengeti and the Kenyan border, holds a formal land certificate that guarantees the village legal statutory property rights (Ndoinyo 2002; Ojalammi 2006). In addition, the whole Loliondo division is also classified, according to the definition given in the Village Land Act of 1999, as "village land" (Tanzania Natural Resource Forum, Maliasili Initiatives 2011).[1] This classification conflicts with the new Wildlife Conservation

[1]The land law of Tanzania differentiates between three categories of land: "reserved land", which consists of all land which is under natural protection, "village land", which is land under the administration of villages, and "general land" which is a residual category since it "means all public land which is not reserved land or village land and includes unoccupied or unused village land" (URT 1998).

Act of 2009 since the act says that no land within a GCA should be classified as village land (URT 2009b).

In the case of Loliondo, the amendments of the Wildlife Conservation Act of 2009 have proved difficult to implement mainly due to the fact that small-scale farmers conducting mainly pastoralism but also farming inhabit the entire division/GCA.

9.2.3 *Redrawing of Boundaries: The District Land Use Framework Plan*

In order to implement the new Wildlife Conservation Act of 2009 in Loliondo and to appease the land conflict in the area, the Tanzanian government proposed a "District Land Use Framework Plan (2012–2030)" in 2010 (URT 2010). The plan covers the Ngorongoro district to which the Loliondo division/GCA belongs. It suggests the separation of the 400,000 ha large Loliondo division into a 134,712.11 ha strip alongside the Serengeti, which would be designated as the new Loliondo GCA and a 265,287.89 ha area, which would be demarcated as village land (ibid., Fig. 9.1 shows the approximate border of the new Loliondo GCA). In the new Loliondo GCA, the Wildlife Conservation Act of 2009 would be applied, which prohibits land uses such as grazing and crop cultivation. Thus, if the Land Use Framework Plan was to be implemented the villages alongside the Serengeti would lose a significant amount of their village land since the inhabitants would no longer be allowed to enter the new Loliondo GCA.

The separation of the Loliondo division is not a new idea. An area similar to the currently proposed Loliondo GCA was recommended in the Serengeti Regional Conservation Strategy (SRCS) in 1985 as a buffer zone for the Serengeti, but was not implemented (Gardner 2012; Ngoitiko et al. 2010). The idea of the new Loliondo GCA proposed in the Framework Plan of 2010 was put on hold, since the councilors of the District Council General Assembly rejected the plan in 2011. The councilors claimed that the plan was not performed in a participatory manner as intended by the guidelines of land-use planning (interview with a government official who participated in the conducting of the District Land Use Framework, 30.10.2012, for the guidelines of land-use planning see URT 2007).

9.2.4 *Redrawing of Boundaries: Putting a Plan into Practice*

In 2013, the idea of the government, outlined in the Framework Plan of 2010, was put back on the agenda. In a press release on March 26, 2013, Hon. Khamis Sued Kagasheki, the then minister for Natural Resources and Tourism, announced that the redrawing of the boundaries of the Loliondo GCA will be implemented (URT; Ministry of Natural Resources and Tourism 7.4.2013). Supposedly due to the increasing media reports and international protests against the government plans[2], the Tanzanian government frames their decision in a second press release as a "unique history of land-grabbing from [the Tanzanian government] itself" (ibid.: 2). With the perspective that the whole Loliondo division is (only) demarcated as GCA and the local population living "illegally" in the area the government argues that they provide "land to the growing landless population in the area" by redrawing the boundaries of the GCA to the strip alongside the Serengeti (ibid.)

Several civil society organizations strongly oppose the way the government framed the issue. They emphasize that the whole Loliondo division is not only demarcate as GCA, but also as village land. Taking the perspective that the whole Loliondo division is village land, the government seeks to demarcate part of the village land as GCA even though the land falls under administration of the particular villages of the area. Thus, civil society organizations claim that the Tanzanian government is not giving land to the local population but "grabbing […] land falling within village boundaries" (Pastoralists Indigenous NGO's Forum et al. 2013: 3).

In addition, the civil society organizations argue that the decision for the new Loliondo GCA is based on the aim to secure the revenues gained through the hunting company OBC (ibid.). With the new Wildlife Conservation Act of 2009, the Loliondo GCA could be allocated exclusively to OBC without any interference with the land-use activities of the local population. Meanwhile, the government claim that they have to allocate the area alongside the Serengeti as GCA (with restricted land-use regulations) due to the "significantly importance [of this area] to the entire Serengeti and Ngorongoro ecosystem" (URT, Ministry of Natural Resources and Tourism 7.4.2013: 2). In particular, the press release of the government highlights the function of the area as a crucial breeding area for wildlife, as a corridor for iconic great wildebeest migration and as a crucial water catchment area (ibid.).

The government's announcement to establish the new Loliondo GCA received an extensive media echo (see for instance Kimati 2013; Athumani 2013a; James 2013; Nkwame 2013a; Athumani 2013b; Patinkin 2013) followed by protest of NGOs and affected communities in Loliondo (Kisanga 2013). The conflict abated after the order of the prime minister in September 2013 to abandon the plan to redraw the boundaries of the Loliondo GCA (Nkwame 2013b; Makoye 2013). Nevertheless, the issue erupted again in the media in November 2014. An article in The Guardian states that "Maasai have been ordered to quit their traditional lands by the end of the year" to allow for hunting by OBC (Smith 2014). However, the Tanzanian government dismissed the report of The Guardian (Mbashiru 2014). This

[2]see for instance the online petition of the online activist organization Avaaz which has been initiated in 2012 and has received more than 2.3 million signatures. The latest update of the petition is available under: http://www.avaaz.org/en/maasai_fb_dm_3/ (checked on 27.4.2016)

might be due to the fact that the article argues, similar to the Avaaz petition, that the Maasai will be evicted from the strip alongside the Serengeti in order to allow for hunting by OBC (Smith 2014). The government instead has argued in the past that they will redraw the boundaries of the Loliondo GCA in order to conserve the Greater Serengeti Ecosystem and to resolve the land conflict in the area (The URT; Ministry of Natural Resources and Tourism 7.4.2013; URT 2010). Thus, from the official perspective of the government, they do not intent to evict the pastoralists from the area in order to allow for hunting, but they rather aim to conserve the area.

9.2.5 The Loliondo Land Conflict: A Case Study for Competition over Provisioning Services and Recreational Services in a Nature Conservation Area

In this section's conceptual chapter (Chap. 8), we argue that land competition between different actors does not only occur due to competition for land itself, seen as a territory or property, but also due to the competition for ecosystem services which can be derived from the particular plot of land. In addition we state that synergies and trade-offs, but also mutual loss, exclusion and no interdependency can ensue between different land-based ecosystem services if they are meant to be derived from one piece of land. Competition between actors over land-based ecosystem services can, however, only occur if an increase in the use of one ecosystem service reduces the utilization of another ecosystem service (i.e., trade-off) or results in total loss of the other service (i.e., mutually exclusive).[3]

The Loliondo case can be interpreted as competition between the pastoralists of Loliondo and the hunting company OBC over the use of provisioning services and recreational services in the Loliondo GCA. The competition has led to a distinct conflict, which is reinforced by government policies. These policies include the amendments of the conservation law in favor of the exclusive use of recreational ecosystem services within GCAs followed by the intention to redraw the boundaries of the Loliondo GCA. This argument will be presented in the following in more detail.

The land conflict in Loliondo has been introduced through the granting of a hunting permit to the hunting company OBC for the Loliondo GCA. Indeed, this permit introduced the competition over land-based ecosystem services between OBC and the pastoralists of Loliondo. While the pastoralists seek to derive provisioning services such as feed (range forage) and water in order to obtain food

---

[3]By employing the term use of ecosystem services we do not imply that ecosystem services are provided by ecosystem to humans, they are rather "generated in complex social processes" (Spangenberg et al. 2014, p. 40, see also Ernstson 2013). We apply the terms use, derive and obtain ecosystem services interchangeable in this chapter.

(milk and meat) from the strip of land alongside the eastern border of the Serengeti, OBC aims to obtain recreational ecosystem services by conducting trophy hunting from the same strip of land.[4] The use of provisioning and recreational ecosystem services from this strip of land is difficult if not impossible to reconcile.

Through the amendments of the Wildlife Conservation Act of 2009, a new dimension has been added to the competition between the use of provisioning and recreational services in the Loliondo GCA, which has reinforced the conflict. By prohibiting grazing and crop cultivation within GCAs, the Tanzanian government has defined nature conservation and the use of provisioning services through pastoralism as mutually exclusive. At the same time, the use of recreational services and nature conservation has not been defined as mutually exclusive. Trophy hunting is still permitted within GCAs by the Wildlife Conservation Act of 2009. This has fueled the conflict, because the Tanzanian government is suspected of being not only interested to conserve the strip alongside the eastern border of the Serengeti "for the benefit of the present and future generations of humankind" (URT: Ministry of Natural Resources and Tourism 7.4.2013: 2). Rather, many suspects that the revenues derived from allowing OBC to obtain recreational services from this strip of land also play a significant role (Pastoralists Indigenous NGOs Forum et al. 2013).

The intention of the Tanzanian government to allocate the strip of ~135,000 ha alongside the Serengeti to nature conservation and the use of recreational services (i.e., trophy hunting) and the remaining ~265,000 ha of the current GCA to the inhabitants of the Loliondo division for pastoralism and farming did not resolve the conflict as intended by the government (URT: Ministry of Natural Resources and Tourism 7.4.2013; URT 2010). It has rather intensified the conflict. This is due to the fact that the pastoralists of the area and their advocates argue that by redrawing the boundaries of the Loliondo GCA, the Tanzanian government is taking village land that belongs to the local population (Pastoralists Indigenous NGOs Forum et al. 2013, interviews with Maasai elders in Ololosokwan 26.10.2012, 25.10.2012).

The plan of the government did not resolve the competition and conflict between OBC and the pastoralists, since both are interested in the ecosystem services that can be derived from precisely the strip of land alongside the eastern border of the Serengeti. OBC can derive recreational services from the strip of land because wildlife is abundant in this area. In case of the pastoralists living in the villages adjacent to the Serengeti, the plan of the Tanzanian government will result in the loss of a significant part of their village land to the new Loliondo GCA. This loss concerns not only the land itself, but also the specific provisioning services, which the villagers currently obtain from that land, especially during the dry season.

The next section will focus on these specific ecosystem services the pastoralists gain from the contested land. Taking the village Ololosokwan as an illustrative

[4]Trophy hunting can be also considered as deriving provisioning service since it aims to obtain wild game as a trophy. In the following trophy hunting will be addressed as recreational ecosystem services which is part of the category "Cultural Services" (The Millennium Ecosystem Assessment 2005).

example the section aims to demonstrate how the pastoralists' land claims are directly tied to their grazing system. The section draws on observations, informal talks and official interviews conducted with the local population during an one-month research stay in the village.

9.3 The Strip alongside the Eastern Border of the Serengeti as a Dry Season Grazing Area

The Loliondo division is located within a semiarid savanna ecosystem characterized by a bimodal but generally low rainfall pattern with high inter-annual dynamics (Ojalammi 2006). Precipitation is the limiting climate factor in such ecosystems. Hence, the spatial and seasonal availability of net primary production is dependent on the precipitation patterns (Lieth 1974). Pastoralism with mobility as its key strategy, allows turning sparse, patchy, and transient net primary production into human food such as milk and meat, which is hardly possible with other modes of production within such ecosystems (Homewood 2008; Reid et al. 2008; Scoones 1994; De Jode 2010).

Maasai pastoralists follow a transhumant grazing strategy and are mobile within a certain territory (Cook 2007). Traditionally, the grazing territories of the Maasai were localities (in the local Maasai language Maa: inkutot), which were associated with a certain Maasai section (broadest subdivision of the Maasai society). The localities were usually geographically divided in uplands (osopuko) and lowlands (olpurkel) and utilized for grazing in seasonal rotation; lowlands were used during the rainy season and uplands during the dry season (Goldman 2006). However, the Ujamaa policy, introduced by the Arusha Declaration of 1967, forced the resettlement and particularly sedentarization of Maasai in distinct villages (O'Malley 2000). As a consequence, Maasai had to reorganize their rangeland management within distinct villages. Villages were subdivided into dry season grazing area and rainy season grazing area. Often, the village areas were too small to include upland and lowlands at the same time (interview with Maasai elder in Ololosokwan 17.10.2012). Nevertheless, the customary zoning of the village areas into rainy and dry season grazing areas is not random but follows distinct rationalities, as the example of the village Ololosokwan shows.

The pastoralists of Ololosokwan village demarcated the eastern area of their village as a rainy season grazing area. Herdsmen move during the rainy season with their livestock on different day routes within this area and return back to the permanent homestead (boma) in the evening. During the rainy season, traditionally all livestock is found in this rainy season grazing area. Also, most of the infrastructures such as schools, dispensaries, cattle dips, small shops, and agricultural fields are located in this area.

The western part of the village area that is adjacent to the Serengeti National Park is demarcated as a dry season grazing area (ronjo). During the rainy season, grazing in the dry season grazing area is restricted to allow the replenishment and accumulation of biomass (Reid et al. 2008). With the start of the dry season, the Maasai elders consult and decide when it is necessary to move to the dry season grazing area. Once the elders open ronjo, the herdsmen start to move to the area close to the Serengeti with the majority of the livestock. Here they build temporary homesteads. Some of the calves and sick livestock remain in the rainy season area in addition to several lactating cows, which provide the daily milk for the family members, particularly women, who remain in the permanent homestead in the rainy season area. During the stay in the dry season grazing area, the livestock is often only watered every second day, since this decreases the feed intake and thus conserves the dry season grazing resources (Perrier 1994; Igoe 2006). With the beginning of the heavy rains, the herdsmen return to the rainy season area and the grazing rotation starts again.

The rationale behind the alternation between two grazing areas depending on the season is the possibility to include periods of replenishment. But the ecosystem of the dry season grazing area has usually specific characteristics which allow pastoralists to derive provisioning ecosystem services also during the dry season. In the case of Ololosokwan the pastoralists report that, these characteristics include the comparable longer humidity of the area to the eastern border of the Serengeti National Park, which leads to a later dry up of range forage (interview with Maasai elder in Ololosokwan, 21.10.2012). Another reason to choose the area close to the Serengeti as the dry season grazing area is the availability of water throughout the year. Furthermore, the area is characterized by particularly salty soils (interview with Maasai elder in Ololosokwan, 26.10.2012). This all makes the area next to the Serengeti most valuable for the pastoralists since they can obtain provisioning services such as feed, water, and salt for their livestock also during the dry season, which ensures food security for the pastoralists.

By establishing the new boundaries of the Loliondo GCA and implementing the amendments of the new Wildlife Conservation Act of 2009 to this area, the village Ololosokwan would lose more than half of its village area, including the entire dry season area to the new GCA (Bartels 2013). The following quotes from elders of Ololosokwan illustrate that the pastoralists are especially concerned to be unable to obtain reliable provisioning ecosystem services from their current dry season grazing area (ronjo) if the new GCA is established:

> "When we lose ronjo we will lose a lot of things beside the additional land." (Interview with Maasai elder in Ololosokwan, 26.10.2012). Another elder added: "[…] the villagers can no longer access grass, quality grass, we cannot access water and we cannot access salt licks, salty areas for our animals. This is the place where we used to graze during the dry season and that used to save us during the dry season." (Interview with Maasai elder in Ololosokwan, 11.10.2012).

Indeed, previous work has shown that the current livestock herd cannot be sustained on the reduced village area of Ololosokwan, i.e., if the new Loliondo GCA is established (Bartels 2013). This is precisely the fear of the Maasai of Ololosokwan. Moreover, the Maasai of the whole Loliondo division are in general affected by the plan of the government and not only the Maasai of the villages which are adjacent to the Serengeti. This is due to the fact that the contested strip alongside the Serengeti serves as a drought buffer for all pastoralists of the division such that they migrate to the area in times of drought in order to be able to feed and water their livestock (O'Malley 2000).

It should be noted that the rotational grazing system of Ololosokwan has started to erode. This is indicated by the fact that the dry season grazing area is slowly becoming an area which is used as grazing and settlement area throughout the year. This is a gradual process which is driven by several factors, as reported by the interviewed Maasai: i) the decrease of the size of the available grazing area in the rainy season grazing area due to an increase in permanent structures, ii) an increase in both human and livestock population, and iii) that a part of the village area is already not freely accessible due to the presence of OBC. However, the area to the eastern border of the Serengeti National Park is still essential for the Maasai pastoralists not for accessing land per se but to obtain ecosystem services from the contested area especially during the dry season. Consequently, the impact of the land loss for the Maasai does not correlate in a linear fashion with the size of the lost area (Coughenour 2008). In the words of Hary et al.: "The essential point to be made here is that even a severe reduction in stocking rates on wet-season pasture cannot compensate for the loss of dry-season grazing reserves." (1996: 229).

9.4 Conclusions

This chapter has presented a case study of competition for land-based ecosystem services in northern Tanzania that has developed into a land conflict. The case study illustrates competition between pastoralists and a hunting company over the use of provisioning services (feed and water for livestock) and recreational services (trophy hunting) in a nature conservation area. The case study specifically demonstrates that distinct government policies have fostered the competition, since they made nature conservation by law compatible with recreational services but incompatible with the use of provisioning ecosystem services through pastoralism. Moreover, the intention of the government to redraw the boundaries of the nature conservation area in order to solve the land conflict has rather intensified the conflict. This is due to the fact that the government aims to allocate the contested area exclusively to nature conservation and the use of recreational services. For the pastoralists of the area, this would result in a considerable loss of access to land. But more importantly, pastoralists would not be able to derive provisioning services such as feed, water, and salt from the contested area anymore. These provisioning ecosystem services are obtainable also during the dry season from the contested strip of land and thus ensure that the area can be used as a dry season grazing area. Dry season grazing areas are distinct components of the pastoral grazing system and the livelihood of the Maasai, since they secure the survival of the livestock during this season and thus the food security of the pastoralists. The local food security is, however, jeopardized by the current attempt to exclude the Maasai from the contested area for the purpose of nature conservation and in favor of safeguarding the use of recreational services.

Acknowledgments The author is grateful in particular to Paul Ole Saing'ue and Geofrey Mwanjela for their invaluable contribution to the fieldwork on which this chapter is based. Helmut Haberl and Daniel Müller contributed constructive comments on earlier versions of this chapter. This research has been partly funded by the German Federal Ministry of Education and Research (BMBF) under the (project funding) reference number 01 LN 1316 A.

References

Alexander, C. (1993). The Brigadier's shooting party. *The New York Times*, 13.11.1993. Available online at http://www.nytimes.com/1993/11/13/opinion/the-brigadier-s-shooting-party.html, checked on 15.7.2013.

Athumani, R. (2013a). Loliondo partition abides by law. *Tanzania Daily News*, 2.4.2013. Available online at http://www.dailynews.co.tz/index.php/local-news/16050-loliondo-partition-abides-by-law, checked on 15.7.2013.

Athumani, R. (2013b). Tanzania: State sticks to its guns on Maasai Land. *Tanzania Daily News*, 10.4.2013. Available online at http://allafrica.com/stories/201304100315.html, checked on 15.7.2013.

Azarya, V. (2004). Globalization and International tourism in developing countries. Marginality as a commercial commodity. *Current Sociology, 52*(6), 949–967.

Baldus, R., Kibonde, B., & Siege, L. (2003). Seeking conservation partnerships in the Selous game reserve. *Tanzania. Conservation Partnership in Africa, 13*(1), 50–61.

Bartels, L. E. (2013). Socio-ecological impacts of land grabbing for nature conservation on a pastoral community: A HANPP-based case study in Ololosokwan Village, Northern Tanzania. *Social Ecology Working Paper 149*. Vienna. Available online at https://www.uni-klu.ac.at/socec/downloads/WP149_webversion.pdf, checked on 15.7.2013

Benjaminsen, T. A., & Bryceson, I. (2012). Conservation, green/blue grabbing and accumulation by dispossession in Tanzania. *Journal of Peasant Studies, 39*(2), 335–355.

Benjaminsen, T. A., Bryceson, I., Maganga, F., & Refseth, T. (2011). Conservation and land grabbing in Tanzania. *Paper presented at the International Conference on land Grabbing*, 6–8 April 2011, University of Sussex. Available online at http://www.iss.nl/fileadmin/ASSETS/iss/Documents/Conference_papers/LDPI/20_Tor_A._Benjaminsen_-_Ian_Bryceson_-_Faustin_Maganga_-_Tonje_Refseth.pdf, checked on 1.11.2013.

Brockington, D. (2002). *Fortress conservation: The preservation of the Mkomazi game reserve, Tanzania*. Bloomington: Indiana University Press.

Brockington, D. (2008). Preserving the New Tanzania: Conservation and land use change. *The International Journal of African Historical Studies, 41*(3), 557–579.

Brockington, D. (1999). Conservation, displacement and livelihoods. The consequences of the eviction for pastoralists moved from the Mkomazi Game Reserve, Tanzania. *Nomadic People, 3*(2), 74–96.

Cook, A. E. (2007). *Subdividing the Savanna: The ecology of change in Northern Tanzania*. Dissertation. University of North Carolina, Chapel Hill. Available online at http://cee.unc.edu/people/graduate-students/theses/Cooke_PhD.pdf, checked on 1.11.2013.

Coughenour, M. B. (2008). Causes and consequences of herbivore movement in landscape ecosystems. In K. A. Galvin, S. Reid, R. H. Behnke Jr, & T. N. Hobbs (Eds.), *Fragmentation in semi-arid and arid landscapes. Consequences for human and natural landscapes* (pp. 45–91). Dordrecht: Springer.

De Jode, H. (2010). *Modern and mobile. The future of livestock production in Africa's drylands*. International Institute for Environment & Development; Sahel International UK.

Ernstson, H. (2013). The social production of ecosystem services. A framework for studying environmental justice and ecological complexity in urbanized landscapes. *Landscape and Urban Planning, 109*(1), 7–17.

Exner, A., Bartels, L. E., Windhaber, M., Fritz, S., See, L., Politti, E., & Hochleithner, S. (2015). Constructing landscapes of value. Capitalist investment for the acquisition of marginal or unused land—The case of Tanzania. *Land Use Policy, 42*, 652–663.

FAO (ed.). 2005. *Livestock sector brief*. United Republic of Tanzania, Available online at http://www.fao.org/ag/againfo/resources/en/publications/sector_briefs/lsb_TZA.pdf checked on 10.7.2012.

Gardner, B. (2012). Tourism and the politics of the global land grab in Tanzania: Markets, appropriation and recognition. *Journal of Peasant Studies, 39*(2), 377–402.

Geisler, C., & de Sousa, R. (2001). From refuge to refugee: The African case. *Public Administration and Development, 21*(2), 159–170.

Goldman, M. J. (2006). *Sharing pastures, building dialogues: Maasai and wildlife conservation in nothern Tanzania*. Dissertation. University of Wisconsin-Madison, Wisconsin.

Gibson, A. (2011). *Sustainability of Conservation Efforts in Tanzania*. Management of Game Controlled Areas in Tanzania; Challenges & Prospects: A Case Study of Kilombero Game Controlled Area. Saarbrücken: Lambert Academic Publishing.

Hary, I., Schwartz, H.-J., Pielert, V. H. C., & Mosler, C. (1996). Land degradation in African pastoral systems and the destocking controversy. *Ecological Modelling, 86*, 227–233.

Homewood, K. (2008). *Ecology of African pastoralist societies*. Oxford, Athens, Pretoria: James Currey, Ohio University Press, Unisa Press.

Igoe, J. (2006). Ecosystem dynamics and institutional inertia: A discussion of landscape conservation in Northern Tanzania. In M. Jayalaxshmi & B. Andreas (eds.), *Savannas and dry forest. Linking people with nature* (pp. 77–103). Aldershot: Ashgate.

James, B. (2013). Activists say plans to split Loliondo a brazen attempt to snatch local land. *The Citizen*, 5.4.2013. Available online at http://africa.widmi.com/index.php/tanzania/the-citizen/tanzanian-news/1523-activists-say-plans-to-split-loliondo-a-brazen-attempt-to-snatch-local-land, checked on 15.7.2013.

Kimati, B. (2013). Loliondo land dispute in perspective: Change, resentment from stakeholders. *Tanzania Daily News*, 20.4.2013. Available online at http://allafrica.com/stories/201304220687.html, checked on 15.7.2013.

Kisanga, D. (2013). Loliondo Maasai elders want to meet President Kikwete. *IPP-Media*, 17.5.2013. Available online at http://www.ippmedia.com/frontend/?l=54801, checked on 15.7.2013.

Lieth, H. (1974). Modeling the primary productivity of the world. In H. Lieth & R. H. Whittaker (eds.), *Primary productivity of the biosphere* (pp. 237–263). Berlin-Heidelberg-New York: Springer (Ecological Studies, 14).

Lindsey, P., Roulet, P., & Romanach, S. (2007). Economic and conservation significance of the trophy hunting industry in sub-Saharan Africa. *Biological Conservation, 134*(4), 455–469.

Makoye, K. (2013). Tanzania allows Maasai Herders to stay in disputed wildlife corridor. *AlertNet*, 2.10.2013. Available online at http://allafrica.com/stories/201310021621.html?viewall=1, checked on 8.11.2013.

Mbashiru, K. (2014). Nyalandu: Maasai will stay. *The Citizen*, 21.11.2014. Available online at http://www.thecitizen.co.tz/News/Nyalandu–Maasai-will-stay/-/1840392/2530184/-/k315yc/-/index.html, checked on 23.11.2014.

The Millennium Ecosystem Assessment (2005). *Ecosystems and human well-being*. Synthesis. Washington, DC: Island Press.

Ndoinyo, Y. E. (2002). *Natural resources managment in Ololosokwan Village, Ngorongoro district, Tanzania*. Bachelor Thesis. University of Dar es Salaam, Dar es Salaam.

Nelson, F., Ole Makko, S. (2003). Communities, conservation, and conflicts in the Tanzanian Serengeti. Preserving rights to gain benefits. In B. Child & M. W. Lyman (eds.), *Natural resources as community assets. Lessons from two continents. Turning natural ressources into assets*. Sand County Foundation (Wisconsin), The Aspern Institute (Washington), pp. 121–145. Available online at http://www.sandcounty.net/assets/chapters/assets_chapter_5.pdf, checked on 26.2.2013.

Nelson, F., Sulle, E., & Lekaita, E. (2012). Land grabbing and political transformation in Tanzania. *Paper presented to the Global Land Grabbing II Confrerence*. Cornell University, Ithaca NY, October 17–19, 2012. Available online at http://www.cornell-landproject.org/download/landgrab2012papers/nelson.pdf, checked on 28.2.2012.

Neumann, R. P. (1998). *Imposing wilderness. Struggles over Livelihood and nature preservation in Africa*. Berkeley, Los Angeles, London: University of California Press.

Ngoitiko, M., Sinandei, M., Meitaya, P., & Nelson, F. (2010). Pastoral activists: Negotiating power imbalances in the Tanzanian Serengeti. In Nelson, F. (ed.), *Community rights, conservation and contested land. The politics of Natural Resource governance in Africa* (pp. 269–289). London: Earthscan.

Nkwame, M. (2013a). Tanzania: CCM moves to solve Loliondo land dispute. *Tanzania Daily News*, 5.4.2013. Available online at http://www.dailynews.co.tz/index.php/local-news/16147-ccm-moves-to-solve-loliondo-land-dispute, checked on 15.7.2013.

Nkwame, M. (2013b). Tanzania: PM ends Loliondo long-running land conflict. *Tanzania Daily News*, 25.9.2013. Available online at http://allafrica.com/stories/201309250290.html, checked on 8.11.2013.

Ojalammi, S. (2006). *Contested lands: Land disputes in semi-arid parts of Northern Tanzania. Case studies of the Loliondo and sale divisions in the Ngorongoro District*. Dissertation. University Helsinki, Helsinki. Department of Geography. Available online at http://ethesis.helsinki.fi/julkaisut/mat/maant/vk/ojalammi/conteste.pdf, checked on 27.2.2013.

O'Malley, M. E. (2000). *Cattle and cultivation: Changing land use and labor patterns in pastoral Maasai livelihoods, Loliondo Division, Ngorongoro District, Tanzania*. Dissertation. University of Colerado, Colorado. Department of Anthropology.

Pastoralists Indigenous NGOs Forum—PINGOs Forum; Tanzania Land Alliance—(TALA); HakiArdhi; Legal and Human Rights Centre (LHRC); Tanzania Human Rights Defenders; Tanzania Gender Network Program (TGNP); Tanzania Natural Resource Forum (TNRF); Community Research and Development Services (CORDS); Ujamaa Community Resource Team (CRT); Pastoralists Women Council (PWC); Ngorongoro NGOs Network (NGONET); Tanzania Pastoralists Community Forum (TPCF); OSEREMI (2013). Press Statement from Civil Society Organisations in Tanzania regarding Plans by the Government to alinate in Lolidondo, 4.4.2013 Arusha.

Patinkin, J. (2013). Tanzania's Maasai battle game hunters for grazing land. *BBC News*. Available online at http://www.bbc.co.uk/news/world-africa-22155538, checked on 15.7.2013.

Perrier, G. (1994). New directions in range management planning in Africa. In I. Scoones (Ed.), *Living with uncertainty: New directions in pastoral development in Africa* (pp. 47–57). London: Intermediate Technology Publications.

Reid, R. S., Galvin, K. A., & Kruska, R. S. (2008). Global significance of extensive grazing lands and pastoral societies: An introduction. In K. A. Galvin, R. S. Reid, R. H. Behnke Jr, & T. N. Hobbs (eds.), *Fragmentation in semi-arid and arid landscapes. Consequences for human and natural landscapes* (pp. 1–24). Dordrecht: Springer.

Sarwatt, S., & Mollel, E. (2000). *United Republic of Tanzania. Country pasture/forage resources profils*. FAO. Available online at http://www.fao.org/ag/AGP/AGPC/doc/Counprof/tanzania/tanz.htm#1, checked on 23.11.2014.

Scoones, I. (Ed.). (1994). *Living with uncertainty: New directions in pastoral development in Africa*. London: Intermediate Technology Publications.

Skof, A. (2008). The tourism industry. In R. J. Utz (Ed.), *Sustaining and sharing economic growth in Tanzania* (pp. 159–168). Washington, DC: World Bank.

Smith, D. (2014). Tanzania accused of backtracking over sale of Masai's ancestral land. *The Guardian*, 16.11.2014. Available online at http://www.theguardian.com/world/2014/nov/16/tanzania-government-accused-serengeti-sale-maasai-lands, checked on 23.11.2014.

Spangenberg, J. H., Görg, C., Truong, D. T., Tekken, V., Bustamante, J. V., & Settele, J. (2014). Provision of ecosystem services is determined by human agency, not ecosystem functions. Four case studies. *International Journal of Biodiversity Science, Ecosystem Services & Management, 10*(1), 40–53.

Tanzania Natural Resource Forum; Maliasili Initiatives (2011). *Integrating pastoralist livelihoods and wildlife conservation? Optioned for land use and conflict resolution in Loliondo Division, Ngorongoro District.* Available online at http://www.tnrf.org/files/Integrating%20Pastoralist%20Livelihoods%20and%20Wildlife%20Conservation_FINAL_FINAL.pdf, checked on 3.3.2013.

United Republic of Tanzania; Ministry of Natural Resources and Tourism (7.4.2013). Press release. Dar es Salaam.

United Republic of Tanzania (1974). *The wildlife conservation act of 1974.*

United Republic of Tanzania (1998). *Land Act of 1998*

United Republic of Tanzania (2007). *The land use planning act of 2007.*

United Republic of Tanzania (2009a). *Fourth National Report on Implementation of Convention on Biological Diversity (CBD).* Dar es Salaam. Available online at http://www.cbd.int/doc/world/tz/tz-nr-04-en.pdf, checked on 22.9.2013.

United Republic of Tanzania (2009b). *Wildlife conservation act of 2009.*

United Republic of Tanzania (2010). *Draft: District land use framework plan (2010–2030).* With assistance of national land use planning commission, Ngorongoro District Council.

United Republic of Tanzania (2012). *National sample census of agriculture 2007/2008.* Volume III: Livestock Sector—National Report.

Wade, D. J., Mwasaga, B. C., & Eagles, P. F. J. (2001). A history and market analysis of tourism in Tanzania. *Tourism Management, 22*(1), 93–101.

Chapter 10
How the Collapse of the Beef Sector in Post-Soviet Russia Displaced Competition for Ecosystem Services to the Brazilian Amazon

Florian Schierhorn, Alex Kramer Gittelson and Daniel Müller

Abstract The collapse of the Russian livestock sector triggered widespread agricultural land abandonment in post-Soviet Russia. The beef industry declined in particular, and consequently, Russia became heavily dependent on beef imports, from Europe in the 1990s and from Brazil after 2002. Therefore, Russian demand substantially contributed to the growth of the Brazilian beef sector and fostered widespread agricultural land expansion and deforestation in the Brazilian Amazon. The beef trade from Brazil to Russia was associated with substantial environmental costs in terms of carbon emissions and loss of biodiversity. While the abandoned agricultural land in Russia has become an important terrestrial carbon sink that would be largely diminished by re-cultivation, we argue that increasing agricultural output through re-cultivation or the expansion of grazing within Russia may be desirable from a global perspective, if the high environmental costs of production elsewhere are taken into account.

Keywords Land abandonment · Deforestation · Soybean · Trade-off · Carbon sink

F. Schierhorn (✉) · D. Müller
Leibniz Institute of Agricultural Development in Transition Economies (IAMO), Halle, Germany
e-mail: schierhorn@iamo.de

A.K. Gittelson
Fulbright Research Scholarship, International Institute of Education, New York, USA

D. Müller
Department of Geography, Humboldt-Universität zu Berlin, Berlin, Germany

D. Müller
Integrative Research Institute on Transformations of Human-Environment Systems (IRI THESys), Humboldt Universität zu Berlin, Unter den Linden 6, 10099 Berlin, Germany

J. Niewöhner et al. (eds.), *Land Use Competition*,
Human-Environment Interactions, DOI 10.1007/978-3-319-33628-2_10

10.1 Introduction

Demand for food, feed, and bioenergy will continue to grow and increase land competition (Foley et al. 2011; Valin et al. 2014). Moreover, land resources have important functions with regard to biodiversity conservation and climate change mitigation through carbon sequestration (Lal 2004; Tscharntke et al. 2012). Meat and milk productions are particularly land demanding (Wirsenius et al. 2010), and the livestock industry currently occupies approximately 30 % of the planet's ice-free terrestrial surface area or 70–75 % of all agricultural land (Foley et al. 2011; Steinfeld et al. 2006). Global expansion of pasture and feed crops at the expense of forested land accounts for 8 % of total anthropogenic CO_2 emissions (Steinfeld et al. 2006). Consequently, the livestock sector contributes considerably to the rising global land-use scarcity and is intertwined with the increasing competition for global land resources (Lambin and Meyfroidt 2011; Smith et al. 2010).

A number of regions have ample potential for future increases in agricultural output at comparatively low environmental costs (Lambin et al. 2013). For example, the Russian Federation is a global hot spot for untapped agricultural resources (Schierhorn et al. 2014b). In 1991, the dissolution of the Soviet Union (SU) triggered unprecedented socioeconomic, institutional, and environmental changes, including strong rural population decline, curtail of governmental support, land reforms, and widespread agricultural land abandonment (Liefert and Liefert 2012; Prishchepov et al. 2013). Competition for land-based ecosystem services during the era of the SU (Peterson 1993) substantially decreased after 1991. The cropland area in Russia has declined by 41 million hectares since 1991, largely driven by the collapse of the livestock sector (Schierhorn et al. 2014b). As a result, Russia harbors the largest amount worldwide of unused former agricultural lands. The widespread abandoned agricultural land resources suggest ample scope for agricultural expansion in order to produce food, feed, fiber, and fuel, as well as for sequestration of carbon in the abandoned soils (Kurganova et al. 2014; Schierhorn et al. 2013; Vuichard et al. 2009). However, increasing biomass production and mitigating climate change (through carbon sequestration) may often be mutually exclusive. The abandoned former agricultural lands in Russia offer an excellent case study to examine the competition between different land-based ecosystem services that can be produced on these currently unused land resources that arise.

Since 1991, Russia has become one of the largest net importers of livestock products, brought about by fading state support as well as low productivity and technical efficiency in livestock production (Lioubimtseva and Henebry 2012). Meat consumption decreased since 1991, but not as much as production, leading to large insufficiencies in domestic meat production, beef in particular. During the 1990s, Russia became the largest importer of beef from the EU (FAO 2014). Beef production in the EU, in turn, is dependent on Brazilian soy, which provides the protein basis in EU cattle fattening. In that way, land-use and production changes in Russia indirectly accounted for deforestation and loss of ecosystem services in the Brazilian Amazon (Nepstad et al. 2006). In the early 2000s, Brazil emerged as the leading global producer and exporter of beef, selling the highest quantities of its production directly to Russia where domestic production collapsed but demand for beef rebounded in the 2000s. Substantial environmental costs in terms of carbon emission and loss of biodiversity were directly associated with the increasing beef trade from Brazil to Russia.

The Russian example demonstrates the globalization of land-use competition because trade from Brazil to Russia satisfies Russian demand for land-based services with production in distant locations. Such trade links manifest the increasing disconnection between places of production and consumption and, in that way, complicate the understanding of drivers and implications of land competition (see also Chaps. 1, 2 and 8). The collapse of the Russian beef sector and the displacement of its beef production, along with the impacts on land-use and ecosystem services, to the Brazilian Amazon is the topic of this case study. First, we analyze the drivers and implications of the collapse of beef production in Russia after the dissolution of the SU. We focus on beef because feeding cattle is highly land demanding and because the collapse of the Russian beef sector accounts for the bulk of post-Soviet cropland abandonment. The contraction of the Russian livestock sector has contributed to rising global demand for beef, which in turn was satisfied by the expanding beef sector in Brazil and contributed to deforestation of the Brazilian Amazon. We demonstrate losses of ecosystem services due to deforestation in the Brazilian Amazon triggered by expanding beef production through an extensive literature review. We argue in the concluding statements that a sophisticated assessment of the potential for Russia to increase its beef production should account for the environmental costs associated with trade.

10.2 Consumption, Production, and Trade of Beef

10.2.1 High Beef Consumption and Production in the Soviet Union

In the early 1970s, meat consumption of Soviet citizens lagged behind that of Americans and Europeans. However, per capita meat consumption in the SU drastically increased, from 49 kg in 1970 to 72 kg in 1989, almost to the level of Europe (79 kg in 1989, FAO 2014). Meat consumption during the late period of the SU was largely based on beef and pork, and much less on poultry (Fig. 10.1). Beef accounted for almost half of the total meat intake per capita in 1990 (Fig. 10.1). High meat and dairy product consumption in the SU was politically motivated and supported by the state (Osborne and Trueblood 2002). Medical and nutritional establishments created standards for dairy products that called for high levels of protein intake. The SU ensured increasing consumption of high-value livestock products with policies to improve the standard of living (Liefert 2004; Sedik et al. 2003). The substantial state subsidies and rigorous price policy resulted in consumer prices for meat that were far below the real costs of production (Liefert and Liefert 2012). Food and agricultural subsidies reached 11.8 % of GDP in 1990. As

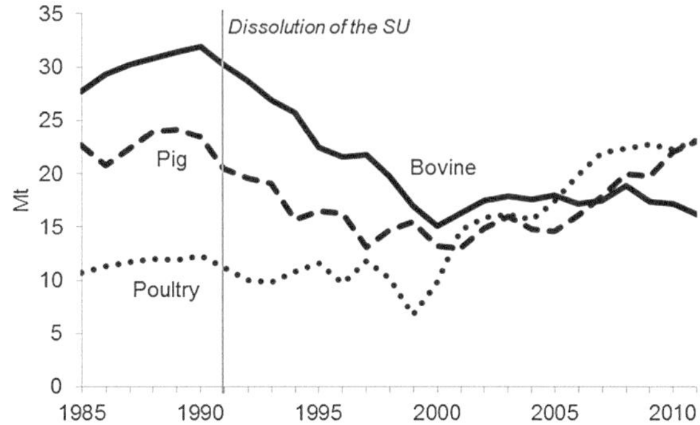

Fig. 10.1 Meat consumption in Russia (Mt/yr, million tons per year). Data taken from USDA (2014)

a result of the low consumer prices, meat consumption in the SU was much higher compared to countries of similar per capita GDP and income (Sedik et al. 2003).

Despite the very high consumption of meat, the SU achieved self-sufficiency in domestic meat production. Cattle numbers increased from 76 to 122 million heads between 1961 and 1987. Almost 10 % of global cattle stock came from the SU in 1987, and in 1989, the SU accounted for 18 % of world meat production (Csaki and Lerman 1992; FAO 2014) and only 3 % of the beef was imported. However, domestic harvests of grain were not sufficient to feed the large livestock population despite the allocation of substantial amounts of agricultural land to fodder and feed production. Therefore, the SU became a major importer of grains for feed toward the end of the socialist period. Grain imports peaked at 49 Mt/yr in 1988 (FAO 2014). The bulk of the total SU cereal imports (29 Mt/yr or 55 %) went to Soviet Russia in 1988 (USDA 2014).

Soviet Russia produced only 70–77 % of its beef consumption domestically (USDA 2014). Long-term contracts for imports from Belarus, Ukraine, and other SU republics filled these production gaps (Csaki and Lerman 1992; USDA 2014). Per capita meat consumption in Soviet Russia was higher than that in the rest of the SU, in terms of both quantity and quality (USDA 2014). In the case of beef, annual imports during the last years of Soviet Russia were higher than annual imports to the Russian Federation (Russia hereafter) in the years following the collapse of the SU (Fig. 10.2), a fact frequently disregarded in the literature. In sum, the enormous

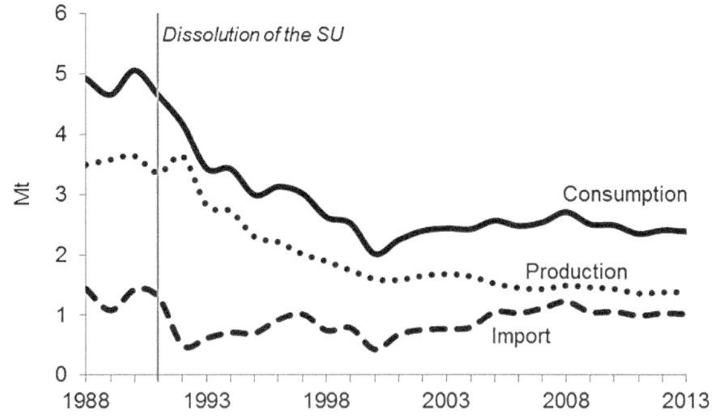

Fig. 10.2 Beef production, consumption, and import in Soviet Russia (1988–1991) and Russia (1992–2014). Mt/yr, million tons per year. Data taken from USDA (2014)

demand for state-subsidized meat in Soviet Russia resulted in large net imports of agricultural products, not only of meat (which mainly originated from Belarus, Ukraine, and other SU republics), but also of fodder grains.

10.2.2 The Collapse of the Beef Sector After 1991

The dissolution of the SU in 1991 triggered drastic changes and divisions in the regional economies. GDP per capita and year decreased from almost 8000 US$ in 1990 to 4000 US$ in 1996, the unemployment rate increased rapidly (Stillman 2006), and the 1992 reforms also caused a short period of hyperinflation (Osborne and Trueblood 2002). Per capita consumer income fell by 47 % in the first year after the dissolution of the SU (Liefert 2004). Along with the collapse in state revenues, subsidies to the agricultural sector decreased from 11.8 % to less than 2 % of GDP between 1990 and 1995 (Osborne and Trueblood 2002). Hence, input and output prices for pork and beef quickly adjusted to the real production costs (Liefert 2004; Staudigel 2011; Wegren 1992).

Because of the economic downturn, real wages decreased and prices increased, leading to a sharp decline of consumption of meat and milk products. Annual per capita beef consumption halved, from 32 kg in 1990 (FAOSTAT data for 1990 are only available for entire SU) to 15 kg in 2000 (FAO 2014). The drop in the consumption of beef was much higher than that of poultry and pork (Fig. 10.1). Beef prices climbed because of a longer duration of the production cycle, resulting in lower feed efficiency of cattle than of poultry and pigs (Hocquette and Chatellier 2011). Meanwhile, consumption of non-meat products increased, particularly that of grain and potatoes, which compensated for the losses of calories from livestock products (Sedik et al. 2003; Staudigel 2011). According to annual measures of per capita caloric intake, sizable decreases in the consumption of animal products failed to compromise food security in Russia, because diets shifted away from meat to plant-based calories. However, concerns regarding food security have persistently been an issue for the political establishment in post-Soviet Russia (Wegren 2011).

After the financial crisis of 1998 and under the new president Vladimir Putin, the Russian economy rebounded thanks to increases in oil and gas exports as well as a rise in world energy prices, which led to increases in export earnings (Stillman 2006; Wegren 2002). The resulting steady growth of GDP and increases in real wages in turn stimulated higher consumption of livestock products. Still, the increase of intake of pork and poultry was much higher than that of beef after 1998 (Fig. 10.1). In the mid-2000s, consumption of poultry and pork meat overstepped beef consumption (Fig. 10.1). The booming poultry sector and increasing consumption of chicken partly contributed to the decline in beef production (Antonova and Zeller 2007).

The number of cattle decreased by 64 %, from 55 million in 1992 to less than 20 million in 2013 (FAO 2014). Worldwide, a similar drastic decline in the number of cattle only occurred in the USA after the mid-1970s (FAO 2014). Beef consumption decreased along with cattle stock (Fig. 10.2). When the effects of the financial crisis subsided, beef consumption slightly rebounded, but low per capita consumption of beef (relative to late SU times) has been persistent. Beef consumption after 1998, however, has dropped less than domestic production (Fig. 10.2). As a result, Russia has become a large importer of beef, which also reflects the lack of competitiveness of the Russian beef production sector in relation to the world market (Liefert 2004; Liefert and Liefert 2012). In 2001, Russia accounted for 19 % of global beef imports (FAO 2014).

10.2.3 Collapse of the Livestock Sector Triggered Land Abandonment

After 1991, the Russian agricultural sector shifted its focus away from meat and fodder grain toward export-oriented crop production (Liefert et al. 2010). Consequently, in the early 2000s, Russia emerged as a leading player in the world grain market; the country was among the top five wheat-exporting countries between 2006 and 2011 (FAO 2014). According to Liefert (2002), the switch from meat to crop production has been an inevitable market-regulated adjustment due to Russia's comparative advantage. Despite this increase in production and crop exports, a significant drop in domestic livestock production—particularly in beef production—caused a massive decline in the demand of fodder crops (Schierhorn et al. 2014b). The contraction of fodder crops and grains other than wheat were large contributors in the decrease in total sowing area in Russia, which dropped from 118 Mha in 1990 to 77 million hectares (Mha) in 2011 (ROSSTAT 2014). This decrease (41 Mha) is equivalent to the entire sowing areas in 2010 of France, Germany, and Spain combined. The abandoned croplands cluster outside the black soil belt in temperate Russia and in the southern steppes along the border of Kazakhstan (Schierhorn et al. 2013). Although abandoned cropland is mostly located in areas where environmental conditions are only moderately suitable for crop production, the large size of abandoned cropland in Russia suggests that agricultural output can be substantially increased (Schierhorn et al. 2014b). However, a large carbon sink developed in the soils and successional vegetation on the cropland that was abandoned soon after the dissolution. The re-cultivation of these lands would lead to large carbon emissions (Kurganova et al. 2014; Schierhorn et al. 2013; Vuichard et al. 2008).

10.2.4 Russia's Beef Imports After 1991

As described earlier, the SU had been almost independent from foreign meat markets, but soon after 1991, the Russian Federation became heavily dependent on meat imports. Globally, Russia and Japan were the largest importers of meat between

1994 and 2009 (except 1999 and 2000 for Russia, Fig. 10.3). The USA was the largest importer of beef between 1992 and 2011 (except for 1995), but Russia ranked third between 1993 and 2003 (after Japan) and second between 2004 and 2011.

During the 1990s, most of the imported beef came from the EU (Fig. 10.4), where beef production was heavily subsidized to reduce surplus on domestic European markets (Antonova and Zeller 2007). Conversely, imports from countries of the former SU ceased after 1991 because of an unfavorable balance of payments, food export restrictions imposed by Russia (Csaki and Lerman 1992), and changes in consumer preferences of the Russian population after many decades of limited food choice during the SU. Meat imports in Russia, particularly of beef and pork, decreased substantially after the severe financial crisis in 1998 (Fig. 10.4), but Europe and the USA provided food aid to Russia, thereby mitigating food insecurity (Liefert 2004).

After the crisis, Russia's economy has improved remarkably. Since 1999, per capita GDP has grown at an average annual rate of 6 % (Liefert 2004), and domestic demand for livestock products increased because of high income elasticity of livestock products (Regmi et al. 2001). However, annual per capita beef consumption after 2000 remained low at approximately 18 kg or only 60 % of the 30 kg of beef consumption during the end of the SU (FAO 2014). Nevertheless,

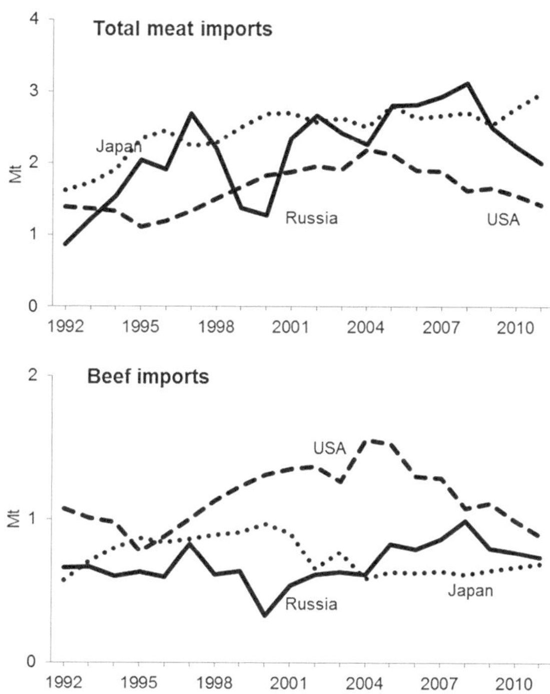

Fig. 10.3 Total meat imports and beef imports for the top 3 meat and beef importing nations. Mt/yr, million tons per year. Data taken from FAO (2014)

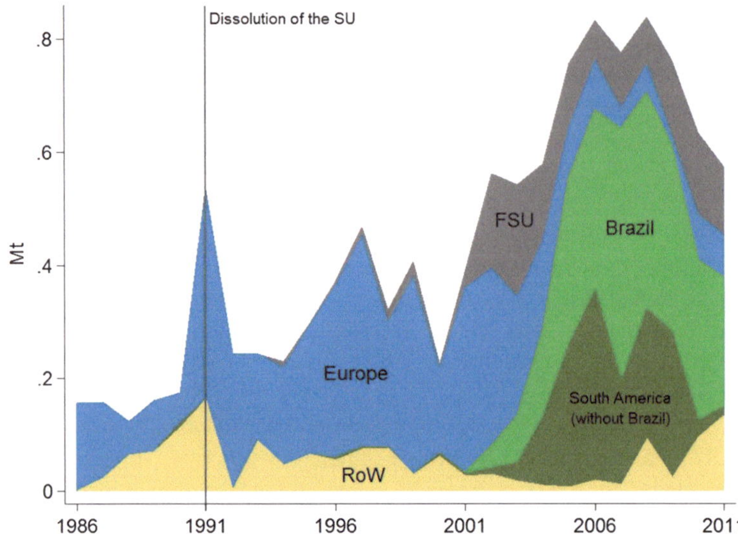

Fig. 10.4 Beef exports to Russia (Mt, million tons). Data taken from FAO (2014). *RoW* rest of the world; *FSU* former Soviet Union excluding Russia

Russian livestock producers could not satisfy increasing domestic demand; hence, the increase in imports was due to the substantial and continuing decline in the Russian beef production (Fig. 10.2). At the same time, beef produced in the EU was banned by Russia, because of sanitary concerns and disease outbreaks, such as bovine spongiform encephalopathy (BSE) in the UK (Dyck and Nelson 2003). As a result, Brazil replaced the EU as the leading exporter of beef to the Russian market.

10.3 The Booming Beef Sector in Brazil

Brazil has ample land and water resources as well as relatively cheap agricultural labor (Nepstad 2005), which has allowed beef production to become one of the most important sectors of Brazilian agribusiness. Today, Brazil's cattle herd is the largest in the world (FAO 2014) and grazing lands occupy approximately 160 Mha of Brazil in 2006 (Cederberg et al. 2011). In 2009, about 8.5 million people in Brazil were directly or indirectly employed in the Brazilian cattle industry, generating approximately 7 % of the total Brazilian gross domestic product (Millen and Arrigoni 2013). However, the Brazilian cattle industry has gained its status relatively recently—annual beef production increased from 6.5 Mt in 2000 to 9.3 Mt in 2007, an increase of 43 %. During this period, annual beef exports soared by 349 %, from 0.5 Mt in 2000 to 2.2 Mt in 2007 (USDA 2014). Brazilian beef exports peaked in 2007, when 21 % of the world beef trade originated from Brazil (FAO 2014).

A number of drivers have fostered the boom of the Brazilian beef sector during the early 2000s: First, the global price for beef almost doubled between December 1998 and December 2002, because of the devaluation of the Brazilian currency (Nepstad et al. 2006; Richards et al. 2012). At the same time, the price of Brazilian beef in dollars decreased, which made Brazilian beef more competitive on international markets (Kaimowitz et al. 2004; Vieira and Traill 2008). Higher beef prices and improved financial position on trade markets were important incentives for farmers to increase production and exports of beef (Kaimowitz et al. 2004). Second, foot-and-mouth disease (FMD) outbreaks in Brazil decreased since the late 1990s, which improved the competitiveness of Brazilian beef in the world meat trade (Kaimowitz et al. 2004). Third, other important beef producers, namely Argentina, Uruguay, North America, and the European Union, were affected by outbreaks of BSE and FMD and had to slaughter high numbers of animals in response (Jin et al. 2004). Beef from Brazil partly compensated for the production shortfalls from these markets (Paulino and Duarte 2014). Europe has also imported larger quantities from Brazil during the early 2000s because European consumers lost confidence in European beef in response to the outbreaks of BSE and FMD in Europe. Fourth, large capital inflows from domestic and foreign investments triggered increases in livestock productivity during the 2000s, with the use of better genetic lines of cattle and improved land management of pastures (Nepstad et al. 2006). Driven by low costs for labor and high productivity, production costs of beef in Brazil are 50 and 60 % lower than those in the USA and Australia, respectively (Paulino and Duarte 2014). Last, the Association of Brazilian Beef Exporters picked up the previously listed structural improvements for beef production and successfully promoted Brazilian beef on international markets (Thomé et al. 2012).

Despite the institutional barriers in the marketing channels for Brazilian beef in Russia (Thomé et al. 2012), Russia imported increasing amounts of Brazilian beef, from 0.04 Mt in 2002 to 0.45 Mt in 2007 (+1025 %). Indeed, Russia was the largest importer of Brazilian meat between 2004 and 2011 (Fig. 10.4). Since 2005, Russia has imported 35–49 % of its total beef imports from Brazil at the expense of imports from Europe and the USA. One advantage of Brazilian beef is its low-fat content, which matches well with Russian consumer preference (Millen and Arrigoni 2013).

10.3.1 Export of Beef and Soy Fostered Widespread Deforestation in the Brazilian Amazon

Since the late 1970s, the Brazilian Amazon has witnessed rapid and large-scale forest conversion, particularly along the "arc of deforestation." From 1970 to 2008, more than 17 % of the Amazon forests have been cleared (Baccini et al. 2012). Deforestation has peaked in the early 2000s, but declined after 2006 (Macedo et al. 2012). Forest carbon density is mostly high in the Amazon forest, which holds about one-tenth of the global carbon in terrestrial ecosystems of tropical regions

(Baccini et al. 2012; Saatchi et al. 2007). The annual net emissions from deforestation were approximately 191 million tons of CO_2 for the period 2000–2007 (McAlpine et al. 2009).

The primary direct cause of Amazonian deforestation has been a growing Amazonian cattle population for beef production, which triggered widespread pasture expansion in the Brazilian Amazon. From 1990 to 2007, Amazonian cattle increased from 22 million to 74 million head (Kaimowitz et al. 2004), thanks to maintaining an FMD-free status, which contributed to an increase in beef exports (McAlpine et al. 2009; Nepstad et al. 2006, see Sect. 10.3). In 2008, approximately one-third of all Brazilian beef exports originated from the Amazon (McAlpine et al. 2009) and the cattle industry was responsible for more than two-thirds of annual deforestation in the Brazilian Amazon (Nepstad et al. 2006).

Soybean expansion is another prominent cause of Amazonian deforestation. Since the mid-1990s, Brazilian production and export of soybean oil and soybean meal increased substantially, driven by the increasing global demand for high-protein animal feed and refined cooking oil, particularly in the EU and China (FAO 2014). In 2001, the EU banned animal-based protein fodder in response to Europe's widespread BSE outbreaks (Martin 2014). The resulting deficit of protein was filled mainly by Brazilian soy (Nepstad et al. 2006). Between 2000 and 2010, Brazil's soybean exports increased by 125 % and area harvested for soybean increased from 17.7 Mha in 1990 to 46.2 Mha in 2010 (+260 %). Soybean expansion concentrated initially in the Cerrado but more recently included the Amazon as well (Pacheco 2012). One-third of all Brazilian soy is produced in the state of Mato Grosso, the main agricultural hot spot of Brazil. Only relatively small direct conversion of Amazon forests to soybean planting occurred, but soybean production expanded to land previously cleared for pasture. Thus, soybean expansion displaced deforestation for expansion of cattle pastures further into the Amazon rainforest (Arima et al. 2011; Barona et al. 2010).

10.3.2 The Contribution of Russian Beef Imports to Deforestation in the Brazilian Amazon

Recently, a number of studies addressed the increasing global displacement of land use, i.e., when production and consumption of products are spatially separated (Karstensen et al. 2013; Lathuillière et al. 2014; Meyfroidt et al. 2013; Zaks et al. 2009). An excellent example for such displacement is the beef-related teleconnection between Brazil and Russia. During the 2000s, Russia became the largest importer of Brazilian beef and thus indirectly contributed to a sizable portion of deforestation due to the expansion of pastures in the Amazon. Therefore, environmental costs in terms of carbon emission and loss of biodiversity are embodied in the beef trade from Brazil to Russia. Karstensen et al. (2013) quantified CO_2 emissions from Brazilian deforestation along the supply chain to all countries importing agricultural commodities from Brazil. They found that Russia has become by far the world's largest importer of CO_2 emissions embodied in Brazilian beef by the mid-2000s. From 2004 to 2010, Russia annually accounted for 15–20 Mt CO_2 of the total carbon emissions from deforestation embodied in beef products. However, the CO_2 emissions embodied in trade to Russia that were used by Karstensen et al. (2013) are likely too low because their global bilateral trade data fail to account for the illegal imports of meat to Russia, accounting for approximately one-third of all meat consumption in 2004 (Antonova and Zeller 2007). Moreover, livestock-related greenhouse gas (GHG) emissions include not only CO_2 emission from land-use change but also emission associated with enteric fermentation, manure management, and fossil fuel use that need to be considered (Caro et al. 2014; Cederberg et al. 2011). In fact, it was found that Russia became the largest net importer of non-CO_2 emission embodied in traded beef during the period 1996–2010. In 2010, 2.7 Mt of CO_2-eq embodied in Brazilian beef that are not related to land-use change was exported to Russia (Caro et al. 2014).

Russia also imports soybean products for feeding its livestock (FAO 2014); however, such imports are fairly low because the livestock industry to date is only weakly developed. Between 2008 and 2010, Russia imported on average 400,000 tons of soybean from Brazil, substantially less than the 16 million tons of soybeans imported by China from Brazil since 2009 (FAO 2014). Therefore, the CO_2 emissions associated with direct trade of soy to Russia are relatively small (Karstensen et al. 2013; Lathuillière et al. 2014). However, the high quantities of imported beef from third countries after the collapse of the SU also had indirect consequences for the area cultivated with soybeans since soybeans or soybean meal are included in modern beef cow diets, particularly in the EU (Simon and Wiegmann 2009). Europe was the largest supplier of beef for Russia until 2002 (Fig. 10.4), and Brazil was the largest exporter of soybean to the European market during the 1990s until the mid-2000s (Lathuillière et al. 2014). Consequently, 15–27 % of total carbon emissions from deforestation linked to soybean production in Brazil was exported to Europe between 1992 and 2002 (Karstensen et al. 2013). This is a conservative estimate because Karstensen et al. (2013) did not account for the indirect effects of land-use change, such as through the displacement of cattle ranchers due to the expansion of soybean production.

10.3.3 Complex Trade-Offs Between Beef Production and Carbon Sequestration

Concerns over the high dependency on imports of food products are a prime economic incentive for the Russian government to support the domestic livestock sector. According to the food security doctrine, signed by Russian President Dmitry Medvedev in 2010, the Russian government aims to produce 85 % of its meat and milk products domestically by 2020, which would require an increase in domestic production by 15 % between 2014 and 2020 (Wegren 2014). Consequently, the

Russian government recently implemented massive import regulations and bans along with increasing state support to increase livestock production. In 2013, the Russian government subsidized the livestock sector with 1.3 billion USD (Wegren 2014). In 2014, Russia banned food imports in retaliation for Western sanctions after the annexation of Crimea by Russia. Together with the depreciation of the Russian ruble in 2014, this contributed to the rapid rise of domestic prices for imported food products. The Russian agricultural sector including the beef industry is thus under increasing pressure to increase its production.

Higher agricultural yields are one promising avenue to increase production, particularly because yield gaps in Russia remain large (Schierhorn et al. 2014a). Moreover, unused or underused agricultural lands in Russia suggest a large scope for increasing domestic food production and to reduce dependency on imports, particularly beef. However, abandoned agricultural land in Russia also offers ecosystem services other than food provision, particularly sequestering carbon in vegetation and soils (Kurganova et al. 2014; Schierhorn et al. 2013; Vuichard et al. 2008). Indeed, abandoned cropland in Russia has become one of the most important terrestrial carbon sinks. The annual carbon sequestration rate compensates for about 4 % of the CO_2 release due to global land-use change (Kurganova et al. 2014). If re-cultivated for production of food, feed, or energy, the abandoned croplands will become a major source for carbon emissions (Poeplau et al. 2011; Shevliakova et al. 2009). Re-cultivation of the abandoned land to increase beef production will hence entail a critical trade-off between provision of food and carbon sequestration (Fig. 10.5, see Chap. 8).

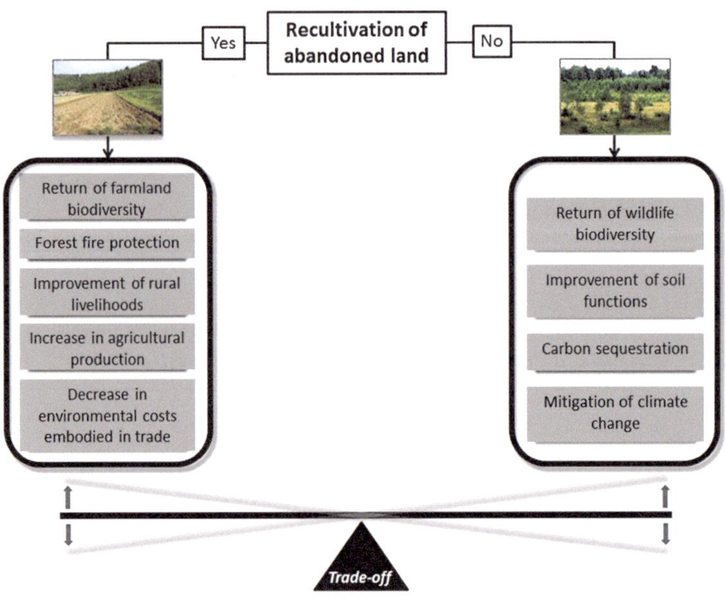

Fig. 10.5 Conceptualization of important trade-offs between re-cultivation of abandoned agricultural land and leaving these lands abandoned

Roughly 10–15 % of abandoned croplands have already been reverted to forest, particularly in temperate regions of Russia (Hansen et al. 2013; Sieber et al. 2013). Consequently, environmental and economic costs to reuse abandoned land are high and reforested lands may stay unused. However, the conversion of abandoned land that has not been reforested yet or will never be reforested due to biogeographic constraints can also lead to carbon emissions, particularly of soil carbon.

The land use before abandonment and the land use after re-cultivation affect the carbon emissions of re-cultivation (Vuichard et al. 2009). For example, the strength of a carbon sink in response to land abandonment depends on former crop management practices such as fertilizer use and land-use legacies (Vuichard et al. 2008); carbon emissions are generally lower if abandoned land is converted to pastures than to cropland (Poeplau et al. 2011). Consequently, the land-related carbon costs to produce beef depend on the livestock production systems and their feeding efficiency (traditional pastoral systems vs. intensified livestock production). Intensified livestock systems, such as those that predominate in the EU, are based less on open-range feeding and more on concentrated feed inputs, such as from soybeans (Bouwman et al. 2005). Moreover, length of time for abandoned lands and geophysical characteristics (soil and climate) control the carbon sequestration rate and the amount of carbon loss due to land re-cultivation (Schierhorn et al. 2013). For example, carbon sequestration capacity and permanence are, inter alia, related to soil clay content (Poeplau et al. 2011).

Therefore, substantial trade-offs need to be taken into consideration if Russia is to reuse its abandoned land with the aim of increasing beef production (Fig. 10.5). The trade-off between reusing abandoned land and beef production versus carbon sequestration becomes even more complex if the carbon emissions associated with the beef imports are compared with the carbon emissions associated with beef production in Russia. The dissolution of the SU and the ensuing collapse of the livestock sector triggered one of the most important natural restoration projects in human history. A globally relevant carbon sink developed in the soils and successional vegetation on the agricultural land that was abandoned after the dissolution. However, livestock production and land competition were increasingly displaced, particularly to Brazil. Therefore, a key message of this article is that beef production in Brazil must be considered as the reference system because increasing production in Russia would replace imports from Brazil. Comprehensive studies should thus estimate the life cycle carbon emissions per unit of beef accumulated along the whole value chain from land-use change and livestock production in Brazil to transportation to the Russian consumer (Davis et al. 2014).

Comprehensive studies should also account for carbon costs of beef that can be subject to change over time. In the early 2000s, the booming beef sector accounted for most of the high deforestation, so carbon emissions from the beef trade between Brazil and Russia were substantial (see above). Later, Brazilian farmers intensified beef, which contributed to the sharp drop in deforestation (Lathuillière et al. 2014). Consequently, the carbon price of Brazilian beef likely decreased. The life cycle GHG emissions of beef production not including deforestation and inherent in enteric fermentation, manure management, and fossil fuel use are still relatively high in Brazil because cattle production systems are mostly extensive (Caro et al. 2014). However, there are clear signals that the cattle production systems will intensify (Millen and Arrigoni 2013) and, thus, that emission intensity of beef production may decrease.

Of course, the trade-offs in practice go beyond the food versus carbon nexus because of a range of other factors that may hinder or support the re-cultivation of abandoned agricultural land (Meyfroidt et al. 2016). For example, improvement of rural livelihoods and a renaissance of cultural landscapes are strong arguments for the re-cultivation of abandoned land, despite inherent carbon costs (Fig. 10.5). Moreover, land cultivation can be justified by ecological motives because land abandonment can reduce water availability (Rey Benayas 2007), increase wildfire risk (Moreira and Russo 2007), or lead to soil erosion or salinization (Penov 2004). Abandoned land may later also form new habitats for wildlife biodiversity that has been shown to benefit from the collapse of the SU (Bragina et al. 2015).

10.4 Conclusion

The dissolution of the SU triggered widespread land abandonment, the collapse of the Russian livestock sector, and high dependency on food imports. Russia became one of the largest importers of beef, mainly from Europe in the 1990s and then from Brazil after 2002. The beef trade to Russia, particularly from the Brazilian Amazon, led to substantial environmental costs in terms of carbon emissions and habitat losses. Studying the environmental costs inherent in displacement of land use via trade is vital to better understand an increasingly connected global land-use system and competition for land-based ecosystem services across different spatial scales. Moreover, evaluating the effects of changes in agricultural production in one place and the resulting effects on competition for ecosystem services in another place enables a better understanding of the trade-offs inherent in agricultural production and an identification of sustainable pathways to necessary production increases. Our analysis examined the abandoned land resources in Russia to demonstrate the trade-offs between beef production and carbon sequestration and allowing us to disentangle the impact of distant drivers on land-use change. The re-cultivation of many abandoned lands in Russia will cause large carbon emissions at the local level, but when factoring in the high carbon costs of beef production in Brazil, re-cultivating abandoned lands in Russia may have substantially lower emissions and be socially more desirable from a global perspective.

Acknowledgements We are grateful for the financial support of the Leibniz Association's "Pakt für Forschung" and the German Federal Ministry of Education and Research (BMBF) (Code01 LL0901A).

References

Antonova, M., & Zeller, M. (2007). *A time series analysis of the beef supply response in Russia: Implications for agricultural sector development policies*. Joint IAAE-104th EAAE Seminar, Budapest, Hungary.

Arima, E. Y., Richards, P., Walker, R., & Caldas, M. M. (2011). Statistical confirmation of indirect land use change in the Brazilian Amazon. *Environmental Research Letters, 6*(2), 024010.

Baccini, A., Goetz, S., Walker, W., Laporte, N., Sun, M., Sulla-Menashe, D., et al. (2012). Estimated carbon dioxide emissions from tropical deforestation improved by carbon-density maps. *Nature Climate Change, 2*, 182–185.

Barona, E., Ramankutty, N., Hyman, G., & Coomes, O. T. (2010). The role of pasture and soybean in deforestation of the Brazilian Amazon. *Environmental Research Letters, 5*(2), 024002.

Bouwman, A. F., Van der Hoek, K. W., Eickhout, B., & Soenario, I. (2005). Exploring changes in world ruminant production systems. *Agricultural Systems, 84*(2), 121–153.

Bragina, E. V., Ives, A. R., Pidgeon, A. M., Kuemmerle, T., Baskin, L.M., & Gubar, Y.P., et al. (2015). Rapid declines of large mammal populations after the collapse of the Soviet Union. *Conservation Biology, 29*(3), 844–853.

Caro, D., LoPresti, A., Davis, S. J., Bastianoni, S., & Caldeira, K. (2014). CH_4 and N_2O emissions embodied in international trade of meat. *Environmental Research Letters, 9*(11), 114005.

Cederberg, C., Persson, U. M., Neovius, K., Molander, S., & Clift, R. (2011). Including carbon emissions from deforestation in the carbon footprint of Brazilian beef. *Environmental Science and Technology, 45*(5), 1773–1779.

Csaki, C., & Lerman, Z. (1992). Land reform and farm sector restructuring in the former Soviet Union and Russia. *Aula*, 7–22.

Davis, S. J., Burney, J. A., Pongratz, J., & Caldeira, K. (2014). Methods for attributing land-use emissions to products. *Carbon Management 5*(2), 233–245.

Dyck, J. H., & Nelson, K. E. (2003). Structure of the global markets for meat. *United States Department of Agriculture, Economic Research Service*.

FAO (2014). FAOSTAT data. In: *Food and Agriculture Organization of the United Nations*.

Foley, J. A., Ramankutty, N., Brauman, K. A., Cassidy, E.S., Gerber, J. S., Johnston, M., Mueller, N. D., O'Connell, C., Ray, D. K., West, P. C., Balzer, C., Bennett, E. M., Carpenter, S. R., Hill, J., Monfreda, C., Polasky, S., Rockström, J., Sheehan, J., Siebert, S., Tilman, G. D., Zaks, D. P. M. (2011). Solutions for a cultivated planet. *Nature, 478*, 337–342.

Hansen, M. C., Potapov, P. V., Moore, R., Hancher, M., Turubanova, S. A., Tyukavina, A., Thau, D., Stehman, S. V., Goetz, S. J., Loveland, T. R., Kommareddy, A., Egorov, A., Chini, L., Justice, C. O., Townshend, J. R. G. (2013). High-resolution global maps of 21st-century forest cover change. *Science, 342*, 850–853.

Hocquette, J.-F., & Chatellier, V. (2011). Prospects for the European beef sector over the next 30 years. *Animal Frontiers, 1*, 20–28.

Jin, H. J., Skripnitchenko, A., & Koo, W. W. (2004). *The effects of the BSE outbreak in the United States on the beef and cattle industry*. Center for Agricultural Policy and Trade Studies, Department of Agribusiness and Applied Economics, North Dakota State University.

Kaimowitz, D., Mertens, B., Wunder, S., & Pacheco, P. (2004). *Hamburger connection fuels Amazon destruction*. Bangor, Indonesia: Center for International Forest Research.

Karstensen, J., Peters, G. P., & Andrew, R. M. (2013). Attribution of CO_2 emissions from Brazilian deforestation to consumers between 1990 and 2010. *Environmental Research Letters, 8*, 024005.

Kurganova, I., Lopes de Gerenyu, V., Six, J., & Kuzyakov, Y. (2014). Carbon cost of collective farming collapse in Russia. *Global Change Biology, 20*, 938–947.

Lal, R. (2004). Soil carbon sequestration impacts on global climate change and food security. *Science, 304*, 1623–1626.

Lambin, E. F., Gibbs, H. K., Ferreira, L., Grau, H. R., Mayaux, P., Meyfroidt, P., Morton, D. C., Rudel, T. K., Gasparri, I., Munger, J. (2013). Estimating the world's potentially available cropland using a bottom-up approach. *Global Environmental Change, 23*, 892–901.

Lambin, E. F., & Meyfroidt, P. (2011). Global land use change, economic globalization, and the looming land scarcity. *Proceedings of the National Academy of Sciences, 108*, 3465–3472.

Lathuillière, M. J., Johnson, M. S., Galford, G. L., & Couto, E. G. (2014). Environmental footprints show China and Europe's evolving resource appropriation for soybean production in Mato Grosso, Brazil. *Environmental Research Letters, 9*, 074001.

Liefert, W. M. (2002). Comparative (dis?) advantage in Russian agriculture. *American Journal of Agricultural Economics, 84*(3), 762–767.

Liefert, W. (2004). *Food security in Russia: Economic growth and rising incomes are reducing insecurity*. Food Security Assessment/GFA-15/May.

Liefert, W., Liefert, O., Vocke, G., & Allen, E. (2010). *Former Soviet Union region to play larger role in meeting world wheat needs*. Amber Waves: U.S. Department of Agriculture, Economic Research Service.

Liefert, W. M., & Liefert, O. (2012). Russian agriculture during transition: Performance, global impact, and outlook. *Applied Economic Perspectives and Policy, 34*, 37–75.

Lioubimtseva, E., & Henebry, G. (2012). Grain production trends in Russia, Ukraine and Kazakhstan: New opportunities in an increasingly unstable world? *Frontiers of Earth Science, 6*, 157–166.

Macedo, M. N., DeFries, R. S., Morton, D. C., Stickler, C. M., Galford, G. L., & Shimabukuro, Y. E. (2012). Decoupling of deforestation and soy production in the southern Amazon during the late 2000s. In *Proceedings of the National Academy of Sciences, 109*, 1341–1346.

Martin, N. (2014). What is the way forward for protein supply? The European perspective. *OCL, 21*, D403.

McAlpine, C. A., Etter, A., Fearnside, P. M., Seabrook, L., & Laurance, W. F. (2009). Increasing world consumption of beef as a driver of regional and global change: A call for policy action based on evidence from Queensland (Australia), Colombia and Brazil. *Global Environmental Change, 19*, 21–33.

Meyfroidt, P., Lambin, E. F., Erb, K.-H., & Hertel, T. W. (2013). Globalization of land use: Distant drivers of land change and geographic displacement of land use. *Current Opinion in Environmental Sustainability, 5*, 438–444.

Meyfroidt, P., Schierhorn, F., Prishchepov, A. V., Müller, D., Kuemmerle, T. (2016). Drivers, constraints and trade-offs associated with recultivating abandoned cropland in Russia, Ukraine and Kazakhstan. *Global Environmental Change, 37*(1).

Millen, D. D., & Arrigoni, M. D. B. (2013). Drivers of change in animal protein production systems: Changes from 'traditional' to 'modern' beef cattle production systems in Brazil. *Animal Frontiers, 3*, 56–60.

Moreira, F., & Russo, D. (2007). Modelling the impact of agricultural abandonment and wildfires on vertebrate diversity in Mediterranean Europe. *Landscape Ecology, 22*, 1461–1476.

Nepstad, D. C. (2005). Governing the world's forests. *Conserving Biodiversity*, 37–52.

Nepstad, D. C., Stickler, C. M., & Almeida, O. T. (2006). Globalization of the amazon soy and beef industries: Opportunities for conservation. *Conservation Biology, 20*, 1595–1603.

Osborne, S., & Trueblood, M. A. (2002). *Agricultural productivity and efficiency in Russia and Ukraine*. Agricultural Economic Report.

Pacheco, P. (2012). *Soybean and oil palm expansion in South America: A review of main trends and implications*. CIFOR Working Paper no. 90.

Paulino, P., & Duarte, M. (2014). Brazilian beef production. *Beef cattle production and trade, 107*.

Penov, I. (2004). The use of irrigation water in Bulgaria's Plovdiv region during transition. *Environmental Management, 34*, 304–313.

Peterson, D. J. (1993). *Troubled lands: The legacy of Soviet environmental destruction*. Boulder, CO (United States): Westview Press.

Poeplau, C., Don, A., Vesterdal, L., Leifeld, J., Van Wesemael, B. A. S., Schumacher, J., & Gensior, A. (2011). Temporal dynamics of soil organic carbon after land-use change in the temperate zone—carbon response functions as a model approach. *Global Change Biology, 17*, 2415–2427.

Prishchepov, A. V., Müller, D., Dubinin, M., Baumann, M., & Radeloff, V. C. (2013). Determinants of agricultural land abandonment in post-Soviet European Russia. *Land Use Policy, 30*, 873–884.

Regmi, A., Deepak, M., Seale Jr, J. L., & Bernstein, J. (2001). *Cross-country analysis of food consumption patterns. Changing structure of global food consumption and trade,* 14–22.

Rey Benayas, J. (2007). Abandonment of agricultural land: An overview of drivers and consequences. *CAB reviews: Perspectives in agriculture, veterinary science, nutrition and natural resources, 2.*

Richards, P. D., Myers, R. J., Swinton, S. M., & Walker, R. T. (2012). Exchange rates, soybean supply response, and deforestation in South America. *Global Environmental Change, 22*, 454–462.

ROSSTAT 2014. Regions of Russia. Socio-economic Indicators. Russian Federal Service of State Statistics, Moscow, Russia. Available from: http://www.gks.ru (in Russian).

Saatchi, S. S., Houghton, R. A., Dos Santos AlvalÁ, R. C., Soares, J. V., & Yu, Y. (2007). Distribution of aboveground live biomass in the Amazon basin. *Global Change Biology, 13*, 816–837.

Schierhorn, F., Faramarzi, M., Prishchepov, A. V., Koch, F. J., & Müller, D. (2014a). Quantifying yield gaps in wheat production in Russia. *Environmental Research Letters, 9*(8), 084017.

Schierhorn, F., Müller, D., Beringer, T., Prishchepov, A. V., Kuemmerle, T., & Balmann, A. (2013). Post-Soviet cropland abandonment and carbon sequestration in European Russia, Ukraine, and Belarus. *Global Biogeochemical Cycles, 27*(4), 1175–1185.

Schierhorn, F., Müller, D., Prishchepov, A. V., Faramarzi, M., & Balmann, A. (2014b). The potential of Russia to increase its wheat production through cropland expansion and intensification. *Global Food Security, 3*(3), 133–141.

Sedik, D. J., Sotnikov, S., & Wiesmann, D. (2003). *Food security in the Russian Federation.* Food & Agriculture Org.

Shevliakova, E., Pacala, S. W., Malyshev, S., Hurtt, G. C., Milly, P. C. D., Caspersen, J. P., Sentman, L. T., Fisk, J. P., Wirth, C., & Crevoisier, C. (2009). Carbon cycling under 300 years of land use change: Importance of the secondary vegetation sink. *Global Biogeochemical Cycles, 23.*

Sieber, A., Kuemmerle, T., Prishchepov, A. V., Wendland, K. J., Baumann, M., Radeloff, V. C., Baskin, L. M., Hostert, P. (2013). Landsat-based mapping of post-Soviet landuse change to assess the effectiveness of the Oksky and Mordovsky protected areas in European Russia. *Remote Sensing of Environment, 133*, 38–51.

Simon, S., & Wiegmann, K. (2009). Modelling sustainable bioenergy potentials from agriculture for Germany and Eastern European countries. *Biomass and Bioenergy, 33*, 603–609.

Smith, P., Gregory, P. J., Van Vuuren, D. P., Obersteiner, M., Havlík, P., Rounsevell, M., Woods, J., Stehfest, E. et al. (2010). *Competition for land. Philosophical Transactions of the Royal Society B: Biological Sciences, 365*, 2941–2957.

Staudigel, M. (2011). How (much) do food prices contribute to obesity in Russia? *Economics & Human Biology, 9*, 133–147.

Steinfeld, H., Gerber, P., Wassenaar, T., Castel, V., Rosales, M., & Haan, C. D. (2006). *Livestock's long shadow: Environmental issues and options.* Rome: Food and Agriculture Organization of the United Nations (FAO).

Stillman, S. (2006). Health and nutrition in Eastern Europe and the former Soviet Union during the decade of transition: A review of the literature. *Economics & Human Biology, 4*, 104–146.

Thomé, K. M., Vieira, L. M., & dos Santos, A. C. (2012). International marketing channels for Brazilian beef: Comparison between Russia and the United Kingdom. *Journal of East-West Business, 18*, 301–320.

Tscharntke, T., Clough, Y., Wanger, T. C., Jackson, L., Motzke, I., Perfecto, I., Vandermeer, J. & Whitbread, A. (2012). Global food security, biodiversity conservation and the future of agricultural intensification. *Biological Conservation, 151*, 53–59.

USDA. (2014). *Production, supply and distribution online*. Washington DC, USA: US Department of Agriculture, Foreign Agricultural Service.

Valin, H., Sands, R. D., van der Mensbrugghe, D., Nelson, G. C., Ahammad, H., Blanc, E., et al. (2014). The future of food demand: understanding differences in global economic models. *Agricultural Economics, 45*, 51–67.

Vieira, L. M., & Traill, W. B. (2008). Trust and governance of global value chains: The case of a Brazilian beef processor. *British Food Journal, 110*, 460–473.

Vuichard, N., Ciais, P., Belelli, L., Smith, P., & Valentini, R. (2008). Carbon sequestration due to the abandonment of agriculture in the former USSR since 1990. *Global Biogeochemical Cycles, 22*.

Vuichard, N., Ciais, P., & Wolf, A. (2009). Soil carbon sequestration or biofuel production: New land-use opportunities for mitigating climate over abandoned Soviet farmlands. *Environmental Science and Technology, 43*, 8678–8683.

Wegren, S. K. (1992). Dilemmas of agrarian reform in the Soviet Union. *Europe-Asia Studies, 44*, 3–36.

Wegren, S. K. (2002). Russian agrarian policy under Putin. *Post-Soviet Geography & Economics, 43*, 26–40.

Wegren, S. (2011). Food security and Russia's 2010 drought. *Eurasian Geography and Economics, 52*, 140–156.

Wegren, S. K. (2014). Human capital and Russia's agricultural future. *Post-Communist Economies, 26*, 537–554.

Wirsenius, S., Azar, C., & Berndes, G. (2010). How much land is needed for global food production under scenarios of dietary changes and livestock productivity increases in 2030? *Agricultural Systems, 103*, 621–638.

Zaks, D. P. M., Barford, C. C., Ramankutty, N., & Foley, J. A. (2009). Producer and consumer responsibility for greenhouse gas emissions from agricultural production—A perspective from the Brazilian Amazon. *Environmental Research Letters, 4*, 044010.

Chapter 11
Of Trees and Sheep: Trade-Offs and Synergies in Farmland Afforestation in the Scottish Uplands

Marvin Beckert, Pete Smith and Stephen Chapman

Abstract Afforestation projects are viewed as potentially effective measures for carbon sequestration and therefore climate change mitigation. Increasing demands for bioenergy products also require land for woodland plantations. Much of the land in temperate regions suitable for afforestation is used for agriculture and consequently afforestation of farmland is frequently proposed. Land owners are commonly reluctant to sacrifice fertile land for purposes other than food and feed production. In Scotland's uplands, grazed pastures are a common land use that could be put under pressure by demands for woodland planting. This chapter explores how farm woodland planting for carbon sequestration and biofuel production affects livestock output. Sheep productivity, soil carbon and tree biomass data from a hill farm in Aberdeenshire, Scotland, are used to estimate the carbon balance and agricultural output of farm woodland plots and silvopastoral systems. The conceptual ideas of trade-off curves are illustrated with the example of 'trees versus sheep'. Planting trees on pastureland can have varying impact on sheep depending on density and planting system. Within the concepts of trade-offs and synergies, it is furthermore discussed whether agroforestry systems such as silvopasture can ameliorate trade-offs between woodland and agriculture by enhancing agricultural productivity while also providing carbon benefits. The concepts presented show that there is great potential for integrating agriculture and forestry to achieve environmental benefits without compromising productivity.

Keywords Food · Feed · Land use competition · Agroforestry · Agricultural · Productivity

M. Beckert (✉) · P. Smith
Institute of Biological and Environmental Sciences, University of Aberdeen, Aberdeen, Scotland, UK
e-mail: r01mrb14@abdn.ac.uk

M. Beckert · P. Smith · S. Chapman
The James Hutton Institute, Craigiebuckler, Aberdeen, Scotland, UK

J. Niewöhner et al. (eds.), *Land Use Competition*,
Human-Environment Interactions, DOI 10.1007/978-3-319-33628-2_11

11.1 Introduction

Land suitable for agricultural production is limited. The Institute for Sustainable Leadership at the University of Cambridge compiled a report stating that in the UK a 'significant shortage of farmland' is to be expected by 2030 (McGrath 2014). Besides new housing and infrastructure developments that take land away from agricultural production, increasing demands for woodland for timber, bioenergy or carbon can put agricultural land under pressure (Moran et al. 2008).

Early grants to promote farm afforestation in the UK were developed to encourage reduction of agricultural productivity (Morgan-Davies et al. 2008). This occurred in the 1980s when overproduction caused by the common agricultural policy (CAP) of the European Union required measures that promised diversified farm income from other activities than food production. More recently, woodland expansion has been put on the political agenda and Scotland will need to afforest more farmland to reach woodland expansion targets (Morgan-Davies et al. 2008). So far, 17 % of Scotland is covered with woodland and an additional 10,000 hectares of new woodland per year for the period 2012–2022 are proposed to be established. The land available in Scotland for such woodland expansion is limited (Fig. 11.1). The vast peat areas of the uplands are of limited suitability for afforestation as they are valuable ecosystems in themselves and tree planting may lead to carbon emissions from soil disturbance (Hargreaves et al. 2003). The most suitable land for forest planting is land that is also most suitable for livestock production (Campbell et al. 2013).

Agri-environmental schemes to integrate agriculture into regional planning, environment and economy have gained importance in the UK since devastating diseases such as BSE and Foot and Mouth Disease, and the development of an economic crisis in the agricultural sector in recent decades (Marsden and Sonnino 2008). In 1988, a farm diversification scheme came into place to bring rural communities away from overdependence on agriculture (Ilberry and Stiell 1991). It was therefore proposed by the Woodland Expansion Advisory Group (WEAG

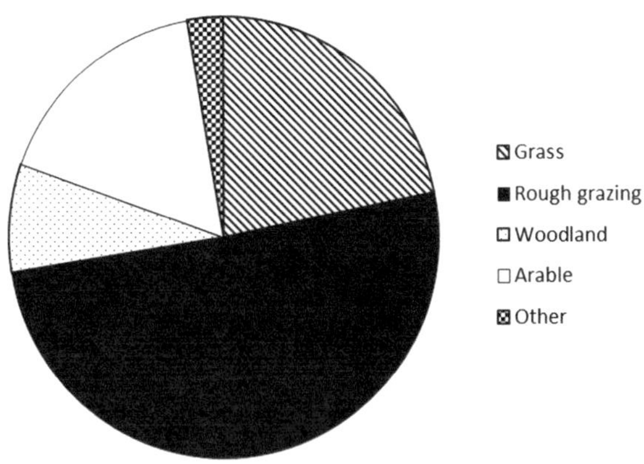

Fig. 11.1 Proportions of land use in agricultural areas in Scotland in 2014. Data adapted from 'Scottish Agricultural Statistics 1982 to 2014'. 'Rough grazing' is often carbon rich soil that is unsuitable for afforestation and 'arable' is considered too valuable. 'Grass' is mostly improved pasture and often deemed suitable for conversion

2012) to integrate woodland with other land uses such as farming. This aligns with the government strategy to reduce conflict and promote synergy between agricultural and forestry land uses. The woodland expansion advisory group recognizes that 'grazing land has significant potential for creation of … woodland'. However, this should be achieved in ways that seek to 'avoid adverse impacts on local patterns of agriculture and enhance the agricultural and environmental value of the remaining unplanted land' (WEAG 2012). Farmland afforestation can be managed in several ways with different objectives and outcomes; woodlands can be planted in dedicated plots excluded from grazing, with focus on timber or bioenergy production. Another option is to integrate trees and pasture on the same piece of land. This land use is referred to as silvopasture, a form of agroforestry.

Udawatta and Jose (2012) point out that silvopasture can sequester more carbon than forests and pasture separately as they combine carbon capture mechanisms of both ecosystems. For example, Sharrow and Ismail (2004) found that a silvopastoral system in Oregon had accumulated more carbon per unit area than both forests and pasture and Haile et al. (2010) indicated that silvopastoral systems can enhance carbon stabilisation in soils. Consequently, agroforestry has gained increasing attention as a potential mitigation measure against climate change (Tobias 2011; Schoeneberger et al. 2012; Mbow et al. 2014). It has also been suggested that agroforestry systems connect ecosystems and can increase biodiversity at the landscape level (Jose 2012). Richards and Méndez (2014) report synergies between biodiversity conservation and carbon sequestration in coffee-based agroforestry systems.

The degree of implementation, however, depends on the decisions made by land managers. Particularly farmers in Europe are often critical towards agroforestry (Nerlich et al. 2013), Cubbage et al. (2012) point out that farmers often perceive the competition between trees and forage as a disadvantage of silvopasture. Hawke (1991) investigated the interactions between pasture growth, livestock and Pinus radiata tree growth in New Zealand and found a decrease in pasture productivity with tree establishment on pasture. At 400 stems per hectare, the pasture yield is reported to be only 12 % of open pasture after 10 years. This was attributed to competition for resources between trees and pasture. Similarly, Kallenbach et al. (2006) found a 20 % reduction in forage production of a 7–8-year-old pine silvopasture, but no negative impact on beef heifer weight gain was observed. Conversely, research conducted in Scotland suggests that in dry summers silvopastoral trees are beneficial for pasture growth, because of soil moisture retention (Sibbald et al. 1991).

In this chapter, woodland creation on agricultural grasslands is viewed as an example to show the competition, and trade-offs and synergies, that can occur on a single piece of land, and the drivers and pressures that act on the land use. The case of a farm in north-east Scotland that has been experimentally planted with woodland plots and silvopasture is examined.

11.2 Livestock Production and Carbon Sequestration in Glensaugh, North-East Scotland

Case study data stem from the ‘Glensaugh’ Agricultural Research Station owned by the James Hutton Institute around 56 km south-west of Aberdeen, Scotland. The site is classified as severely disadvantaged for agricultural production. The farm includes a silvopasture experiment that was established in spring 1988 as part of a silvopastoral network across the UK. The experiment contains three tree species, namely Scots pine (Pinus sylvestris), hybrid larch (Larix eurolepis) and sycamore (Acer pseudoplantanus) at densities of 100, 200, 400 trees ha^{-1} for the silvopastoral treatments and 2,500 trees ha^{-1} woodland control plots. Additionally, pasture plots without trees act as control treatments. The silvopastoral and pasture control plots are ca. 0.5–0.7 ha in size and the woodland plots ~ 0.1 ha (Beckert et al. 2015).

The purpose of the trial was to monitor the impact of silvopastoral systems on sheep output and also the potential of the widely spaced trees for timber production. Timber from the silvopasture trees was intended to offset financial losses in sheep output. Sibbald et al. (2001) present data for the sycamore treatment and Sibbald (1996) for hybrid larch for the first 6 years after planting (Table 11.1), while Sibbald (2006) discusses the output after 12 years for all tree species without presenting detailed data. Sibbald (2006) report no significant reduction of agricultural output after 12 years for sycamore and Scots pine, but a 20 % reduction in hybrid larch stands. Now, ca. 27 years after planting, the trend of declining forage availability in the hybrid larch stands seems to have continued. Much of the area is now covered in a dense litter layer. Also the pasture under sycamore has visibly suffered and cannot provide adequate grazing anymore. In contrast to those two species, the pasture in the Scots pine silvopasture is still grazed and seems to have deteriorated very little. This suggests that trade-offs between trees and pasture growth and therefore livestock production ultimately leads to a decline in pasture growth, but under Scots pine the two manage to coexist even after canopy closure.

Initially, trees of the silvopastoral treatments were intended for timber products. However, wood quality from widely spaced trees is largely viewed as unsuitable for high-grade timber (Balandier and Dupraz 1998). Mäkinen and Hein (2006) examined differences in wood properties of trees planted at 350, 700 and 1,600 trees per hectare densities and found that the widely spaced trees had most knots in the stem wood which lowers their suitability as timber wood. However, in recent decades it has become clear that increasing the biomass of a land-use system can be

Table 11.1 Livestock output 6-12 years after planting of 400 tree ha^{-1} silvopastures (data adapted from Sibbald et al. 2001 and Sibbald 1996)

Tree species	Stocking density (sheep ha^{-1})
Sycamore	14.6
Hybrid larch	14.3
Pasture	14.0

a valuable energy (bioenergy) and environmental asset. Carbon sequestration, especially, is identified as a crucial ecosystem service for climate regulation.

Beckert et al. (2015) assessed the carbon stocks of the pasture, 400 trees ha^{-1} silvopastoral plots and the 2,500 trees ha^{-1} forest plots 24 years after planting. It was found that both woodland establishment strategies increased total carbon stock. This is a logical consequence of increased woody biomass. But interestingly, the soil carbon stock had also slightly increased 24 years after planting. Even though soil carbon stock differences were small across all treatments and not statistically different, it shows that a commonly assumed trade-off in grassland afforestation does not seem to take place here; increasing biomass carbon is not offset by losses in soil carbon.

The agricultural and carbon data sets from the same site show how farmland afforestation can be managed to minimize trade-offs and increase synergies. If we first look at the carbon sequestration potential of the two systems, we see that, theoretically, farm woodland plots should provide the highest carbon sequestration potential (Table 11.2). However, it should be noted that in order to achieve high levels of carbon storage through woodland plots, a considerable amount of pasture land has to be set aside. Here, a clear linear trade-off of carbon with livestock production can be observed. The higher the proportion of land that is taken out of agricultural production by forest planting, the less sheep the remaining pasture can carry. The resulting carbon stocks of farm woodland–pasture systems can simply be estimated as the difference between the total carbon stocks per hectare of woodland and pasture adjusted to the area they take up, respectively. Table 11.2 shows exemplary data from Beckert et al. (2015) for scenarios with 100, 50 and 10 % woodland, respectively. Increasing the proportion of woodland comes at a cost of land available for animal grazing if the carrying capacity is not to be exceeded.

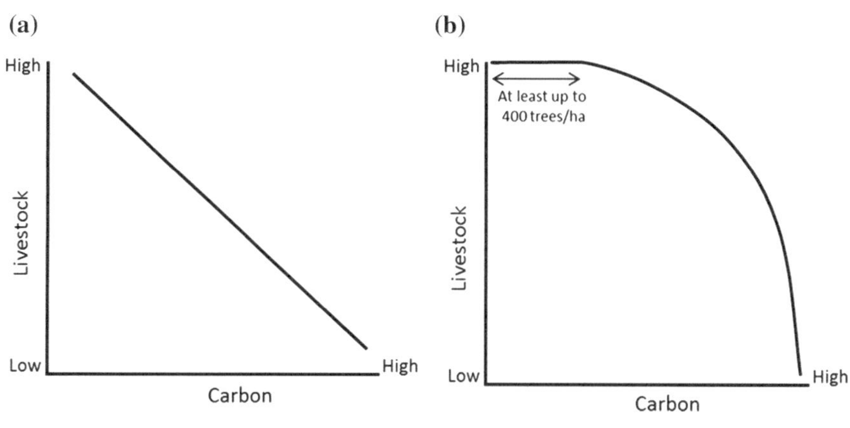

Fig. 11.2 Carbon versus livestock in a farm woodland **a** and silvopastoral system **b**. The farm woodland system shows a linear trade-off between livestock and carbon, whereas the silvopastoral system can support both simultaneously to a certain extent

Table 11.2 Comparison of carbon stocks (Megagrams per hectare, Mg ha^{-1}) in silvopasture, woodland and partially afforested pasture (50 % and 10 % woodland, respectively)

Tree species	Silvopasture (C Mg ha^{-1})	100 % woodland (C Mg ha^{-1})	50 % afforested pasture (C Mg ha^{-1})	10 % afforested pasture (C Mg ha^{-1})
Scots pine	166	237	151	83
Hybrid larch	176	246	156	84
Sycamore	104	153	109	75

Silvopasture trees are at a density of 400 trees ha^{-1} and woodland and afforested pasture at 2500 trees ha^{-1}. Data from Beckert et al. (2015)

Carrying capacity of pastures is commonly expressed as the stocking density a given area can support. Whenever the available area is reduced, the number of animals that can be supported decreases. In order to maintain ideal stocking rates, the total number of livestock must therefore be reduced if significant carbon sequestration, bioenergy or timber with 2,500 trees per hectare is a land management goal. Figure 11.2a shows a simplified representation of how carbon gain comes at a cost of livestock output in these non-mixed woodland–pasture systems. Table 11.1 shows that in the 400 trees/ha silvopastoral configuration the stocking density could be maintained at similar, or even slightly higher, stocking rates compared to pasture plots. This indicates that silvopasture might indeed be a non-interdependent, if not even synergistic, land use that can provide carbon benefits without compromising agricultural productivity to a great extent.

Also, the carbon data for silvopasture support its promise as a land-use system that can serve multiple purposes effectively. As Table 11.2 suggests, even if 50 % of grazed pasture was to be converted into woodland, silvopasture of the same total area has higher or very similar total carbon stocks (e.g. Scots pine silvopasture stores 166 Mg carbon ha^{-1} and an area of 50 % pasture and 50 % woodland of the same species stores 151 Mg carbon ha^{-1}). The carbon data used for these calculations comprise both biomass and soil carbon data. Beckert et al. (2015) report that the coniferous silvopastoral treatments had greater woody biomass per tree and also greater (Scots pine) or very similar (hybrid larch) soil carbon stocks compared to their respective woodland counterparts. Silvopasture of all species had stabilized carbon contents similar to pasture and greater than the woodland treatments. Therefore, both soil carbon stocks and stabilization data suggest that the silvopastures in this study may combine the carbon sequestration mechanisms of forest and pasture ecosystems and provide an overall higher soil carbon sequestration potential than pasture and woodland separately. Given that Scots pine silvopasture combines the carbon sequestration benefits of woodland with pasture growth that still supports sheep grazing even after canopy closure, this species shows particular promise as a working silvopastoral system in Scotland.

11.3 Implications for Land-Use Competition

The different land uses in this case chapter can exemplify differences between the needs of the global community and those of local land users, government and industry. The global community raises demands for greenhouse gas mitigation, which can be achieved by tree planting, the timber industry demands wood, which can be best produced from well managed woodlots, and the farmer requires sufficient livestock output to meet consumer demand for meat, dairy and wool products. Rather than direct competitors for land, in this case study the final decision is made by the land manager, often a farmer or farm manager, but pressures from both the markets and policies/regulations can influence the decision-making processes in deciding the importance of the different direct and indirect ecosystem services that can be gained from the land (Fig. 11.3).

In the conceptual introduction to this section (Chap. 8), land-use competition is classified as competition for the services from the land rather than the property itself. This is what manifests in rural land use, where provisioning and non-provisioning ecosystem services are factors that may compete and need to be integrated to reduce such competition. The conceptual chapter explains competition and the associated trade-offs, among others, in economic and ecological terms. Livestock production is a provisioning service provided by pasture; pasture is relatively easy to maintain and the land use provides steady income. Introducing trees into the farmland is, at first,

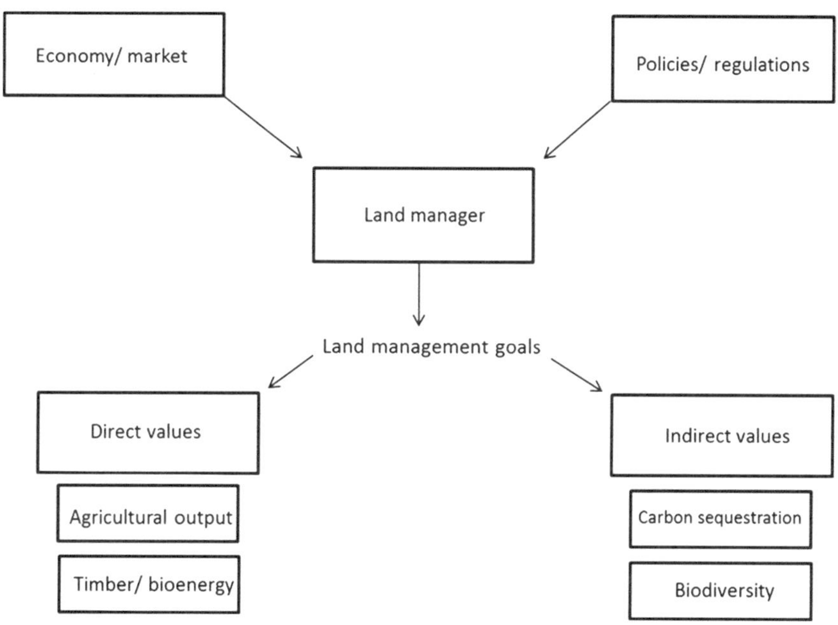

Fig. 11.3 Shows how the economy and regulations influence land management decisions by rural land managers for direct and indirect ecosystem services from the land

an additional cost factor with a direct negative impact on farm financial output. Financial returns from woodland planting would in the first years of establishment come from grants and incentives which depend on an outside actor, most likely national or supranational government bodies. Economic benefits from the wood itself will only be realised after years for bioenergy, or decades for timber. When large proportions of land are taken out of agricultural production, any land-use change to woodland becomes unfeasible for the land owner.

Crabtree et al. (2001) discuss possible incentives for farmland afforestation and point out that farmers are only likely to plant trees when financial outputs from forestry exceed agricultural outputs. The silvopastoral system has the ability to diversify into woodland, while still maintaining agricultural production (Schoeneberger 2009), at least in the first decade after planting. Switching to silvopasture involves less risk than conversion to forest land as farm income is not reduced immediately. However, silvopastoral trees must initially be protected against browsing, unlike dedicated forest plots that are fully fenced off, which is an additional cost factor.

Whether trees are introduced into farmland in dedicated plots or as silvopasture, the case study data suggest that trade-offs between agricultural production and woodland products take place to some degree. The silvopastoral system could buffer the impact of woodland introduction better, while still providing significant carbon benefits. The land use might therefore be non-interdependent, i.e., both services can exist simultaneously, or even be synergistic. According to the definition proposed in Müller et al. (Chap. 8), for services to be synergistic, one service must increase the productivity of the other. In terms of agricultural and carbon output, this would mean that the system can produce more livestock with increasing biomass and vice versa. Increasing the woody biomass seems to have negligible effects on livestock outputs in the first years after planting, but the trend is towards decrease rather than increase. In other circumstances, however, trees on pasture can potentially increase the output, by improving shelter and therefore welfare and survival of the animals. A recent review found that shelter improves welfare and productivity of cattle (Van laer et al. 2014). Also, Pollard (2006) point out that shelter can improve lamb survival and Schütz et al. (2014) found that shade reduced panting in cows in New Zealand.

11.4 Conclusion

This case study showed that agricultural output and carbon sequestration can be combined on the same piece of land. Trade-offs exist between the competing ecosystem services of agricultural productivity and woodland creation for carbon, timber or bioenergy. The data indicate that silvopasture with some tree species, such as Scots pine, promises to minimise the trade-offs and increase overall land productivity by supporting similar animal numbers as pasture while providing carbon stocks that can only be attained if large proportions of land are converted into pure woodland. While farm woodland plots are an example of how one ecosystem service is compromised for the benefit of another, silvopasture can potentially fit the definition of a non-interdependent or even synergistic land use.

References

Balandier, P., & Dupraz, C. (1998). Growth of widely spaced trees. A case study from young agroforestry plantations in france. *Agroforestry Systems, 43*, 151–167.

Beckert, M. R., Smith, P., Lilly, A., & Chapman, S. J. (2015). Soil and tree biomass carbon sequestration potential of silvopastoral and woodland-pasture systems in North East Scotland. *Agroforestry Systems*. doi:10.1007/s10457-015-9860-4.

Campbell, C. D., Lilly, A., Towers, W., Chapman, S. J., Werritty, A., & Hanley, N. (2013). Land use and a low-carbon society. *Earth and environmental science transactions of the royal society of edinburgh, 103*, 165–173.

Crabtree, B., Chalmers, N., & Eiser, D. (2001). Voluntary incentive schemes for farm forestry: Uptake, policy effectiveness and employment impacts. *Forestry, 74*, 455–465.

Cubbage, F., Balmelli, G., Bussoni, A., et al. (2012). Comparing silvopastoral systems and prospects in eight regions of the world. *Agroforestry Systems, 86*, 303–314.

Haile, S. G., Nair, V. D., & Nair, P. K. R. (2010). Contribution of trees to carbon storage in soils of silvopastoral systems in Florida, USA. *Global Change Biology, 16*, 427–438.

Hargreaves, K. J., Milne, R., & Cannell, M. G. R. (2003). Carbon balance of afforested peatland in Scotland. *Forestry, 76*, 299–317.

Hawke, M. F. (1991). Pasture production and animal performance under pine agroforestry in New Zealand. *Forest Ecology and Management, 45*, 109–118.

Ilberry, B., & Stiell, B. (1991). Uptake of the farm diversification grant scheme in england. *Geography, 76*, 259–263.

Jose, S. (2012). Agroforestry for conserving and enhancing biodiversity. *Agroforestry Systems, 85*, 1–8.

Kallenbach, R. L., Kerley, M. S., & Bishop-Hurley, G. J. (2006). Cumulative forage production, forage quality and livestock performance from an annual ryegrass and cereal rye mixture in a Pine Walnut Silvopasture. *Agroforestry Systems, 66*, 43–53.

Marsden, T., & Sonnino, R. (2008). Rural development and the regional state: Denying multifunctional agriculture in the UK. *Journal of Rural Studies, 24*, 422–431.

Mäkinen, H., & Hein, S. (2006). Effect of wide spacing on increment and branch properties of young Norway spruce. *European Journal of Forest Research, 125*, 239–248.

Mbow, C., Smith, P., Skole, D., Duguma, L., & Bustamante, M. (2014). Achieving mitigation and adaptation to climate change through sustainable agroforestry practices in Africa. *Current Opinion in Environmental Sustainability, 6*, 8–14.

McGrath. (2014). UK faces significant shortage of farmland by 2030. http://www.bbc.com/news/science-environment-28003435. Accessed May 29, 2015.

Moran, D., Macleod, M., Wall, E., Eory, V., Pajot, G., Matthews, R., et al. (2008). *UK Marginal Abatement Cost Curves for agriculture and land use, land-use change and forestry sectors out to 2022, with qualitative analysis of options to 2050*. London: Final Report to the Committee on Climate Change.

Morgan-Davies, C., Waterhouse, A., Pollock, M. L., & Holland, J. P. (2008). Integrating hill sheep production and newly established native woodland: Achieving sustainability through multiple land use in Scotland. *International Journal of Agricultural Sustainability, 6*, 133–147.

Nerlich, K., Graeff-Hönninger, S., & Claupein, W. (2013). Agroforestry in Europe: A review of the disappearance of traditional systems and development of modern agroforestry practices, with emphasis on experiences in Germany. *Agroforestry Systems, 87*, 1211.

Pollard, J. C. (2006). Shelter for lambing sheep in New Zealand: A review. *New Zealand Journal of Agricultural Research, 49*, 395–404.

Richards, M. B., & Méndez, V. E. (2014). Interactions between carbon sequestration and shade tree diversity in a smallholder coffee cooperative in El Salvador. *Conservation Biology, 28*, 489–497.

Scottish Agricultural Statistics 1982 to 2014, the Scottish Government. http://www.gov.scot/Topics/Statistics/Browse/Agriculture-Fisheries/agritopics/LandUseAll. Accessed May 29, 2015.

Schoeneberger, M., Bentrup, G., De Gooijer, H., Soolanayakanahally, R., Sauer, T., Brandle, J., et al. (2012). Branching out: Agroforestry as a climate change mitigation and adaptation tool for agriculture. *Journal of Soil and Water Conservation, 67*, 128A–136A.

Schoeneberger, M. M. (2009). Agroforestry: Working trees for sequestering carbon on agricultural lands. *Agroforestry Systems, 75*, 27–37.

Schütz, K. E., Cox, N. R., & Tucker, C. B. (2014). A field study of the behavioral and physiological effects of varying amounts of shade for lactating cows at pasture. *Journal of Dairy Science, 97*, 3599–3605.

Sibbald, A.R., Griffiths, J.H., & Elston, D.A. (1991). The effects of the presence of widely spaced conifers on under-storey herbage production in the U.K. Forest Ecology and Management 45, pp. 71–77.

Sibbald. (1996). Silvopastoral Systems on temperate sown pasture: A personal perspective. In: Etienne (ed.) *Western European Silvopastoral Systems*. Institute National de la recherche agronomique.

Sibbald, A. R., Eason, W. R., Mcadam, J. H., & Hislop, A. M. (2001). The establishment phase of a silvopastoral national network experiment in the UK. *Agroforestry Systems, 53*, 39–53.

Sibbald 2006. Silvopastoral Agroforestry: A Land Use for the Future. Scottish Forestry, 60, pp. 4–7.

Tobias, P. (2011). Capitalizing on the carbon sequestration potential of agroforestry in Germany's agricultural landscapes: Realigning the climate change mitigation and landscape conservation agendas. *Landscape Research, 36*, 435–454.

Van laer, E., Moons, C.P.H., Sonck, B., & Tuyttens, F.A.M. (2014). Importance of outdoor shelter for cattle in temperate climates. Livestock Science 159, pp. 87–101.

Woodland expansion advisory group 2012. *Final report* http://scotland.forestry.gov.uk/images/corporate/pdf/WEAGFinalReport.pdf. Accessed May 29, 2015.

Chapter 12
Land Use Competition Related to Woody Biomass Production on Arable Land in Germany

Jana Zscheischler, Nadin Gaasch, David Butler Manning and Thomas Weith

Abstract As land is a limited resource, diverging demands drive competition between land uses. The pressure on arable land is increasing. In Germany, about 80 ha of arable land are converted daily (2009–2012), mainly to housing and transport uses. A small portion is also lost each year to reforestation and to ecological compensation measures. The boom in biomass production for energetic purposes in particular is serving to heighten the pressure on arable land and is increasingly competing for land with food and feed production. Over the last decade, the demand for woodfuel has also increased remarkably. Germany aims to cover 14 % of its heat demand with renewable energy by the year 2020, with wood to play a leading role. To meet this target, novel systems of land use will need to be added to the current land use spectrum in order to increase the wood supply. The cultivation of fast-growing trees, such as poplar and willow, represents one means to address the increasing demand while, at the same time, fostering the provision of a range of beneficial ecosystem services. Systems that mix annual crop production and fast-growing trees on the same site constitute an especially promising pathway to achieve synergies between different land use types and to reduce land use competition. This chapter explores the benefits, potentials, limitations and challenges of cropping systems using fast-growing trees in Germany. Special emphasis is placed on the capacity of agroforestry systems to mitigate land use competition by combining wood with food production and conservation on the same plot. Trade-offs and synergies with ecosystem services such as food production, aesthetic values and other regulating and supporting services are discussed at different spatial and temporal scales. Finally, some conclusions for sustainable land management are drawn.

J. Zscheischler (✉) · N. Gaasch · T. Weith
Institute of Socio-Economics, Leibniz Centre for Agricultural Landscape Research, Müncheberg, Germany
e-mail: Jana.Zscheischler@zalf.de

D.B. Manning
Department of Forest Sciences, Dresden University of Technology, Dresden, Germany

T. Weith
Institute of Geography, University of Potsdam, Potsdam, Germany

J. Niewöhner et al. (eds.), *Land Use Competition*,
Human-Environment Interactions, DOI 10.1007/978-3-319-33628-2_12

Keywords Ecosystem services · Food production · Trade-off · Agroforestry · Scales

12.1 Drivers of Land Use Change in Germany

As land is a limited resource, differing and often diverging demands drive competition between land uses and, in specific situations, lead to land use conflicts. The demands, and thus the societal pressure, placed on arable land are increasing. In Germany, about 80 ha of arable land is converted daily (2009–2012), mainly to housing and transport uses (Fig. 12.1). A small portion is also lost each year to forest, which, once established, is protected and essentially cannot be converted back to agricultural land. Whereas the forest area increased in Germany between the years 2004 and 2013, from 29.8 to 30.3 % of the surface area, the area of agricultural land declined by about half a million hectares (from 53.0 to 52.1 %) during this period.[1] Additionally, the designation of ecological focus areas, areas for the implementation of compensation measures and conservation areas also drive land use change on arable land and can lead to additional pressure on the remaining arable area. A further driver of land use change is energy policy, which states that the energy demand should increasingly be covered by renewable resources.

Germany is characterised by high primary energy consumption, a correspondingly high energy dependency and an ecological footprint per capita more than twice the global average (Ewing et al. 2008). In recent years, Germany has become a world leader in innovation in the field of renewable energies and is currently striving to transform its energy system to a higher share of renewables (Energiewende). Although Germany is a densely populated and highly industrialised country, large agricultural land resources still occupy nearly 50 % of the surface area. This is one reason why the production of biomass for energetic purposes is an important cornerstone of the energy transformation process in Germany. The production of annual energy crops like maize for biogas is still booming,[2] promoted by the German Renewable Energies Act (Erneuerbare Energien Gesetz, EEG). This development is proceeding uninterrupted in spite of scientific, political and even societal discussions about the negative long-term ecological effects of the large maize monocultures that have resulted from the sharp rise of biogas. Parts of society have also expressed their concerns with regard to the aesthetics of the increasingly monocultural landscapes.

---

[1]https://www.destatis.de/EN/FactsFigures/EconomicSectors/AgricultureForestryFisheries/LandUse/Tables/Areas.html.

[2]www.fnr.de/fileadmin/fnr/images/daten-undfakten/2014/Abb_108_2014_Maisanbau_tabellarisch.zip.

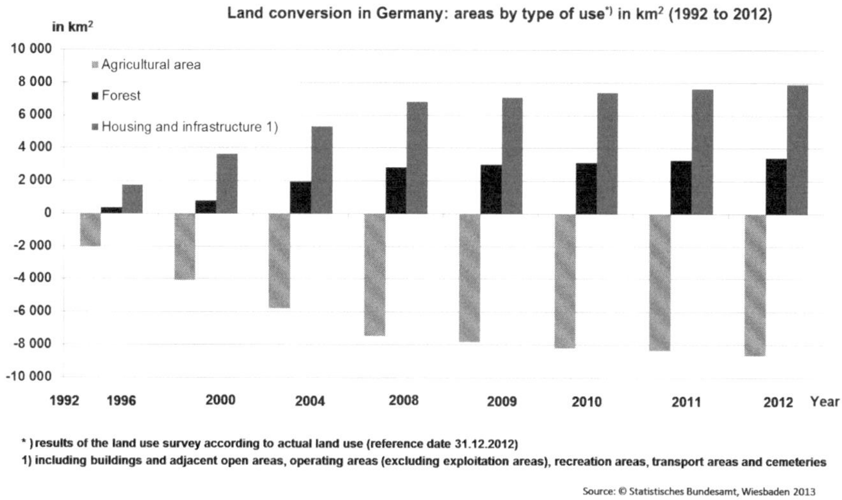

Fig. 12.1 Development of the land area occupied by agriculture, forestry and housing and infrastructure in Germany from 1992 to 2012 (own graph based on data from: Federal Statistics Office, Germany) https://www.destatis.de/DE/Publikationen/Thematisch/LandForstwirtschaft/Flaechennutzung/BodenflaechennutzungPDF_2030510.pdf?__blob=publicationFile

The energy transformation process, therefore, is another source of pressure on arable land. In the year 2014, a total of 2.1 million ha of arable land in Germany were being cultivated with crops grown for energy. This equated to 12.5 % of the area of agricultural land (FNR 2014).[3] Alongside the boom in the production of annual energy crops in Germany, the demand for woodfuel has also increased dramatically over the last decade (Fig. 12.2). Due to its widespread availability, favourable combustion characteristics and high energy density, wood is one of the most important energy sources worldwide (IEA 2013). Even in highly industrialised Germany, where for many decades wood played almost no role in the provision of energy, the government's aim to cover 14 % of Germany's heat demand with renewable energy by the year 2020 (BMU/BMELV 2010) has meant that the energetic use of wood has undergone an unheralded revival. At present, 16 million German[4] households use wood for heating—with an upward trend (Mantau 2012). The introduction of new technologies (e.g. biomass-to-liquid, gasification) and expanding material and chemical wood applications are expected to increase consumption even further. The policy and practice of sustainable forest management in Germany means, however, that the national forest resource will not be sufficient to satisfy this growing demand for wood. Subject to a continuation of recent developments, an annual deficit of approximately 30 million m^3 is forecast for the German timber market by the year 2020 (Thrän et al. 2009). Over the same period,

---

[3]http://mediathek.fnr.de/media/downloadable/files/samples/a/b/abb_696.jpg.

[4]http://bioenergie.fnr.de/bioenergie/festbrennstoffe/.

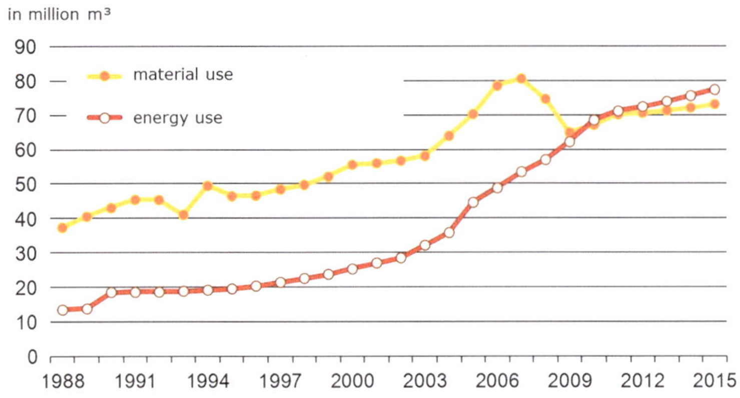

Fig. 12.2 Development of material and energy use of wood in Germany from 1988 to 2010 and projected for the years 2011 to 2015 (*source* Mantau 2012)

the annual deficit for all EU member states will amount to about 400 million m^3 (Hetsch et al. 2007; Mantau 2010). Whereas initially harvest residues and thinning material were the main sources of woodfuel, higher quality roundwood is increasingly required to fuel-combined heat and power plants.

In 2014, bioenergy contributed 12.2 % to total heat consumption in Germany, 87 % of this derived from bioenergy stemming from biogenic solid fuels—primarily wood (FNR 2015). Wood, therefore, is central to the German renewable energy supply and in order to reach the 14 % target in 2020, novel systems of land use will need to be added to the current land use spectrum in order to increase the wood supply. An alternative to increasing the wood supply nationally is to increase wood imports to Germany. This approach is not considered further here as it poses numerous difficulties with respect to the sustainability of the sources, the transport effort and the continued reliance on non-domestic sources for energy supply.

However, the cultivation of fast-growing trees in Germany, such as poplar, willow and black locust, constitutes one option to address the increasing demand while, at the same time, fostering the provision of a range of beneficial ecosystem services such as soil protection or an enhanced habitat diversity (e.g. Bemmann and Knust 2010; Busch and Lamersdorf 2010; Dimitriou et al. 2011, Weih and Dimitriou 2012).

The aim of this chapter is to highlight the possible effects of the production of woody biomass on arable land in terms of both reducing and increasing the competition between land-based ecosystem services in Germany. Features of the two cultivation systems, short rotation coppice and agroforestry systems, are introduced before trade-offs and synergies with other ecosystem services such as food, aesthetic values and regulating and supporting services are discussed. Finally, challenges facing the implementation are presented and implications for sustainable land management are outlined.

12.2 Features of Agroforestry Systems and Short Rotation Coppice

When producing woodfuel, requirements of trees common to forestry such as straight shafts, small crowns and even, narrow growth rings become redundant. The focus is solely on the achievement of high biomass yields, fast juvenile growth, good coppicing ability, easy reproduction, a capacity for growth at high stocking densities and a low vulnerability to pests and diseases (Schildbach et al. 2009). In central Europe, these criteria are best fulfilled by the tree genera poplar (Populus spp.) and willow (Salix spp.), and the species black locust (Robinia pseudoacacia). Other species approved for cultivation on arable land in Germany are birch, oak, alder and ash. In certain other European countries, additional species such as Eucalyptus, hazel, lime and sycamore are also planted as energy crops.

There are two cultivation systems for the production of woody biomass on arable land currently under discussion: short rotation coppice and agroforestry systems. Both comprise fast-growing tree species and clones with a high capacity for vegetative regeneration. The trees are coppiced, i.e., cut back at intervals usually of between three to five years so that several new shoots emerge, thereby producing the next crop from the same stools.

Short rotation coppice is usually established as small plantations (1–3 ha) and at high stocking densities (8000–12,000 trees per ha). Agroforestry systems, by contrast, combine agricultural and silvocultural land use (cf. Chap. 11). At least two plant species are combined on the same plot, one a woody perennial and the other a food or fodder species. Agroforestry systems are a traditional form of land use which persisted in many regions of Germany until the beginning of the twentieth century (King 1989) (e.g. pruned and pollarded trees, orchard meadows, wooded meadows). Over the course of the industrialisation and intensification of farming, trees began to disappear from agricultural land, as did knowledge of traditional agroforestry systems amongst farmers. Today, the modern agroforestry systems most widely discussed are predominantly so-called alley cropping systems. Trees are established as hedge-like strips. In between the tree rows, conventional arable crops are cultivated or grassland is maintained. The distances between the rows can vary and will often be adapted to the agricultural equipment available. These systems can be readily accommodated within the context of modern agriculture and the prevailing cultivation practices.

However, the implementation of fast-growing trees as coppice plantations or as hedge-like strips may affect the delivery of ecosystem services from land and therefore alter (reduce or increase) land use competition, as conceptualised by Müller et al. (Chap. 8). The nature of these impacts and their implications are highlighted in the following.

12.2.1 Interactions with Food and Annual Biomass Crops

The issue of land use competition attained particular prominence during the so-called tortilla crisis in the years 2007 and 2008—a corn price spike hitting most visibly Mexico's poorest people. In the period since then, the debate on the use of biomass on agricultural land as renewable energy has heated up markedly, often summarised simplistically as a conflict between food and fuel.

Given the limitation on the total area of land available, one might argue in general terms that the cultivation of fast-growing trees on arable land always results in a reduction of the area of land that can be used for the provision of other goods such as food or fodder. This oversimplification, however, does not allow for conclusions to be drawn with regard to a changed degree of land use competition because spatial scales and the corresponding demands must also be taken into account.

The issue of food security is receiving increasing awareness and interest globally (e.g. FAO 2013) due to concerns over the impacts of forecasted demographic and climatic changes, as well as increasing standards of living and associated changes in consumption patterns. As of now, food availability is secured in Germany; however, subsidised exports of agricultural surplus had and still have many negative effects on local markets and food systems in developing countries. At the moment, there is no evidence that the cultivation of woody biomass on arable land in Germany will cause supply gaps with essential food products in the near future.

The question that remains, however, is the extent to which food and fodder production can be substituted by energy crops before supply becomes an issue. By focusing the production of woody crops on arable land currently used for annual energy crops such as maize, which already cover a large share of agricultural land area in Germany (see above) and involve environmental risks such as soil erosion, an increasing competition can be largely avoided.

However, short rotation coppices and agroforestry systems have different impacts. While the implementation of short rotation coppice results in a higher overall supply of wood for energy and can, therefore, contribute to reducing pressure on forest resources, agroforestry systems contribute less wood but involve a wider range of produce and complex effects; for example, the hedge-like tree strips can be either positive or negative with respect to yield. The shade created by the trees reduces the light available to the accompanying crop and there is also competition for water, nutrients and surface soil. At the same time, however, the crop is sheltered and the litter produced by the trees leads to an increased supply of nitrogen and other nutrients (Ong et al. 1991; Van Noordwijk and Lusiana 2000).

For this reason, when contemplating the suitability of agroforestry systems, the 'land equivalent ratio' (LER; Mead and Willey 1980) is often calculated to determine whether the implementation of agroforestry systems is likely to increase yields at a certain location. From a theoretical point of view, agroforestry systems may be better able to exploit a site than simple cropping because combined cultures can better allocate the vertical space. The LER is defined as the relationship between the 'area needed under sole cropping to the area of intercropping at the same management level to obtain a particular yield' (van der Werf et al. 2007). Grünewald et al. (2007) investigated agroforestry systems with fast-growing trees in former mining landscapes in Germany and found no significant difference between the productivity of an agroforestry system and monocropping. Beyond mere yield, however, the agroforestry systems were characterised by marked (non-monetary) social and ecological benefits (ibid.). For this reason, Nair (1990) argued that the simple calculation of LER is an insufficient means to evaluate agroforestry systems as it does not reflect their sustainability.

12.2.2 Interactions with Regulating and Supporting Ecosystem Services

Some critics claim that short rotation coppice is a land use far removed from nature, represents a strong intervention in the historical cultural landscape and also results in low biodiversity. Often these crops are monocultures and with their cultivation completely new types of ecosystems arise. However, one might argue that the cultivation of short rotation coppice in Germany in almost all situations takes place on arable land previously used intensively for the production of annual crops. Compared to the cultivation of typical agricultural crops, and especially energy crops such as maize, a well-considered implementation of fast-growing trees in the agricultural landscape may result in environmental benefits (Schubert et al. 2009; Don et al. 2011). These include an extended soil regeneration period, a greatly reduced need for artificial fertilisers and pesticides, less disturbance to the natural environment due to reduced management operations, greater structural diversity in the landscape and generally lower energy input in production (Bemmann and Große 2011). The cultivation of trees on arable land can impact positively on the soil by preventing soil and water erosion and increasing soil organic matter. Trees generally contribute to new soil formation (Young 1997; Reeg and Grünewald 2009), while nitrogen fixing trees (e.g. black locust) accumulate nitrogen in the soil. In landscapes characterised by a sensitive hydrological balance (wetlands or very arid areas), the higher rates of water use by tree crops and the capacity to access groundwater must be taken into consideration.

Hedges and the tree rows employed in agroforestry systems also have a considerable influence on the microclimate. Protection against wind can enhance the yields of crops situated beyond the shelter provided by the trees. Several tests employing different tree species and tree strips under diverse climatic conditions revealed that wind speeds on the leeside can be reduced by up to 60 %, resulting in lower rates of evapotranspiration of as much as 30 % and a rise in soil moisture of up to 20 %. Soil temperature has also been found to be higher (e.g. Pollard et al. 1974) and precipitation increased. The rates of infiltration increase by braking wind speeds and by promoting water drainage. As a consequence of this, agroforestry systems can also contribute to preventing the leaching of nutrients from soils. The woodfuel produced contributes to replacing fossil fuels and, as a consequence, to furthering national and global climate protection goals. Under some circumstances, short rotation coppice plantations might also serve as effective carbon sinks (Quinkenstein et al. 2009).

12.2.3 Interactions with Biodiversity

Biodiversity is closely related to ecosystem services, and most studies on the linkages between biodiversity attributes and ecosystem services showed positive correlations (Harrison et al. 2014). The effects of woody biomass production on biodiversity are controversially discussed, as these may be both positive and negative. Proponents of short rotation coppice argue positive impacts on biodiversity or at least reduced pressure on biodiversity, compared to other agricultural crops (Fry and Slater 2009; Dimitriou et al. 2011; Baum 2012). A notably higher number of plant species have been observed in German short rotation coppice plantations than on neighbouring conventional arable land, albeit with differences over time (Schmidt and Glaser 2009; Grunert and Wilhelm 2013a). However, as these plants are predominantly generalist species, the higher number of plant species found in short rotation coppice is not necessarily an indicator of a higher 'quality' of the species spectrum. Studies of fauna have also shown that rare species are seldom found in short rotation coppice (Baum et al. 2009; Schulz et al. 2009). Longer rotation periods result in denser canopies that are increasingly impervious to light and result in the loss of species of the open landscape. Other species, however, benefit from the biotope corridor effect. Additionally, agroforestry systems enable enriched habit structures due to fringe structures while exerting less pressure on species of open landscapes.

12.2.4 Interaction with 'Cultural Services'

From a landscape aesthetics perspective, short rotation coppice may also have contrasting effects. The impact may be negative in densely wooded regions where sight axes become obscured (Boll et al. 2015). By contrast, in very open, intensively used agricultural landscapes the structure given to the landscape may have a positive aesthetic effect, as well as proving beneficial with respect to wind and water erosion in these exposed landscapes (Grunert and Wilhelm 2013b).

As has been shown, the implementation of agroforestry systems and short rotation coppice may serve to increase competition between the provisioning services of arable land, such as with food production. Nonetheless, agroforestry systems and short rotation coppice also include the capacity for numerous synergies such as increasing the wood supply in combination with greater provision of several supporting and regulating ecosystem services. Although the synergetic effects with respect to biodiversity require further study and, from a landscape aesthetics perspective, the specific regional conditions should be taken into account, both agroforestry and short rotation coppice can produce beneficial effects on certain sites. Ecological assessments (Finch et al. 2009) generally rate short rotation coppice somewhere between forest and arable land on a scale of land use intensity. However, the ecological value of short rotation coppice and agroforestry systems depends very much on the design and site-specific conditions.

12.3 Synergies or Trade-Offs? The Selection of Suitable Sites Is the Key

The way, in which competition for land plays out and synergies or trade-offs arise, depends to a large extent on the form of cultivation chosen for a particular site. The optimisation of land use towards sustainable land management requires a smart, mixed cultivation system on most sites, seeking to avoid trade-offs and mutual losses.

As a starting point, it seems reasonable to plant trees on sites where the conditions are unfavourable for agricultural production. Such 'marginal' sites often necessitate higher management costs and energy input. Some authors have proposed the implementation of short rotation coppice on sites of low soil quality (Murach et al. 2009; Vetter et al. 2009) because here trees unlike arable crops still have access to groundwater and the long rotation periods reduce management effort and costs. Other proponents of short rotation coppice, however, emphasise that on such sites yield expectations are low and that, in order to obtain the greatest benefits from short rotation coppice, good agricultural land should also be used. In either case, implementing short rotation coppice on arable land implies a trade-off between food and biomass production on a local scale and from a short-term perspective.

Another possibility is the implementation of short rotation coppice in post-mining landscapes (Böcker 2013) and on contaminated sites (Bungart and Hüttl 2001; Dietzsch 2011). The former are characterised by a low humus content, nutrient deficiency, low biological activity, acidification and salinisation, and correspondingly low productivity. The cultivation of food plants on either site type is often not feasible due to the low yields and the pollution of the soils. Short rotation coppice and agroforestry systems (intercropped with energy plants) support the recultivation and redevelopment of devastated areas and contribute to the supply of bioenergy, albeit only in a small way. In such landscapes, there are no trade-offs between the provision of food or biomass. Equally, the wood yields produced will be correspondingly low.

Dupraz et al. (2005) presented a synergistic approach to the implementation of agroforestry systems on a local scale. In this approach, potential target sites are derived from the subset of available arable land, potential growth areas for trees and sites prone to environmental degradation such as erosion and the leaching of nutrients (Fig. 12.3).

Following this concept, the preferred locations for agroforestry systems are sites with very good soils (see also Vetter et al. 2009), such as the 'Magdeburger Börde' and the 'Hildesheimer Börde' in Germany. Due to their high productivity, forest and structural landscape elements have been almost completely lost in these prime agricultural locations (Fig. 12.4). With a view to sustainable land use, the

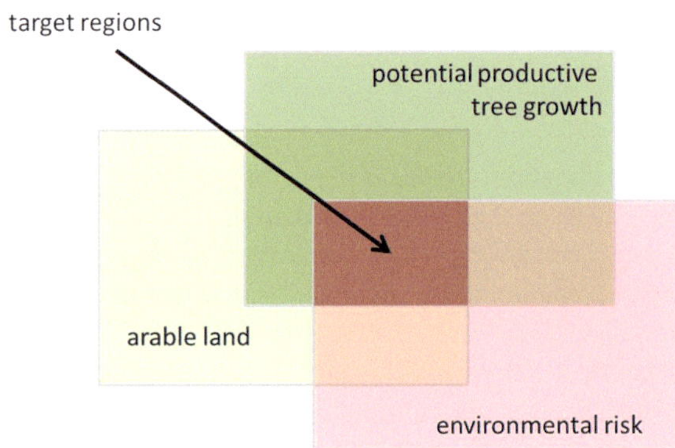

Fig. 12.3 Concept for the identification of target regions for the implementation of agroforestry systems (Dupraz et al. 2005)

Fig. 12.4 Satellite image of land use close to Bördeland. The region possesses the most productive soils in Germany and is characterised by a featureless and almost treeless agricultural landscape (*source* Google Earth)

Table 12.1 Allocation of agroforestry systems and/or short rotation coppice according to the prevailing landscape characteristics

	Forest-poor landscape	Forest-rich landscape
Poorly structured landscape	AFS + SRC	AFS
Highly structured landscape	SRC	–

AFS agroforestry system; *SRC* short rotation coppice

establishment of tree strips within these landscapes could enhance their ecological and aesthetic value and protect the productive soils from wind and water erosion. Unfortunately, these are the very sites where farmers have no interest in integrating trees due to perceived negative impacts in terms of cultivation efficiency and on profits from arable farming. In these areas, positive long-term ecological effects compete with short-term economic interests.

Another option in the search for target sites is consideration of the specific field size and shape (Busch and Meixner 2015). Small, irregularly shaped plots incur high production costs. Therefore, these plots are especially suitable for the installation of short rotation coppice cultivated at longer rotation lengths. Tree strips established along the water courses in agricultural landscapes can serve as buffer strips to protect against the leaching of nutrients (Bärwolff et al. 2014). Some areas of abandoned grassland in Germany's low mountain ranges could also be managed as short rotation coppice rather than being left to natural succession.

Whereas at the local scale the focus is on specific site conditions, at a regional scale, landscape characteristics play an increasingly important role. While the cultivation of agroforestry systems and short rotation coppice in forest-rich and well-structured landscapes would serve to 'close' the landscape, with corresponding trade-offs with respect to the hydrological balance, biodiversity conservation and landscape aesthetics, the incorporation into featureless and almost treeless agricultural landscapes (Fig. 12.4) could result in several synergistic effects (Table 12.1).

Estimates of the potential to produce woody biomass on arable land (e.g. Aust 2012; Zscheischler et al. forthcoming) indicate that there are clear limits to the capacity. These estimates suggest that the 14 % renewable heat target adopted by Germany is achievable based on an increased woodfuel supply but that this is at the very limit of what can be produced domestically when competition with food supply, nature conservation and landscape aesthetics are taken into consideration. Nevertheless, short rotation coppice and agroforestry systems offer the potential for more sustainable land use when established carefully, and targeting the most suitable sites and regions. As yet, however, their successful implementation in practice faces many challenges.

12.4 Challenges to the Implementation of Agroforestry Systems and Short Rotation Coppice

The implementation and multiplication of woody biomass production on arable land in the form of agroforestry systems and short rotation coppice has been slow to date and faces a number of challenges. The latest estimates suggest that solid fuel biomass is cultivated on approximately 9000 ha of arable land in Germany (FNR 2014). This figure falls well short of the great expectations of less than a decade ago (e.g. BMU 2008). The reasons for the slow uptake in Germany are varied (Bemmann et al. 2010) and reflect similar trends observed elsewhere (Alker et al. 2005). Although familiarity with the crop has improved in recent times, acceptance by farmers remains quite low. This has various intangible reasons, such as an aversion by farmers to trees on agricultural land, but also more practical causes, namely the long-term binding of the land, the inability to react to developments on the markets for agricultural produce and uncertainty over the profitability of wood chips combined with a lack of familiarity with the wood market.

Probably the largest non-technical barriers to the expansion of short rotation coppice in Germany at present relate to policy. National targets have been set with regard to energy savings and the contribution of renewable sources of energy to final energy consumption (BMWi and BMU 2010), yet the pathways to achieve these goals remain vague. Consequently, the political debates about implementing the mainly sectoral goals are quite unfocused. In 2012, renewable energies (biomass, biofuels, waste materials, solar thermal, geothermal, photovoltaic, water and wind) contributed 12.3 % to final energy consumption in Germany. Biomass accounts for 62 % of this and 51.8 % of the total primarily in the form of wood (BMWi 2013). In spite of this substantial contribution, wood plays almost no role in the national political debate on energy and successive versions of the renewable energies act have promoted the widespread cultivation of maize and oilseed rape, but have done little to advance the cultivation of wood on the limited agricultural land available. One effect of this has been to exacerbate intensive, unsustainable agriculture while tainting 'biomass' as a whole in the public consciousness. Another issue in this regard is the overwhelming focus of the 'energy discussion' on electricity, largely ignoring the equally important issue of heat. Given the characteristics of wood and the technology currently widely available, wood remains primarily suited to the provision of heat, ideally as part of regional, decentralised heating systems.

A phenomenon that has been especially evident since 2012 has been a wait-and-see approach adopted by farmers while the European Union's common agricultural policy (CAP) is reformed. Originally intended for introduction in 2013, the reform process has still not been completed at the time of writing. Relatively large increases in the area of short rotation coppice cultivated in Germany were recorded between 2010 and 2012, but with reforms looming farmers have been unwilling to take any risks with new crops since 2012, choosing instead to wait out the reform process. Whereas the 'greening' of the CAP was initially seen as an opportunity to advance short rotation coppice, the restrictions on its inclusion in the ecological focus areas indicated in the announcements by the European Commission to date would appear to relegate short rotation coppice to an irrelevance in the context of greening. Additionally, the long, drawn out nature of the process of policy reform itself has had negative implications for novel land uses generally (cf. Alker et al. 2005). Increasingly the growing number of reforms to legislation at European, national and state level is serving to undermine landowner confidence in their ability to make long-term planning decisions.

12.5 The Implications for Sustainable Land Management

Agroforestry systems and short rotation coppice should not be regarded as being either positive or negative per se. Sustainable implementation depends, amongst other factors, on the choice of suitable sites and on the trade-offs between different uses of the land. Only when cultivated on appropriate sites synergistic effects can be achieved and the effects of land use competition be mitigated. A number of challenges exist with respect to the governance of sustainable land use decisions.

12.5.1 Assessing the Intensity of Land Use Competition

In this chapter, we regarded land use competition from a very general and simplified perspective. Reduced to its simplest level, it would appear obvious that woody biomass production on arable land increases competition with other agricultural commodities such as food. Such discussion of increased competition between land uses has grown since the years 2007 and 2008. During the so-called tortilla crisis in Mexico, corn prices underwent a fourfold rise due, according to certain analysts, to increasing bioethanol production in the USA. However, others cited speculation and import dependency as causal factors, too. This clearly shows that dependencies between covering the demand for produce and land use are not always directly related. According to the FAO (2013), current worldwide food production provides 5000 calories per capita per day, which is in theory sufficient to cover global food demand. In reality, however, an estimated 805 million people are chronically undernourished. This illustrates that food security is not simply a question of land use on a global scale but of distribution, poverty and access as well. Demographic and climate change may result in additional land-based demands in future, exacerbating land use competition between food and bioenergy. One major challenge with respect to the achievement of sustainable production will be to assess the intensity of land use competition and its real effect on meeting needs such as the demand for food.

12.5.2 Integrating Ecological with Economic and Societal Demands

Currently, the cultivation of woody biomass on arable land does not compete with food production in Germany rather it offers certain economic and societal advantages and beneficial synergetic effects with regard to supporting and regulating services. Agroforestry systems and short rotation coppice can contribute to regional added value, particularly in rural areas. Regional material and energy loops can be completed by using the wood produced locally in district heating systems. Short rotation coppice in combination with local heating systems can—after initial investment—widen the range of economic options open for some municipalities across Germany by reducing energy expenses paid to providers outside the region. Regional supply chains can contribute, to some extent, to the independence of municipalities from markets and provide security of supply and stable costs. In addition to supplying wood for energy, fast-growing trees can also provide the raw material necessary for certain material applications. Research has shown that small dimension poplar and willow wood can be used in the manufacture of high quality thermoplastic composites (Krause et al. 2015) and particle boards (Friese et al. 2015). In accordance with the concept of cascade use, these materials are being produced in such a way—for example, with appropriate glues—that they can be used for the provision of energy at the end of the product life cycle.

12.5.3 Integrating Knowledge, Sectoral Viewpoints and Values

As land use competition occurs between actors with diverse interests and demands on different scales (see also Chaps. 2 and 8), the sustainable management of land faces many challenges with respect to integrating these competing interests and new conceptual approaches. In addition, 'sustainability' is a normative concept and values play a guiding role when taking action (Wiesmann 1998). Science has been assigned a rationalising role and has been charged with providing more 'socially robust knowledge' to tackle complex, real-world problems. In consequence the emerging approach of transdisciplinary research should be discussed or applied going beyond scientific disciplines and sectoral viewpoints (e.g. Pohl 2010; Zscheischler and Rogga 2015).

Up to now, science and policy have focused mainly on assessments of the potential to supply renewable energy: how much energy can be supplied by biomass in Germany up to the year 2020? Hypothetical options are paramount, whereas practical research focusing on implementation and interactions with other land uses is rare. A stronger focus in science on problem orientation is needed to foster a sustainable use of ecosystem services. The implementation of short rotation coppice and agroforestry systems in a sustainable way will only be possible if the relevant actors operating at different scales develop an awareness of the problems and are disposed to generate, share and put into practice new knowledge (cp. Bock et al. 2009). This implies continuous collaboration between representatives of science, policy, administration, the private sector, associations, foundations and civil society.

Ultimately, competition between land-based ecosystem services cannot only be explained by diverging interests. Competition and conflicts are also caused by deficits and inadequate forms of land use governance (e.g. Davoudi et al. 2008; Hubacek et al. 2010). This also applies to the rare examples of implemented short rotation coppice and agroforestry systems in Germany. Institutional barriers such as sectoral knowledge, bureaucracy, lobbying and path dependencies, strictly separated spatial responsibilities as well as diverse temporal scales can also promote unsustainable land use and competition (cf. Young 2002—problem of scale, fit, interplay). Shared responsibilities as well as inter-sectoral and inter-organisational cooperation can reveal synergies between different land uses and support the implementation of a more multifunctional land use. At the very least, a mutual compatibility in the sense of a coexistence of different types of land use and ecosystem services can be achieved. Research into the institutional settings and governance in the context of sustainable land management must be furthered. Some lessons can be learned from research into integrated water resource management, which increasingly focusses on measures minimising the institutional barriers that lead to land use competition (e.g. Moss 2012; Moss and Newig 2010).

As decision-making in the realm of natural resource use is fraught with complex interactions and uncertainties, an integrative adaptive management approach is required. Knowledge integration and social learning processes play a central role as 'policy-making is an iterative experiment acknowledging uncertainty' (Costanza et al. 1998).

12.5.4 Managing the Demand

The contribution to the total energy demand that can be achieved sustainably by woody biomass grown on arable land remains limited. One promising approach involves steering not only the supply but also the demand. Managing the demand represents a problem-oriented approach to addressing the growing gap between the rising demand for wood in Germany and the stagnant supply (e.g. Strbac 2008; Didden and D'haeseleer 2003; Wilhite et al. 2000). Energy saving is one example of a demand-focused perspective making up an indispensable cornerstone of the attempted transformation of the energy system in Germany. According to the European directives on energy end-use efficiency and energy services (EC2006/EU2012), the member states of the European Union are expected to achieve the indicative target of at least 9 % energy savings by the year 2016. The overall aim is to reduce the European Union's primary energy consumption by 20 % by 2020. Germany has targeted a reduction of the national primary energy consumption of 20 % compared to 2008 by

2020 and 50 % by 2050 (NEEAP 2007/2011/2014). A greater emphasis on the need to save energy, and the raw materials that can be used to provide energy, is required at all levels of society, with particular scope for enhanced, industry-specific savings throughout the various wood-based sectors.

Balancing the competing ecosystem services by fostering a more integrated and transdisciplinary perspective is a central cornerstone of sustainable land use. Sustainable land management focuses not only on the ecological and economic advantages, or disadvantages, of short rotation coppice and agroforestry, but on all aspects in a comprehensive and a trans-sectoral way. It must be viewed as a highly complex sphere of activity considering the broad range of ecosystem services and linked human–environment interactions. This understanding goes beyond the common definition of sustainable land management often employed in discussions of various aspects of land use (participation, ownership, reforestation) in developing countries (e.g. The World Bank 2006). The link between the ecosystem services and human–environment interactions is attracting increasing attention at the global scientific level, as represented by the Global Land Project,[5] as well as at European level, represented by the EU Research and Innovation programme Horizon 2020. At political level, an integrated view of land use is gradually being advanced, for example, under the UN's Millennium Development Goals and Habitat III programme. In Germany, the 'Sustainable Land Management' research programme, launched by the German Federal Ministry for Education and Research in 2008 (BMBF 2008), addresses in particular problem-oriented transfer strategies and concrete implementation measures.[6] These developments make clear the need for integrated multiscale perspectives in order to specifically harness global as well as regional trade-offs between the different types of land use. The necessarily broad nature of the discussion of land use is an indicator of the diversity of the challenges currently facing both science and practice and demonstrates the importance of the ongoing research in this field.

References

Alker, G., Bruton, C., & Richards, K. (2005). Full-scale implementation of SRC-systems. Assessment of technical and non-technical barriers. IEA Bioenergy Task 30. *High Priority Area, 3*, 29.

Aust, C., (2012). *Abschätzung der nationalen und regionalen Biomassepotentiale von Kurzumtriebsplantagen auf landwirtschaftlichen Flächen in Deutschland.* Dissertation. Fakultät für Forst- und Umweltwissenschaften, Albert-Ludwigs-Universität Freiburg im Breisgau. 167 S. Online. http://www.freidok.uni-freiburg.de/volltexte/8630/pdf/Dissertation_Aust_KUP_Potentiale.pdf. Accessed 26 November 2014.

[5]Global Land Project, 2005: Science Plan and Implementation Strategy: http://www.globallandproject.org/arquivos/report_53.pdf.

[6]More information about the 'Sustainable Land Management' programme: http://nachhaltiges-landmanagement.de/en/home.

Bärwolff, M., Reinhold, G., Fürstenau, C., Graf, T., Jung, L. & Vetter, A. (2014). *Gewässerrandstreifen als Kurzumtriebsplantagen oder Agroforstsysteme*. Umweltbundesamt, Texte 94/2013. Online. http://www.umweltbundesamt.de/sites/default/files/medien/378/publikationen/texte_94_2013_gewaesserrandstreifen_als_kurzumtriebsplantagen_1.pdf. Accessed 26 November 2014.

Baum, S., Weih, M., Busch, G., Kroiher, F. & Bolte, A. (2009). The impact of short rotation coppice plantations on phytdiversity. vTI *Agriculture and Forestry Research* (Vol. 59, Issue 3, pp. 163–170). Online: http://www.ti.bund.de/media/publikationen/landbauforschung/Landbauforschung_Vol59_3.pdf

Baum, S., (2012). *Phytodiversity in short rotation coppice plantations*. Dissertation. Georg-August-Universität Göttingen. Online: https://ediss.uni-goettingen.de/bitstream/handle/118-58/00-1735-0000-000D-F05A-1/baum.pdf?sequence=1. Accessed 26 November 2014.

Bemmann, A. & Große, W. (2011). Effiziente Landnutzung—ein Beitrag zur Zukunftssicherung. Vision der Professur für Forst- und Holzwirtschaft Osteuropas. In S. Bonn, J. Erler, S. Herzog (eds.), *Tharandt 2011—200 Jahren Ideen für die Zukunft*. Forstwissenschaftliche Beiträge Tharandt/Contributions to Forest Sciences, Beiheft 12, pp. 16–37.

Bemmann, A., & Knust, C. (Eds.). (2010). *Agrowood—Kurzumtriebsplantagen in Deutschland und europäische Perspektiven*. Berlin: Weißensee Verlag. (340 pp).

Bemmann, A., Nahm, M., Brodbeck, F., & Sauter, U. H. (2010). Holz aus Kurzumtriebsplantagen: Hemmnisse und Chancen. *Forstarchiv, 81*(6), 246–254.

BMBF—German Federal Ministry of Education and Research (2008). *Announcement. Regulations for the 'Sustainable land management' funding measure*. Online: http://www.bmbf.de/en/furtherance/13138.php. Accessed 18 November 2014.

BMU—Federal Ministry for the Environment, Nature Conservation and Nuclear Safety (2008). *Leitstudie 2008. Weiterentwicklung der Ausbaustrategie Erneuerbare Energien vor dem Hintergrund der aktuellen Klimaschutzziele Deutschlands und Europas*. Bonn, 196 pp.

BMU/BMELV—Federal Ministry for the Environment, Nature Conservation and Nuclear Safety and Federal Ministry of Food, Agriculture and Consumer Protection (2010). *National biomass action plan for Germany*.

BMWi—Federal Ministry for Economic Affairs and Energy, 2007/2011/2014: *1st, 2nd and 3rd National Energy and Efficiency Plan of the Federal Republic of Germany* (NEEAP).

BMWi—Federal Ministry for Economic Affairs and Energy (2013). *Erneuerbare Energien im Jahr 2013*. Erste vorläufige Daten zur Entwicklung der erneuerbaren Energien in Deutschland auf der Grundlage der Angaben der Arbeitsgruppe Erneuerbare Energien-Statistik (AGEE-Stat). http://www.bmwi.de/BMWi/Redaktion/PDF/A/agee-stat-bericht-ee-2013,property=pdf,bereich=bmwi2012,sprache=de,rwb=true.pdf. Accessed 14 November 2014.

BMWi/BMU—Federal Ministry for Economic Affairs and Energy and Federal Ministry for the Environment, Nature Conservation and Nuclear Safety (2010). *Energiekonzept für eine umweltschonende, zuverlässige und bezahlbare Energieversorgung*. Berlin, 36 pp.

Bock, S., Hinzen, A., & Libbe, J. (2009). Nachhaltiges Flächenmanagement – in der Praxis erfolgreich kommunizieren. In S. Bock, A. Hinzen, & J. Libbe (Eds.), *Nachhaltiges Flächenmanagement – in der Praxis erfolgreich kommunizieren*. Berlin: Deutsches Institut für Urbanistik.

Böcker, L. (2013). Kurzumtriebsplantagen auf Rekultivierungsflächen. In A. Bemmann, D. Butler Manning (eds.), *Energieholzplantagen in der Landwirtschaft* pp. 155–158. Agrimedia Verlag, Clenze.

Boll, T., Haaren, C. V. & Rode, M. (2015). The effects of short rotation coppice on recreation and the visual landscape—Results of a survey using 3D visualisations for scenarios in different landscape types. In D. Butler Manning, A. Bemmann, M. Bredemeier, N. Lamersdorf, C. Ammer (eds.), *Bioenergy from dendromass for the sustainable development of rural areas*. Weinheim: Wiley-VCH Verlag (In preparation).

Bungart, R., & Hüttl, R. F. (2001). Production of biomass for energy in post-mining landscapes and nutrient dynamics. *Biomass and Bioenergy, 20*(3), 181–187.

Busch, G., & Lamersdorf, N. (2010). *Kurzumtriebsplantagen – Handlungsempfehlungen zur naturverträglichen Produktion von Energieholz in der Landwirtschaft*. Ergebnisse aus dem Projekt Novalis. Osnabrück, Deutsche Bundesstiftung Umwelt (DBU). Online: https://www.dbu.de/OPAC/fp/Kurzumtriebsplantagen.pdf. Accessed 26 November 2014.

Busch, G., & Meixner, C. (2015). A spatially explicit approach to the identification of sites suitable for woody biomass systems based on site factors and field geometry—A case study for the Göttingen district. In D. Butler Manning, A. Bemmann, M. Bredemeier, N. Lamersdorf, C. Ammer (eds.), *Bioenergy from dendromass for the sustainable development of rural areas*. Weinheim: Wiley-VCH Verlag (In preparation).

Costanza, R., Andrade, F., Antunes, P., Van Den Belt, M., Boersma, D., Boesch, D. F., et al. (1998). Principles for sustainable governance of the oceans. *Science, 281*(5374), 198–199.

Davoudi, S., Evans, N., Governa, F., & Santangelo, M. (2008). Territorial governance in the making. *Approaches, Methodologies, Practices. Boletin de la A.G.E.N., 46*, 33–52.

Didden, M. H., & D'haeseleer, W. D. (2003). Demand side management in a competitive European market: Who should be responsible for its implementation? *Energy Policy, 31*, 1307–1314.

Dietzsch, A. (2011). Nutzung kontaminierter Böden. Abschlussbericht des Projektes 'Anbau von Energiepflanzen für die nachhaltige, ressourcenschonende und klimaverträgliche Rohstoffabsicherung zur Erzeugung von Strom/Wärme und synthetischen Biokraftstoffen der zweiten Generation'. *Schriftenreihe des LfULG*, Heft 19.

Dimitriou, I., Baum, C., Baum, S., Busch, G., Schulz, U., & Köhn, J., et al. (2011). Quantifying environmental effects of short rotation Coppice (SRC) on Biodiversity, Soil and Water. *IEA Bioenergy*, Task 43. Online: http://ieabioenergytask43.org/wp-content/uploads/2013/09/IEA_Bioenergy_-Task43_TR2011-01.pdf. Accessed 26 November 2014

Don, A., Osborne, B., Hastings, A., Skiba, U., Carter, M. S., Drewer, J., et al. (2011). Land-use change to bioenergy production in Europe: Implications for the greenhouse gas balance and soil carbon. *GCB Bioenergy*. doi:10.1111/j.1757-1707.2011.01116.x.

Dupraz C., Burgess P., Gavaland A., Graves A., Herzog F., & Incoll L., et al. (2005). *Synthesis of the Silvoarable Agroforestry for Europe project*. INRA-UMR System Editions, Montpellier, 254 p. Online: http://www1.montpellier.inra.fr/safe/english/results/final-report/SAFE%20Final%20Synthesis%20Report.pdf

European Community (2006). *Directive on energy end-use efficiency and energy services*. 2006/32/EC.

European Union (2012). *Directive on energy efficiency*. 2012/27/EU.

Ewing, B., Goldfinger, S., Wackernagel, M., Stechbart, M., Rizk, S. M., Reed, A., & Kitzes, J. (2008). *The ecological footprint atlas 2008*. Oakland: Global Footprint Network.

FAO. (2013). *Statistical yearbook 2013. World food and agriculture*. Rome: FAO.

Federal Statistics Office, Germany https://www.destatis.de/EN/FactsFigures/Economic-Sectors/AgricultureForestryFisheries/LandUse/LandUse.html. Accessed 18 November 2014.

Finch, J. W., Karp, A., McCabe, D. P. M., Nixon, S., Riche, A. B. & Whitmore, A. P. (2009). Miscanthus, short-rotation coppice and the historic environment. *English heritage*. Online: http://nora.nerc.ac.uk/7566/1/EngHerit_Report_final.pdf. Accessed 26 November 2014.

FNR – Fachagentur für Nachwachsende Rohstoffe (2014). *Anbau nachwachsender Rohstoffe in Deutschland*. http://mediathek.fnr.de/media/downloadable/files/samples/r/z/rz_fnr4_0302_grafik_nawaro_anbau_de_neu.jpg. Accessed 14 November 2014.

FNR (2015). Wärmeverbrauch aus erneuerbaren Energien. https://mediathek.fnr.de/grafiken/daten-und-fakten/bioenergie/warme-aus-biomasse.html. Accessed 28 April 2016.

Friese, F., Euring, M., & Kharazipour, A. (2015). Particle boards from newly exploited wood sources as a starting point for cascade utilisation. In D. Butler Manning, A. Bemmann, M. Bredemeier, N. Lamersdorf & C. Ammer (eds.), *Bioenergy from dendromass for the sustainable development of rural areas*. Weinheim: Wiley-VCH Verlag (In preparation).

Fry, D., & Slater, F. (2009). *The biodiversity of short rotation willow coppice in the Welsh landscape*. A report to the Institute of Biological, Environmental and Rural Sciences, Aberystwyth University for EU Project 'Willows for Wales.' Online: https://www.aber.ac.uk/en/media/departmental/ibers/research/willowforwales/Biodiversity-of-src-coppice-in-the-Welsh-Landscape.pdf. Accessed 26 November 2014.

Gruenewald, H., Brandt, B. K. V., Schneider, B. U., Bens, O., Kendzia, G., & Hüttl, R. F. (2007). Agroforestry systems for the production of woody biomass for energy transformation purposes. *Ecological Engineering, 29*(4), 319–328.

Grunert, M., & Wilhelm, E.-G. (2013a). Kurzumtriebsplantagen und Biodiversität. In A. Bemmann, D. Butler Manning (eds.), *Energieholzplantagen in der Landwirtschaft* (pp. 139–144). Clenze: Agrimedia Verlag.

Grunert, M., & Wilhelm, E.-G. (2013b). Bodenschutz vor Wind- und Wassererosion mit Kurzumtriebsplantagen. In A. Bemmann, D. Butler Manning (eds.) *Energieholzplantagen in der Landwirtschaft* (pp. 48–52). Clenze: Agrimedia Verlag.

Harrison, P. A., Berry, P. M., Simpson, G., Haslett, J. R., Blicharska, M., Bucur, M., & Turkelboom, F. (2014). Linkages between biodiversity attributes and ecosystem services: A systematic review. *Ecosystem Services, 9*, 191–203.

Hetsch, S., Steierer, F., & Prins, C. (2007). *Wood resources availability and demands, part 2: Future wood flows in the forest and energy sector, European Countries in 2010 and 2020*. Genf: UNECE.

Hubacek, K., Fraser, E., & Thapa, S. (2010). Land-use governance. In P. A. O'Hara (ed.), *International encyclopedia of public policy—Governance in a global age*, Vol. 4, Social, Environmental and Corporate Governance (pp. 419–428). GPERU: Perth, Online: http://pohara.homestead.com/Encyclopedia/Volume-4.pdf. Accessed 18 November 2014.

IEA—International Energy Agency. (2013). *IEA statistics—renewable information 2013*. Paris: IEA.

King, K. F. S. (1989). The history of agroforestry. In P. K. R. Nair (ed.), *Agroforestry systems in the tropics*. Forestry Sciences (Vol. 31, pp. 3–11). Dordrecht: Kluwer Academic Publishers.

Krause, K. C., Müller, M., Militz, H., & Krause, A. (2015). Converting wood from short rotation coppice and low value beech wood into thermoplastic composites (WPC). In D. Butler Manning, A. Bemmann, M. Bredemeier, N. Lamersdorf & C. Ammer (eds.), *Bioenergy from dendromass for the sustainable development of rural areas*. Weinheim: Wiley-VCH Verlag (In preparation).

Mantau, U. (2010). Is there enough wood for Europe? 19–34. In *EUwood—Final report*. Hamburg/Germany, June 2010, 160.

Mantau, U. (2012). Holzrohstoffbilanz Deutschland, Entwicklungen und Szenarien des Holzaufkommens und der Holzverwendung 1987 bis 2015. Hamburg, 65.

Mead, R., & Willey, R. W. (1980). The concept of a 'land equivalent ratio' and advantages in yields from intercropping. *Experimental Agriculture, 16*, 217–228. doi:10.1017/S0014479700010978.

Moss, T., & Newig, J. (2010). Multi-level water governance and problems of scale. Setting the stage for a broader debate. *Environmental Management, 46*, 1–6. doi:10.1007/s00267-010-9531-1.

Moss, T. (2012). Spatial Fit, from Panacea to practice: Implementing the EU water framework directive. *Ecology and Society, 17*(3), 2.

Murach, D., Hartmann, H., Murn, Y., Schultze, M., Ali, W., & Röhle, H. (2009). Standortbasierte Leistungsschätzung in Agrarholzbeständen in Brandenburg und Sachsen. In T. Reeg, A. Bemmann, W. Konold, D. Murach, & H. Spiecker (Eds.), *Anbau und Nutzung von Bäumen auf landwirtschaftlichen Flächen*. Weinheim: Wiley-VCH Verlag.

Nair, P. K. R. (1990). The prospects for agroforestry in the tropics. *World Bank technical paper*, ISSN 0253-7494; no. 131.

Ong, C. K., Corlett, J. E., Singh, R. P., & Black, C. R. (1991). Above and below ground interactions in agroforestry systems. *Forest Ecology and Management, 45*, 45–57.

Pohl, C. (2010). From transdisciplinarity to transdisciplinary research. *Journal of Engineering & Science, 1*(1), 74–83.

Pollard, E., Hooper, M. D., & Moore, N. W. (1974). *Hedges*. London: Collins.

Qinkenstein, A., Jochheim, H., Schneider, B.-U., & Hüttl, R. F. (2009). Modellierung des Kohlenstofhaushalts von Pappel-Kurzumtriebsplantagen in Brandenburg. In T. Reeg, A. Bemmann, W. Konold, D. Murach, & H. Spiecker (Eds.), *Anbau und Nutzung von Bäumen auf landwirtschaftlichen Flächen*. Weinheim: WILEY-VCH Verlag.

Reeg, T., & Grünewald, H. (2009). Überblick über den Stand der Forschung zu Agroforstsystemen in Deutschland. In T. Reeg, A. Bemmann, W. Konold, D. Murach, & H. Spiecker (Eds.), *Anbau und Nutzung von Bäumen auf landwirtschaftlichen Flächen*. Weinheim: WILEY-VCH Verlag.

Schildbach, M., Grünewald, H., Wolf, H., & Schneider, B.-U. (2009). Begründung von Kurzumtriebsplantagen: Baumartenwahl und Anlageverfahren. In T. Reeg, A. Bemmann, W. Konold, D. Murach, & H. Spiecker (Eds.), *Anbau und Nutzung von Bäumen auf landwirtschaftlichen Flächen* (pp. 57–72). Weinheim: Wiley-VCH Verlag.

Schmidt, P., & Glaser, T. (2009). Kurzumtriebsplantagen aus Sicht des Naturschutzes. In T. Reeg, A. Bemmann, W. Konold, D. Murach, & H. Spiecker (Eds.), *Anbau und Nutzung von Bäumen auf landwirtschaftlichen Flächen* (pp. 161–170). Weinheim: Wiley-VCH Verlag.

Schubert, R., Schellnhuber, H. J., Buchmann, N., Epiney, A., Grießhammer, R., & Kulessa, M., et al. (2009). *Future bioenergy and sustainable land use*. German Advisory Council on Global Change (WGBU). London, Sterling, Earthscan.

Schulz, U., Brauner, O., & Gruß, H. (2009). Animal diversity on short-rotation coppices—A review. vTI *Agriculture and Forestry Research,* Vol. 59, Issue 3, pp. 171–182. Online: http://www.ti.bund.de/media/publikationen/landbauforschung/Landbauforschung_Vol59_3.pdf

Strbac, G. (2008). Demand side management: Benefits and challenges. *Energy Policy, 36*(12), 4419–4426. doi:10.1016/j.enpol.2008.09.030.

Thrän, D., Edel, M., & Seidenberger, T. (2009). *Identifizierung strategischer Hemmnisse und Entwicklung von Lösungsansätzen zur Reduzierung der Nutzungskonkurrenzen beim weiteren Ausbau der energetischen Biomassenutzung*. 1. Zwischenbericht, Leipzig. Online: http://www.fornebik.bayern.de/allgemein/a_netzwerk_news/pdf_images/100526_DBFZ_Biomassekonkurrenzen.pdf

Van der Werf, W., Keesman, K., Burgess, P., Graves, A., Pilbeam, D., Incoll, L. D., et al. (2007). Yield-SAFE: A parameter-sparse, process-based dynamic model for predicting resource capture, growth, and production in agroforestry systems. *Ecological Engineering, 29*(4), 419–433.

Van Noordwijk, M., & Lusiana, B. (2000). WaNuLCAS 2.0, *Background on a model of water nutrient and light capture in agroforestry systems*. International Centre for Research in Agroforestry (ICRAF), Bogor, Indonesia, p. 186.

Vetter, A., Bärwolff, M., & Biertümpfel, A. (2009). *Energieholz aus Plantagen oder Agroforstsystemen – eine vergleichende Betrachtung*. Dornburg. Forum Agroforstsysteme in Jena. Online: http://www.tll.de/ainfo/pdf/afs/afs18_09.pdf

Weih, M., & Dimitriou, I. (eds.). (2012). SRC and the environment. *BioEnergy Research*, 5(3). doi:10.1007/s12155-012-9230-2

Wiesmann, U. Sustainable Regional Development in Rural Africa: Conceptual Framework and Case Studies from Kenya; Geograhica Bernensia and Geograhipical Society of Berne: Bern, Switzerland, 1998; Volume A14.

Wilhite, H., Shove, E., Lutzenhiser, L., & Kempton, W. (2000). The legacy of twenty years of energy demand management: We know more about individual behaviour but next to nothing about demand. In E. Jochem, J. Sathaye, & D. Bouille (Eds.), *Society, behaviour, and climate change mitigation* (pp. 109–126). Netherlands: Springer.

The World Bank (2006). *Sustainable land management. Challenges, opportunities, and trade-offs*. Washington

Young, A. (1997). Agroforestry for soil management (No. Ed. 2). CAB international.
Young, O. (2002). *The institutional dimensions of environmental change. Fit, interplay, and scale*. Cambridge: The MIT Press.
Zscheischler, J., & Rogga, S. (2015). Transdisciplinarity in land use science—A review of concepts, empirical findings and current practices. *Futures, 65*, 22–48.
Zscheischler, J., Uckert, G., & Gaasch, N. (forthcoming) *Potentials of energy wood production on arable land in Germany related to land use competition*.

Chapter 13
Land-Use Competition in the South American Chaco

Matthias Baumann, María Piquer-Rodríguez, Verena Fehlenberg, Gregorio Gavier Pizarro and Tobias Kuemmerle

Abstract The dry forests of the Chaco in South America are under great land conversion pressure, mostly for establishing pastures and soybean fields. Taking recent estimates into account, the rates at which forests are disappearing are similar to those of the Amazon, but compared to the Amazon, the Chaco remains fairly understudied. The land transformations during the past decades went along with a substantial change in the type of actors dominating the Chaco landscape. In this chapter, we discuss the land changes in the Chaco during the past 30 years with emphasis on the actors driving these changes, as outlined in the conceptual chapter of this book section. In the Chaco, it appears that the competition over land is a competition in which the actors are endowed with varying degrees of power, resulting in highly unbalanced competition. This chapter highlights these differences and discusses the potential role of the state as an actor in the competition for land that may help to slow down deforestation in the area and guide the Chaco toward more sustainable land-use futures.

Keywords Transformation · Proximate causes · Underlying drivers · Deforestation

---

M. Baumann (✉) · M. Piquer-Rodríguez · V. Fehlenberg · T. Kuemmerle
Geography Department, Humboldt-Universität zu Berlin, Berlin, Germany
e-mail: matthias.baumann@geo.hu-berlin.de

G. Gavier Pizarro
Grupo de Biodiversidad, Ecología y Gestión Ambiental,
Instituto de Recursos Biológicos (CIRN), Instituto Nacional
de Tecnología Agropecuaria (INTA), CNIAINTA, Nicolas Repetto
y De los Reseros, Buenos Aires, Argentina

T. Kuemmerle
Integrative Research Institute Transformations of Human-Environment
Systems (IRI Thesys), Humboldt-Universität zu Berlin, Berlin, Germany

J. Niewöhner et al. (eds.), *Land Use Competition*,
Human-Environment Interactions, DOI 10.1007/978-3-319-33628-2_13

13.1 Introduction

Deforestation has been of major global concern during the last decades since its identification as one of the major drivers of global environmental change. Its consequences for biodiversity are also profound, mostly because habitat for many species is lost when natural forests are converted into human-dominated landscapes. Deforestation also causes substantial carbon emissions and thus also contributes to global climate change. Although deforestation at the global level has been decreasing since 2000, it is still alarmingly high (Hansen et al. 2013). Between 2000 and 2010, the loss of natural forests amounted to 5.2 million ha per year globally. Despite the particular attention of conservation efforts on countries in South America, Southeast Asia, and Africa, which harbor globally valuable forests, these regions still suffer the largest net losses of forests.

In South America, the Amazon region has historically been a focus region for scientists. Deforestation rates here have been high, and the region plays a significant role in the context of global biodiversity, indigenous populations, and carbon budgets.

Interestingly, and also surprisingly, the South American Chaco has been largely overlooked in the literature. For example, using the same search terms in the ISI Web of Science, the number of studies focusing on the Amazon outnumber the studies in the Chaco by more than a factor of ten (Fig. 13.1). However, the South American Chaco is among the largest dry forest areas worldwide, and those forests

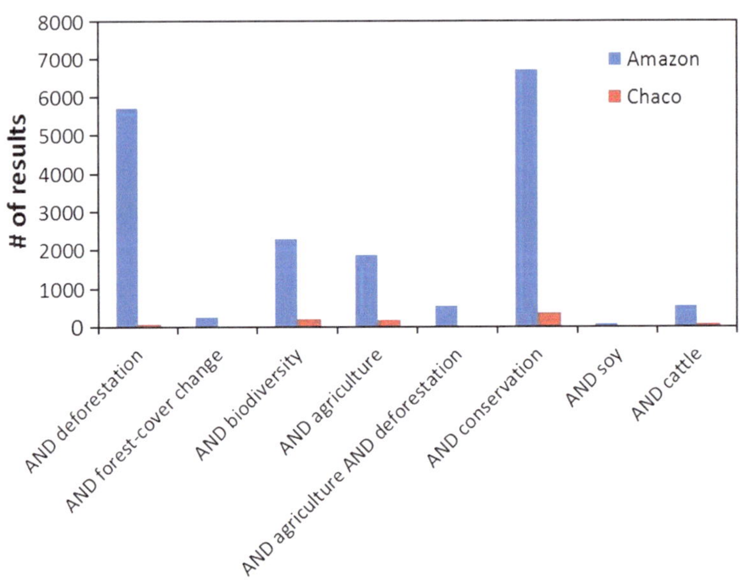

Fig. 13.1 Search results in the ISI Web of Science for Chaco and Amazon in combination with a number of terms used in relation to land-use competition (search conducted on December 10, 2015)

have been disappearing rapidly during the past decades (Hansen et al. 2013). Taking global estimates from the past 10 years, the Chaco has had rates of deforestation comparable to those of the Amazon. Moreover, the dry forests of the Chaco are rich in biodiversity.

The reasons for the ongoing deforestation in the Chaco are manifold and will be discussed in this chapter with a focus on the different actors competing for land. Most of the forests of the Chaco are disappearing in favor of agricultural production, mostly to fulfill increasing global demand in agricultural products resulting from population growth, diets richer in animal products, and the increasing use of agricultural products for biofuel production. This chapter sheds light on the land-use changes that have been shaping today's landscapes of the Chaco. It will discuss the different degrees of power actors possess with regard to land-use decisions and describe the competition over land in this region.

13.1.1 The Chaco

The Chaco is a large, dry forest region covering about 1.1 Mkm^2 located in Argentina (60 % of the Chaco), Bolivia (11 %), Paraguay (27 %), and Brazil (2 %) (Olson et al. 2001, Fig. 13.2). The climate is semiarid and highly seasonal, with a distinct dry season from May to September and a warm, wet season from November to April. The mean annual temperature is about 22 °C, with an average monthly maximum of 28 °C. Annual precipitation ranges from 1200 mm in the east (wet Chaco) to 450 mm in the west (dry Chaco). Elevation varies marginally except for the west and southwest of the Chaco where more hilly terrain prevails. The natural vegetation in the Chaco is a direct result of climate and topography and consists mostly of closed dry forest, open woodlands, shrublands, and palm savannas, which represent a globally significant carbon pool (Gasparri et al. 2008). Forests are the most characteristic vegetation formation and are typically dominated by species of the genera Schinopsis and Aspidosperma ("quebrachos"), Prosopisflexuosa, and shrubs of Larrea sp. (Prado 1993). The Chaco also hosts animal species of importance for local livelihoods and nature conservation. The rich biodiversity of the Chaco includes 145 mammal species (12 endemic), 409 birds (7), 54 reptiles (17), 34 amphibians (8), and more than 80 plant genera (3400 species, of which 46 are trees and with a total of 400 endemic plant species) (Bucher and Huszar 1999).

The diverse biodiversity worth hunting or gathering for their meat and skin is the reason why this region is named "Chaco." The word "Chaco" derives from the Quechua expression "chacu" which means "hunting place," and this is probably how the region was perceived by native settlers. Around 5000 years ago, native tribes from the Pampas (the "pampidos"), who were originally hunters, moved to the Chaco region probably looking for new hunting territories. Other communities inhabiting the Chaco came from the Andes and Amazonia regions. The original nomadic communities of hunters, fishers, and gatherers (estimated to be approximately 300,000 inhabitants) had to move deep into the "wild Chaco" after the onset

Fig. 13.2 The South American Chaco

of the Spanish-criollo invasions in the sixteenth century (Brown et al. 2010). These communities left a rich linguistic inheritance, including the languages of the groups of Zamuco, Gaycuru, Maskoi, and the Matako-Maka. Their agricultural activities were secondary to cattle ranching, and plots were small (below 100 m^2). Until the end of the nineteenth century, the most recent immigrant ethnic groups of the "gringos," "criollos," or "menonitas," largely of Spanish descent, had transformed the land at an unprecedented rate for sugar plantations, cotton, and maize.

13.1.2 Land-Use Change in the Chaco

The land-use pattern and agricultural practices prevailing in the Chaco today do not have much in common with what has been dominating the Chaco's landscape for the past centuries. Traditionally, small-scale subsistence agriculture and extensive cattle ranching characterized land use in the Chaco. Farms were sufficiently large to fulfill a family's own needs and generated only a marginal surplus that could be sold on markets. The predominance of subsistence agriculture resulted in a land-use pattern characterized by small agricultural fields and cattle ranches, with the latter being often referred to as the "puesto system," puesto referring to a particular type of range dwelling (see also Chap. 4).

Today, the picture differs strongly from what has been characterizing the Chaco's landscape for centuries. During the past decades, and in particular during the past 10–15 years, the agricultural system in the Chaco has become dominated by large-scale farms, both for the production of cash crops and for beef production (Altrichter and Basurto 2008; Grau et al. 2005). Agricultural crops today are mostly grown on farms with a size of several 1000 ha. Monocultures are widespread, mostly soy, but also sugarcane, wheat, and corn (maize). Pasture areas are of similar size. This agricultural system has been expanding dramatically in the past, and the expansion has mostly occurred at the expense of natural forest areas. During the past 30 years, almost 20 % of the natural forest area was converted into either croplands or pastures (Vallejos et al. 2015). The speed of deforestation has dramatically accelerated over the past 10 years (Caldas et al. 2015), only interrupted by a two-year slowdown during the economic crisis of the late 2000s (Graesser et al. 2015). In addition to agricultural expansion, agricultural land use has also changed. Many pasture areas have been converted into croplands over the past 15 years, particularly in Argentina (Baumann et al. in review-a) (Fig. 13.3).

The agricultural expansion with the particular focus on soy had substantial positive effects for the economies of the countries, particularly Argentina and Paraguay. In case of Argentina, the focus on soy helped the country recover from the deep economic crisis after 2001 that sent nearly half of the Argentine population into poverty (Leguizamon 2014). Following the introduction of genetically modified variants in 1996, soy production has spiked (Fig. 13.3, left) and the export of soybeans has been rising 8.6 % per year on average (Leguizamon 2014), substantially contributing to making agriculture responsible for more than half of all Argentine export earnings. GM soy is thus often promoted as a success story of how to boost socioeconomic development of poor countries elsewhere (Newell 2009). Likewise, the introduction of new highly productive exotic grasses such as Gatton panic (Panicum maximum) or Buffel grass (Cenchrusciliaris) and the adoption of improved farming techniques and cattle breeds (Vazquez 2013) resulted in productivity increases in cattle production (Ramirez and Laneri 1989; Fig 13.3, right), largely for export purposes (Fatecha 1989). Thus, deforestation and agricultural expansion helped the Chaco countries, particularly Argentina and Paraguay, to strengthen their economies and to recover from economic crises.

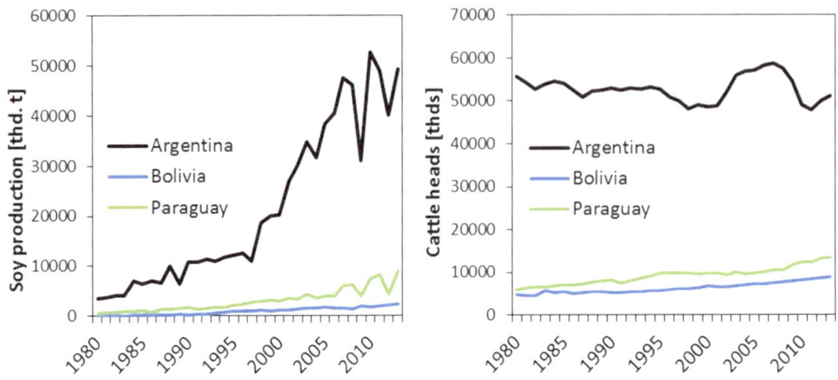

Fig. 13.3 Soy production and number of cattle head between 1980 and 2013 in the three countries that intersect the Chaco. The numbers represent national estimates from the FAO (FAOSTAT 2015)

However, the changes in agricultural practices in the Chaco have also had manifold and substantial consequences for society and the environment of the region, particularly in Argentina. For example, the transition of agricultural practices from traditional farming toward modern, capital intensive agriculture using genetically modified soybean involves the heavy use of herbicides, pesticides, and fertilizers (Gurtler et al. 2007). These chemicals are often used without strict regulations and safety instructions and thus pose substantial health threats to farmers and animals (e.g., for bees, which are important for the pollination process). Such health issues can include skin, lung, and heart diseases of the farmers and of people living close to the farms (Burghardt 2014). Moreover, malformations of newborn babies and early child deaths are increasingly reported and have been associated with the intensive use of herbicides and pesticides (Goldfarb and Zoomers 2013).

In addition, social conflicts have become more frequent. Many conflicts arise as a consequence of land acquisitions by big powerful external investors, irregularities in land titling (e.g., when farmers that have been living and working on the land for several generations are evicted from their land), or the sale of government-owned land to external investors at derisory prices. Further, the highly mechanized agricultural practices with fewer and fewer employees on large farms have been stimulating a rural exodus due to the absence of employment opportunities, leading to decimation of rural communities (Leguizamon 2014).

Besides the socioeconomic implications of the agricultural transition in the Chaco, the environment also experienced (and still experiences) the consequences of the agricultural land changes. Inefficient agricultural practices, resulting either from a lack of knowledge or from a lack of interest in efficient agricultural production, resulted in a loss of nutrients as well as in degradation and soil erosion across large areas (Bucher and Huszar 1999). In addition, the establishment of monocultures has impacted the amount of birds and mammals in the region during the last century (Gavier-Pizarro et al. 2012). Water pollution from the uncontrolled use of pesticides and herbicides is endangering aquatic species. Nutrient and water cycles have been modified due to the intensification of cattle ranching and agricultural production. Overall, the agricultural expansion resulted in some economic development in the Chaco, but there are also substantial negative effects that may outweigh positive developments.

13.2 Drivers of Land-Use Change in the Chaco

Recognizing the dramatic land conversions from natural forest ecosystems into human-dominated agricultural landscapes and their consequences for society, economy, and environment raises the question of the drivers of these land-use changes. Here, we present a selection of dominant drivers that we suggest can account for a significant portion of the observed land-use changes in the Chaco.

13.2.1 Proximate Causes

The most dominant proximate causes of land-use change in the Chaco are agricultural expansion for both crop production and cattle ranching (Fehlenberg 2015). Pasture expansion appears to be the dominant proximate cause of deforestation, as rates of pasture expansion are substantially higher than those of cropland. The picture is becoming more diverse, however, when subdividing the Chaco into the dry Chaco to the west, which is mostly located in Argentina, and the wet Chaco to the east, which is mostly on Paraguayan territory. Analyses of remote sensing data and national statistical information of soybean and cattle production suggest that during the past 12 years the causes diverged substantially between the dry and the wet Chaco. In the wet Chaco, pasture expansion appears to be the most dominant proximate cause of deforestation (Caldas et al. 2015), to the extent that during the past 12 years almost 25 % of natural forests in the wet Chaco were lost to the creation of pastures (Baumann et al. in review-a). The vast majority of these forest-to-pasture conversions were made primarily in areas around Filadelfia (Baumann et al. in review-b). Filadelfia is a Mennonite colony in which traditional farming practices prevailed for a long time (Dana and Dana 2007).

In the dry Chaco, which is to a large proportion on Argentine territory, cropland expansion is the most dominant proximate driver of deforestation (Baumann et al. in review-a; Fehlenberg 2015), although pasture expansion is widespread as well. Across large areas of the Argentine, Chaco soy is grown in monocultures, over 90 % using genetically modified crop varieties (Leguizamon 2014; Reenberg and Fenger 2011). These patterns have been developed in the past despite of partially unfavorable economic conditions including a 2007 Forest Law and high taxes on soy exports (Fehlenberg 2015).

13.2.2 Underlying Drivers

The proximate drivers of land-use change in the Chaco are fueled by and interact with a number of underlying drivers, of which we name only two predominant ones. Similar to many other regions in the world, the underlying drivers are hard to quantify, as they do not originate from the countries within the Chaco region itself but instead have their origin in other places (see also Sect. 13.1). The most important underlying driver of deforestation and cropland expansion in the Chaco is the growing demand for agricultural products, primarily soy. For the last few decades, population growth coupled with increasing affluence that went hand in hand with changing dietary preferences toward more animal protein caused a substantial increase in the demand for animal agriculture. Overall, caloric intake of the world's population is increasing, but also, and this is probably from an environmental point of view more severe, diets are becoming richer in meat. This is problematic because in order to produce more meat, more production of soy as protein-rich animals feed is required. In 2011, 35 % of all crops produced worldwide were used for animal feed which is a highly inefficient use of agricultural products (Reenberg and Fenger 2011) and fundamentally contributed to the increase in demand for agricultural products.

A second dominant underlying driver of the rapid agricultural expansion and the introduction of a system in which monocultures dominate are technological advances, which have revolutionized agricultural practices. In particular, the introduction of genetically modified (GM) soy is a major technological advance that substantially drove and is still driving agricultural expansion in the Chaco. GM soy was introduced in Argentina in the Pampas region in 1996. At present, the vast majority of arable lands are planted with GM soy and that are further expanding (Leguizamon 2014). The GM soy variants are known for the glyphosate resistance, which improves resistance to droughts at reduced pesticide and fuel cost. Other technological advances, including the development of direct tilling techniques or the increasing use of fertilizers and agrochemicals, have fostered the expansion of total soy area in Argentina at rates above those of all other crops (Goldfarb and Zoomers 2013).

The impact of soybean expansion as an underlying driver of deforestation is not immediately visible though. Not entirely understood yet, it is increasingly becoming clearer that forest-to-pasture conversions in the Chaco are often a result of conversions from pastures into soybean elsewhere, through what has been termed leakage effects (Gasparri and le Polain de Waroux 2014). This is particularly true for regions in which forest-to-pasture conversions are dominant, such as in Paraguay but also across wide areas of Argentina. For example, in the Pampas region to the south of the Chaco, many former pasture areas were converted into soybean areas (Graesser et al. 2015), which in turn pushed pasture areas into more marginal areas such as the Argentine Chaco (Manuel-Navarrete et al. 2009; Santarcángelo and Fal 2009). This is likely because converting existing pasture areas into croplands while at the same time deforesting new areas for pastures is overall less costly compared to preparing forestlands for crop production in the Chaco. Similarly, Uruguayan farmers are increasingly selling their land to (foreign) soybean producers and reinvest the revenues by purchasing land in the Paraguayan Chaco to engage in cattle ranching there (Bertello 2008; Gonzales 2013). Moreover, in Paraguay's Atlantic forest region, deforestation for soybean was banned (Huang et al. 2007), which resulted in substantial conversions from pastures into croplands there and the subsequent displacement of pasture expansion into the Paraguayan Chaco where restrictions on deforestation are missing (World Wildlife Fund 2015).[1] All three examples highlight that, although in many areas of the Chaco pasture expansion appears to be the dominant proximate cause of deforestation, the global demand of soybean is actually driving these processes through leakage effects, in many cases across borders. This is, at least in part, a consequence of the change in agricultural actors from local and national land users to internationally operating companies (Gasparri and le Polain de Waroux 2014), which we discuss in the next section.

13.3 Land Competition in the Chaco

The observed land-use changes and their drivers in the Chaco enable us to examine the competition over land that is happening in this region. As highlighted in the conceptual chapter of this book section, competition over land entails a number of strongly interacting factors. In this chapter, we reflect on the competition in the Chaco by applying the theoretical framework of the conceptual chapter on the competition of land-based ecosystem services. Thereby, we pay particular attention to the actors that compete over land, which in case of the Chaco seems to be of substantially different strengths and power with regard to land-use decision making, resulting in an unbalanced competition. Subsequently, there may be clear winners and losers of the competition that can be identified.

The main actors in the competition over land in the Chaco are the actors that actually use the land. These are, on one side, the small landowners including both crop producers and farmers in the puesto system that raise cattle at a small scale. On the other side, the second major group of actors can be characterized as large-scale agribusiness. While historically the small-scale landowners were the biggest group in the Chaco and used most of the agricultural land, during the past decades the landscape of the Chaco has become more and more dominated by large-scale agribusinesses, which often operate simultaneously in different countries (Gasparri and le Polain de Waroux 2014). One important reason for this domination is that large agribusinesses are capable of making substantial investments in technological innovation, which allow them to produce crops at a relatively lower price.

[1]Although particularly in the case of Paraguay, this resulted in the undesirable outcome of accelerated deforestation in the Paraguayan Chaco.

Small-scale farmers, however, lack resources and financial means, often resulting in production costs per unit area that are higher compared to those by agribusinesses. Consequently, small-scale farmers and their products are less competitive on international markets. Over the course of the past few decades, this resulted in the decrease of the number of small-scale farmers. As a consequence, many small-scale farm owners rented out their land to large-scale businesses and shifted to other income opportunities. In other words, this actor type transitioned from a small- or medium-sized producer to an actor renting out land to large-scale agricultural producers on a contractual basis, typically over a relatively short time period of three years (Goldfarb and Zoomers 2013). This process has been widespread in South America in general, and in the Chaco in particular, leading to the concentration of use rights for agricultural land (Goldfarb and Zoomers 2013).

In addition, other actors have been appearing in the Chaco's landscape during the past 30 years. For example, the large-scale agribusinesses are increasingly dominated by so-called sowing pools, formed by groups of international investors that lease land (as well as machinery, seeds, etc.) to produce crops with the goal to maximize profits. Among other reasons, their financial power enables them to act in a similar way as agribusinesses, with the result that small-scale farmers are being marginalized further. Altogether, this development has further advanced the concentration of agricultural activities on fewer but larger farms in the Chaco (Goldfarb and Zoomers 2013). Between 1988 and 2008, the overall number of active agricultural enterprises in Argentina decreased by almost 50 %. Large-scale enterprises and sowing pools increasingly dominate the agricultural sector. Today, only 2 % of all agricultural enterprises farm more than half of the agricultural land in Argentina, whereas 57 % of enterprises only farm 3 % of all agricultural land (Alvarez Kalverkamp 2013; Goldfarb and Zoomers 2013).

Other actors play only a minor role in the competition over land in the Chaco. There are, for example, actors that pursue the goal to conserve natural habitats. In the Chaco, these conservation actors are often nongovernmental organizations, which act entirely from inside of the Chaco but may also be international initiatives. For example, The Nature Conservancy (TNC) has designated a set of priority areas for conservation for the Gran Chaco ("Conservation Portfolio of Priority Areas for Biodiversity"). Experts on Chaco wildlife, conservation, and ecology from Argentina, Paraguay, and Bolivia outlined so-called areas of biodiversity significance (ABS) for each major taxon (birds, amphibians and reptiles, mammals, and vegetation) based on regional knowledge. Based on this ABS, TNC established priority areas taking into account particular threats from current human pressure. These priority areas cover 470,000 km^2 in the entire Chaco, of which 238,000 km^2 are in Argentina (representing 36 % of Argentina), 148,000 km^2 are in Paraguay (49 % of Paraguay), and 83,500 km^2 are in Bolivia (68 % of Bolivia). However, these priority areas do not necessarily imply that land being designated as ABS is under protection. In fact, only a small proportion of the Chaco's land is under some kind of nature protection (less than 65,000 km^2 across the entire Chaco) partly because land acquisition, land management, and land regulation of protected areas entail administrative and bureaucratic hurdles to overcome and require substantial amounts of funds. Overall, conservation actors currently appear to only play a minor role in the competition over land in the Chaco.

This may be one reason why in some cases the state as an actor is in the debate, particularly in Argentina (Goldfarb and Zoomers 2013). At the moment, however, the states of the different countries in the Chaco maintain an ambivalent role, characterized in Argentina, for example, by the existence of procedures of "doubtful legality around administration and handover of land" as well as by a government that supports large-scale farming operations as a policy of rural development (Sili and Soumoulou 2011). Likewise, in Paraguay, the expansion of large-scale cattle farms is one major governmental instrument to promote economic growth (Vazquez 2013), rendering the state as an actor that promotes and supports deforestation at least indirectly. At the same time, the state acts toward reducing deforestation by defining and imposing legal frameworks for enforcing environmental protection. For example, Argentina released the Forest Law in 2007 (Ley de Proteccion ambiental de Bosques nativos 2007) that divides the forests into different zones with the goal to slow down deforestation (Piquer-Rodríguez et al. 2015). Likewise Paraguay released a deforestation moratorium in the Atlantic forests, which essentially stopped deforestation there (World Wildlife Fund 2015). Thus, government policies toward protecting natural forests open the room for discussion of potential mechanisms and interventions that may be effective in doing so.

Protected areas and zoning policies (e.g., the Argentinian Forest Law from 2007) are examples of a restriction-oriented policy mechanism. However, as already noted, protected areas in the Chaco are historically underrepresented. Although being partially successful in slowing down deforestation, the problem with forest zoning policies is that although they impose restrictions on forest conversion and legal forest reserves, such zoning policies are often highly variable and insufficiently enforced (Le Polain de Waroux et al. in review). Nevertheless, given that an expansion of the protected area network for the Chaco currently seems politically unfeasible and therefore unlikely to happen any time soon, the development of effective zoning policies such as the Forest Law seems to be the more promising policy mechanism to slow down deforestation.

Deforestation and agricultural expansion often result in substantial emissions of carbon, which renders deforestation a main contributor to climate change (Houghton 2003; Houghton and Hackler 2001). Payments for ecosystem services (PES) are a mechanism making use of that relationship by paying landowners a compensation for not deforesting their land. This has been promoted as a mechanism to reduce deforestation rates in several world regions with partially good success (Börner et al. 2010; Wendland et al. 2010; Wunder 2007). However, PES schemes are underdeveloped in the Chaco with only a handful of projects in the preparation phase. Further, it is questionable whether the implementation of PES schemes can be successful in the Chaco due to the nature of the actors in the competition for land in the Chaco. Large-scale agricultural businesses produce soy in the Chaco at a comparatively low cost resulting in high net returns, which any payment-oriented conservation tool would have to compete with. Although soy prices have not been increasing substantially during the past years, it is at least questionable whether PES mechanisms can be equipped with sufficient funding to compete with soybean revenues. In sum, given the high revenues from soybean production, the success of payment-oriented tools to protect forest resources in the Chaco seems questionable.

A higher potential for success in conserving natural forests in the Chaco may be seen in policies that incentivize sustainable production in concert with regulation of forest clearing such as the Forest Law in Argentina. The combination of these components may present a promising tool to reduce deforestation while at the same time maintaining production levels. Policy incentives to promote more production could include, for example, the subsidization of the introduction and development of yield-improving technologies (Diaz-Zorita et al. 2002) and the promotion of systems that integrate crop and livestock production (Lemaire et al. 2014). With a simultaneous forest clearing regulation, this would lead to production increases while at the same time reducing deforestation. At the moment, however, the outcomes of such measures are unclear. For example, promoting intensified soybean production may result in conversions of more pastures into croplands, a process which has already been observed in the Chaco (Baumann et al. in review-a) and which may in the short run increase carbon emission related to land-use change due to the pasture-to-cropland conversions (Baumann et al. in review-a). However, in a long-term view, the benefits of reduced deforestation while maintaining production levels may outweigh the disadvantages of having higher short-term emissions.

It is also questionable whether national efforts are enough to curb deforestation rates in the Chaco. Recent studies suggest that the Chaco absorbs substantial deforestation leakages from neighboring countries (Baumann et al. in review-a). Some of these leakages are at least in part a result of strong conservation policies and enforcement of these policies in these countries. For example, Paraguay banned deforestation for cropland expansion in its Atlantic Forests in 2009. This resulted in a conversion from existing pasture areas into croplands and a subsequent displacement of these pastures into the Chaco forests (World Wildlife Fund 2015). Likewise, following stricter conservation policies in the Amazon (Nepstad et al. 2009), many large-scale farmers reinvested their revenues into land acquisitions in the Chaco where they deforested for pasture expansion. Thus, conservation efforts in one place result in deforestation in another place that is subject to less conservation efforts. These shifts and leakage effects are aided by the fact that large-scale agribusinesses are increasingly highly mobile actors within the region and beyond, particularly compared to small-scale local farmers within the traditional puesto system. Agribusinesses have their disposal strategic knowledge, financial resources, and high-level political influence, and their business structures allow them the necessary degree of independence from individual sites and specific local contexts of production. The state as an actor could possibly increase the efficiency of its conservation actions by coordinating and implementing conservation policies internationally in order to regulate the actions of such mobile actors and prevent displacement and unintended outcomes.

Overall, it seems that the competition over land in the Chaco is far from "perfect," as outlined in the introduction to this section (Chap. 8). Instead, the competition in the Chaco is an example, in which large-scale, international actors dominate the land sector, resulting in the land-use changes observed throughout the past decades. It is reasonable to assume that this process will continue in the future and may even accelerate. For example, it is unlikely that the global demand in agricultural products will decrease any time soon. With more people living on the planet, increasing wealth, and, as it appears at the moment, changing diets toward more meat consumption, the global demand for soybean will continue to rise. The current conditions in the Chaco that we outlined suggest that without appropriate land-use planning and regulation, the Chaco will likely absorb a substantial proportion of the additional need for agricultural land to grow soy in a business-as-usual scenario, with all associated environmental and social costs. There are pathways that may contribute toward decreasing deforestation rates, though to be successful will in many cases likely require stronger land-use regulation and enforcement.

References

Altrichter, M., & Basurto, X. (2008). *Effects of land privatisation on the use of common-pool resources of varying mobility in the Argentine Chaco.*

Alvarez Kalverkamp, M. (2013). *Im Reich der Sojabohne: Strukturwandel in der Landwirtschaft des Cono Sur am Beispiel Argentiniens.* Heinrich-Boell-Stiftung, Available in https://www.boell.de/de/oekologie/lateinamerika-argentien-landwirtschaft-agrarpolitik-16374.html.

Baumann, M., Gasparri, I., Piquer-Rodríguez, M., Gavier-Pizarro, G. I., Griffiths, P., Hostert, P., Kuemmerle, T. et al. (in review-a). *Carbon emissions from deforestation and agricultural intensification in the Chaco.*

Baumann, M., Israel, C., Piquer-Rodríguez, M., Gavier-Pizarro, G. I., Volante, J. N., & Kuemmerle, T. (in review-b). *Deforestation and cattle expansion in the Paraguaya Chaco 1987–2012.*

Bertello, F. (2008). Cada vez más productores argentinos van a sembrar soja a Uruguay. Newspaper article in *lanacion.com* from March 4, 2008. [Available in http://www.lanacion.com.ar/992652-cada-vez-mas-productores-argentinos-van-a-sembrar-soja-a-uruguay].

Börner, J., Wunder, S., Wertz-Kanounnikoff, S., Tito, M. R., Pereira, L., & Nascimento, N. (2010). Direct conservation payments in the Brazilian Amazon: Scope and equity implications. *Ecological Economics, 69*, 1272–1282.

Brown, A. D., Foguet, M. J., García Moritán, M., & Maliya, S. (2010). *Bañado de la Estrella* (p. 109). Dinamica fluvial de un espacio compartido: Fundación Pro-Yungas, Tucuman, Argentina.

Bucher, E. H., & Huszar, P. C. (1999). Sustainable management of the Gran Chaco of South America: Ecological promise and economic constraints. *Journal of Environmental Management, 57*, 99–108.

Burghardt, P. (2014). Der Tod kommt mit dem Wind. *Sueddeutsche Zeitung*. http://sz-magazin.sueddeutsche.de/texte/anzeigen/42435/Der-Tod-kommt-mit-dem-Wind.

Caldas, M. M., Goodin, D., Sherwood, S., Campos Krauer, J. M., & Wisely, S. M. (2015). Land-cover change in the Paraguayan Chaco: 2000–2011. *Journal of Land Use Science, 10*, 1–18.

Dana, L.-P., & Dana, T. E. (2007). Collective entrepreneurship in a mennonite community in Paraguay. *Latin American Business Review, 8*, 82–97.

Diaz-Zorita, M., Duarte, G. A., & Grove, J. H. (2002). A review of no-till systems and soil management for sustainable crop production in the subhumid and semiarid Pampas of Argentina. *Soil & Tillage Research, 65*, 1–18.

FAOSTAT 2015. FAOSTAT Database [Online]. Food and agriculture organization of the United Nations. Available in http://faostat3.fao.org/download/F/FO/E [Accessed February 16, 2015].

Fatecha, A. (1989). Present and potential area for agricultural use in the arid Chaco of Paraguay. In M. Hamp & M. A. Tiefert (Eds.), *Agricultural production under semi-arid conditions with special reference to the Paraguayan Chaco: Strategies* (p. 284). Food and Agriculture Development Centre.

Fehlenberg, V. (2015). *Understanding proximate drivers of deforestation in the Gran Chaco: Soybean expansion versus cattle ranching*. MSc-Thesis (p. 51). Berlin: Geography Department, Humboldt-University Berlin.

Gasparri, N. I., Grau, H. R., & Manghi, E. (2008). Carbon pools and emissions from deforestation in extra-tropical forests of Northern Argentina between 1900 and 2005. *Ecosystems, 11*, 1247–1261.

Gasparri, N. I., & le Polain de Waroux, Y. (2014). The coupling of South American soybean and cattle production frontiers: New challenges for conservation policy and land change science. *Conservation Letters,* 8, 290–298.

Gavier-Pizarro, G. I., Calamari, N. C., Thompson, J. J., Canavelli, S. B., Solari, L. M., Decarre, J., et al. (2012). Expansion and intensification of row crop agriculture in the Pampas and Espinal of Argentina can reduce ecosystem service provision by changing avian density. *Agriculture, Ecosystems & Environment, 154*, 44–55.

Goldfarb, L., & Zoomers, A. (2013). The drivers behind the rapid expansion of genetically modified soya production into the Chaco region of Argentina. In Z. Fang (Ed.), *Biofuels—Economy, environment and sustainability* (pp. 73–95). InTech Publisher.

Gonzales, D. (2013). Uruguayos ya explotan el 12, 5 % de las tierras ganaderas en Paraguay. *Ultimahora.com*. Edicion Impresa (February 12, 2013).

Graesser, J., Aide, M., Grau, R., & Ramankutty, N. (2015). Cropland/pastureland dynamics and the slowdown of deforestation in Latin America. *Environmental Research Letters, 10*, 034017.

Grau, H. R., Gasparri, N. I., & Aide, T. M. (2005). Agriculture expansion and deforestation in seasonally dry forests of North-West Argentina. *Environmental Conservation, 32*, 140–148.

Gurtler, R. E., Kitron, U., Cecere, M. C., Segura, E. L., & Cohen, J. E. (2007). Sustainable vector control and management of Chagas disease in the Gran Chaco, Argentina. *Proceedings of the National Academy of Sciences of the United States of America, 104*, 16194–16199.

Hansen, M. C., Potapov, P. V., Moore, R., Hancher, M., Turubanova, S. A., Tyukavina, A., et al. (2013). High-resolution global maps of 21st-century forest cover change. *Science, 342*, 850–853.

Houghton, R. A. (2003). Revised estimates of the annual net flux of carbon to the atmosphere from changes in land use and land management 1850–2000. *Tellus Series B-Chemical and Physical Meteorology, 55*, 378–390.

Houghton, R. A., & Hackler, J. L. (2001). *Carbon flux to the atmosphere from land-use changes: 1850–1990* (p. 74). Washington D.C.: Environmental Science Division, U.S. Department of Energy.

Huang, C. Q., Kim, S., Altstatt, A., Townshend, J. R. G., Davis, P., Song, K., et al. (2007). Rapid loss of Paraguay's Atlantic forest and the status of protected areas—A landsat assessment. *Remote Sensing of Environment, 106*, 460–466.

Le Polain de Waroux, Y., Garrett, R. D., Heilmayr, R., & Lambin, E. F. (in review). *A rush to the margins: Impacts of land use policies on corporate investments in agriculture in the Gran Chaco.*

Leguizamon, A. (2014). Modifying Argentina: GM soy and socio-environmental change. *Geoforum, 53*, 149–160.

Lemaire, G., Franzluebbers, A., Carvalho, P. C. D., & Dedieu, B. (2014). Integrated crop-livestock systems: Strategies to achieve synergy between agricultural production and environmental quality. *Agriculture, Ecosystems & Environment, 190*, 4–8.

Ley de Proteccion ambiental de Bosques nativos (2007). Ley No 26331, primera seccion. Argentina, 3pags.; Available in http://www.infoleg.gov.ar/infolegInternet/anexos/135000–139999/136125/norma.htm.

Manuel-Navarrete, D., Gallopín, G., Blanco, M., Díaz-Zorita, M., Ferraro, D., Herzer, H., et al. (2009). Multi-causal and integrated assessment of sustainability: The case of agriculturization in the Argentine Pampas. *Environment, Development and Sustainability, 11*, 621–638.

Nepstad, D., Soares-Filho, B. S., Merry, F., Lima, A., Moutinho, P., Carter, J., et al. (2009). The end of deforestation in the Brazilian Amazon. *Science, 326*, 1350–1351.

Newell, P. (2009). Bio-Hegemony: the political economy of agricultural biotechnology in Argentina. *Journal of Latin American Studies, 41*, 27–57.

Olson, D. M., Dinerstein, E., Wikramanayake, E. D., Burgess, N. D., Powell, G. V. N., Underwood, E. C., et al. (2001). Terrestrial ecoregions of the worlds: A new map of life on Earth. *BioScience, 51*, 933–938.

Piquer-Rodríguez, M., Torella, S., Gavier-Pizarro, G., Volante, J., Somma, D., Ginzburg, R., Kuemmerle, T. et al. (2015). Effects of past and future land conversions on forest connectivity in the Argentine Chaco. *Landscape Ecology*, 1–17.

Prado, D. E. (1993). What is the gran Chaco vegetation in South America? I. A review: contribution to the study of flora and vegetation of the Chaco, *V. Candollea*, 48, 145–172.

Ramirez, E. G., & Laneri, J. L. (1989). Fodder and feeding of cattle in the Paraguayan Chaco. In M. Hamp, & M. A. Tiefert (Eds.) *Agricultural production under semi-arid conditions with special reference to the Paraguayan Chaco: Strategies and appropriate technologies* (pp. 139–148). DSE Feladelfing.

Reenberg, A., & Fenger, N. A. (2011). Globalizing land use transitions: the soybean acceleration. *Geografisk Tidsskrift-Danish Journal of Geography, 111*, 85–92.

Santarcángelo, J. E., & Fal, J. (2009). *Producción y rentabilidad en la ganadería Argentina: 1980–2006*. Mundo agrario, 10.

Sili, M., & Soumoulou, L. (2011). *The issue of land in Argentina. Conflicts and dynamics of use, holdings and concentration*. Report, International Fund for Agricultural Development (IFAD), 12p.

Vallejos, M., Volante, J. N., Mosciaro, M. J., Vale, L. M., Bustamante, M. L., Paruelo, J. M. et al. (2015). Transformation dynamics of the natural cover in the dry Chaco ecoregion: A plot level geo-database from 1976–2012. *Journal of Arid Environments*.

Vazquez, F. (2013). *Geografia humana del Chaco paraguayo—Transformaciones territoriales y desarollo regional*. Asuncion: Ediciones ADEPO.

Wendland, K. J., Honzák, M., Portela, R., Vitale, B., Rubinoff, S., & Randrianarisoa, J. (2010). Targeting and implementing payments for ecosystem services: Opportunities for bundling biodiversity conservation with carbon and water services in Madagascar. *Ecological Economics, 69*, 2093–2107.

World Wildlife Fund (2015). *Soy and deforestation—The Gran Chaco*. http://wwf.panda.org/what_we_do/footprint/agriculture/soy/soyreport/soy_and_deforestation/the_gran_chaco/ [last access: March 16, 2015].

Wunder, S. (2007). The efficiency of payments for environmental services in tropical conservation. *Conservation Biology, 21*, 48–58.

Part III
Understanding the Interactions of Land and Food Systems

Chapter 14
The Future Is Made. Imagining Feasible Food and Farming Futures in an Unpredictable World

Christian Lauk and Juliana Lutz

Abstract In land system science, the issue of land use competition is often explored in the context of future scenarios of the agro-food system. While land system science shares its research topic with the so-called agro-food studies, there is little communication between these two strands of research. In order to explore the reasons for this communicative divide and how it could be bridged, we first critically examine the ontological foundations of futures studies in land system science, arguing that they (a) tend to have a global and remote, rather than a place-specific perspective, (b) typically consider biophysical constraints via according models and (c) build on an "economistic" understanding of social relations, insofar as complex social relations are represented by an economic model. This makes it difficult to relate the futures studies from land system science to the mostly place-specific, participatory and social science-based perspective that agro-food studies provide. Still, we conclude that while land system science could benefit from a more place-specific and social science perspective, agro-food studies could profit from translating biophysical considerations and scenario thinking into a place-specific and participatory perspective. The chapters in this section show some fruitful approaches, which take up a place-specific perspective, without losing the sight of biophysical constraints as well as cross-scalar interactions.

Keywords Food security · Interdisciplinarity · Place · Biophysical constraints · Participation

14.1 Introduction

Looking at the recent wave of literature which refers to the notion of land use competition, it stands out that a considerable part of this literature is concerned with the future of land use competition. That is, the main focus is not the analysis of

C. Lauk (✉) · J. Lutz
Institute of Social Ecology Vienna (SEC), Alpen-Adria Universität, Klagenfurt, Vienna, Graz, Austria
e-mail: christian.lauk@aau.at

J. Niewöhner et al. (eds.), *Land Use Competition*,
Human-Environment Interactions, DOI 10.1007/978-3-319-33628-2_14

current processes of patterns of land use competition. Rather, land use competition is presented as a future "challenge" against the background of an increasing demand for biomass due to population growth, richer diets and the increasing use of biomass for energy. In these studies, possible, probable or desirable futures are explored, often on the basis of scenarios and models representing land use and the food and energy system.

Berndes et al. (2003), for example, discuss 17 earlier studies on the future contribution of biomass for energy and how they looked at either competition between biomass-based electricity and biofuels, or at the competition between different uses of biomass. Hoogwijk et al. (2003) aim to make the future competition and synergy between different biomass flows more transparent by showing the impact of the future development of crucial parameters such as human diets or crop yields on global bioenergy potentials. In another scenario-based study, Johansson and Azar (2007) analyse the future potential food–fuel competition in the USA, based on an economic optimization model. And Havlik et al. (2011) aim at providing "policy analysis on global issues concerning land use competition between the major land-based production sectors", based on the global equilibrium model GLOBIOM. These are just a few examples in a much longer list of futures studies referring in some way to land use competition (e.g. Field et al. 2008; Johansson and Azar 2007; Lapola et al. 2010; Wirsenius et al. 2010).

Most of these studies on land system futures can be related to a research tradition within environmental science referred to as land system science (Rounsevell et al. 2012; Verburg et al. 2015), land use science (Müller and Munroe 2014) or land change science (Turner et al. 2007). This tradition is characterized not only by land use being its core research topic; it is also dominated by a particular perspective on issues related to land use. Not the studied issue but which questions are asked, how problems are framed, which geographic scale is primarily addressed, which explanatory framework is applied and how it implicitly relates to fundamental epistemological issues such as the role of subjectivity or normativity in research constitute this perspective (cf. Turner and Robbins 2008 for an according comparison between land change science and political ecology).

In fact, there are other research traditions sharing the main topic of analysis with land system science, i.e. land use and the agro-food and energy system, which, however, rarely refer at all to the methods of future studies, i.e. models and scenarios. Within so-called agro-food-studies (Goodman 1999), for example, a bulk of literature focused on the relocalization of food production, distribution, and consumption with social, political, and cultural aspects being at the heart of the inquiry (Sonnino 2013). Most recently, the future of agro-food systems has become a core topic within agro-food studies, focusing on the governance of transition pathways and the role of local food systems in regard to global transitions. This comprises discussions on the up-scaling of local food systems and on the relocalization of food-governance, especially in regard to conflicts on land and the related issue of "land grab" (McMichael 2012; Sonnino 2013).

Is it possible to reconcile those approaches when looking at food and farming futures? To approach this question, we first critically examine the foundations of futures studies referring to land use and more specifically land use competition. We do so by discussing these studies along three important ontological dimensions and how these dimensions are considered in these studies: (a) Relations of space and place, (b) biophysical constraints and (c) social systems, social change and actors. For each of these dimensions, we critically discuss how it is considered in futures studies referring to land use competition. Building on this critique, we then tentatively explore some options for an alternative, place-based approach to futures studies of the agro-food system.

14.2 Land System Futures

14.2.1 Relations of Space and Place

Most futures studies referring to land use competition, in particular the most cited ones, take up a primarily global perspective (Berndes et al. 2003; Golub et al. 2009; Havlik et al. 2011; Hoogwijk et al. 2003; Murphy et al. 2011; Popp et al. 2011; Reilly and Paltsev 2009; Smith et al. 2010; Wirsenius et al. 2010). While there exist some studies referring to the national scale (Chen et al. 2011; Johansson and Azar 2007; Lapola et al. 2010; Rathmann et al. 2010), a place-specific perspective is virtually absent. The “global” appears already as part of the main question, which typically refers to issues such as global potentials of bioenergy or global implications of biofuel targets and this perspective shapes the research derived from it. Place-specific information is not completely absent, as the models used in these studies spatially differentiate material/technological (such as crop yields and cropland areas) on plot levels and economic parameters (such as GDP or costs of production), often on the level of countries or world regions. Nevertheless, not only remains the place-specific information secondary, with the primary aim to answer a global question and to present a global scenario, but the sheer size of the globe also makes it necessary to rigorously abstract and generalize from the specifics of place-bound socio-ecological relationships.

The global perspective adopted by most futures studies referring to land use competition is part of a broader tradition within the environmental sciences and the environmental movement in general, which takes a mainly global perspective on today’s environmental problems, often under the heading of “Global Ecology” or “Global Change” and closely linked to the paradigm of “sustainable development” (Sachs 1993, 1992). The blue planet earth seen from space and photographed by Neil Armstrong in 1972 for the first time became the icon symbolizing this perspective.

The global perspective on the environment is argued on grounds of the global interconnectedness of today’s environmental problems, with the global climate as the paradigmatic case: Not only is the climate itself a system which can only be understood on a global level, but it is also understood as largely irrelevant in terms of climate impacts in which part of the world greenhouse gases are emitted. Concerning the agro-food system, reference to the global is made on two grounds: on the one hand, some environmental impacts of biomass production, in particular those relating to greenhouse gas emissions, impact the global environment. On the other hand, the agro-food system is globally interconnected via trade and investments.

However, as Sachs (1992) notes, a global perspective on "our planet earth" can be quite ambiguous: "On the one hand, 'our' can imply participation and highlight man's dependence on an encompassing reality. On the other hand, it can imply ownership and emphasize man's vocation to master and to run this common property. Consequently, the image of 'our' planet conveys a contradictory message; it can either call for moderation or for megalomania".

Thus, although global interconnections of the agro-food system deserve attention, such a global perspective might also convey an image in which land—and the products produced on it such as bioenergy or food—is seen as a "global resource" (Harvey and Pilgrim 2011). Such a view not only invokes the task to control these "global commons" through corresponding institutions on national and supranational levels, but also resonates with the idea that "biomass potentials" in a certain region are available for sale on a global market.

Accordingly, a primarily global and rather remote view on environmental problems in general and land use in particular tends to overlook many of the more complex, place-specific interrelations. Thus, although many environmental problems relating to land use and the agro-food system are truly global in scale, there are still many issues which require a place-specific understanding of the land use system. On-site and often time-intensive studies, which are able to delve deeper into the social system and concrete practices on the ground, remain paramount for a better understanding of issues such as land use competition. This not only concerns environmental problems related to the biophysical environment, such as soil degradation, water contamination, biodiversity or nitrogen flows, but it also concerns the place-specific and complex social interrelations related to land use and its environmental impacts.

Corresponding to their global perspective, land system studies with a global scope are often financed and the results taken up by national or supranational institutions such as the European Union, with a tendency to approach the management of social-ecological change top-down. Such approaches often produce tensions with grassroots-oriented perspectives. While these also draw on knowledge about global phenomena, they tend to give primacy to a nuanced understanding of place-specific aspects.

14.2.2 Biophysical Constraints

Most futures studies referring to land use competition include a model of the metabolism of the agro-food system. This model accounts for all biomass flows, from the harvest of primary biomass (i.e. crops and roughage), through the conversion of biomass by animals and sometimes industrial conversion technologies such as biofuel production technologies, to the final use of more or less processed biomass by society. In some cases, this biomass accounting model is coupled with a model that accounts for greenhouse gas emissions or other environmental problems caused by agricultural activities and land cover change. The main theoretical foundation of this model, which analyses the biomass metabolism, are the thermodynamic principles.

There are several parameters and related technologies, which determine the modelled biomass metabolism and at the same time—assuming these technologies change over time—allow for a certain degree of freedom. This in particular includes areas used as cropland or for grazing, crop yields and grassland productivities, input–output-efficiencies of livestock and biomass processing technologies, human diets, population size, as well as the exchange of biomass between regions. A consistent representation of the biomass metabolism in relation to these parameters allows to explore the biophysical boundary conditions or the biophysical option space (Erb et al. 2016) of the future agro-food system, which relates to the idea of a "safe operating space", suggested by Rockström et al. (2009).

Global models can be useful in the context of questions such as the size of global or regional bioenergy potentials or the (biophysical) viability of a certain future agro-food system. For example, many advocate an agrarian system less dependent on artificial inputs such as synthetic fertilizers and organic pesticides (e.g. organic farming). However, such an agrarian system might be linked to lower crop yields compared to industrialized farming and is therefore sometimes criticized as not being able to feed the world (Connor 2008). Using a biophysical modelling approach, Erb et al. (2016) showed that a switch to organic farming with lower crop yields is indeed biophysically feasible, yet only if certain conditions, such as diets with a low share of animal products or increasing cropland areas, are met. Land use competition is another topic in which the use of a biophysical model can contribute, as the severity and type of competition depends—not only but also—on the biophysical relation between biomass demand and potential supply of biomass (see e.g. Haberl 2015 for a socio-metabolic perspective on land use competition).

Two major difficulties complicate the modelling of biophysical constraints of the future development of the agro-food system: first, the global or large-scale regional scope of many futures studies discussed above makes it difficult to consider more complex and place-specific biophysical structures and processes of agro-food systems. In many global or large-scale regional models, the biophysical structure of farming systems is largely defined by quantifiable parameters such as cropland and grassland areas, crop yields, grassland productivities and livestock efficiencies. More complex, place-specific interrelations, which often require partly qualitative approaches, are largely masked in these types of models.

Second, although the biophysical system is constrained by certain fundamental and unchangeable natural laws, in particular thermodynamic principles, technology allows for a considerable variation concerning the biophysical feasibility of its future development. The question of how many people the earth can support (Cohen 1996) nicely illustrates this point. As shown by Cohen, estimates of the global carrying capacity differ by more than a factor of 1000, ranging from 0.5 to 1.2 billion people, given by Ehrlich (1971 cited in Cohen 1996: 410) to 1 trillion people, given by Marchetti (1979, cited in Cohen 1996: 413). Both estimates take into consideration biophysical limits, yet they assume very different technologies applied for the production of food—most importantly, in the case of Marchetti, a virtually unlimited energy supply. This is closely linked to the question of land use competition: if a production potential to feed 1000 billion people could be realized in a world of 10 billion people, then the problem of land use competition due to land scarcity would simply disappear.

Obviously, those are the most extreme examples but they illustrate an important point: the biophysical feasibility of futures depends on technologies available and applied in the future, here especially relating to food production. This, in turn, also depends on which technological developments are considered feasible, desirable or even become visible for researchers, stakeholders and politicians, respectively. Farmers themselves constantly develop (new) ways of producing food without systematic documentation. The way land is used and agro-systems are shaped thus relates not only to socio-economic, but also to ethical questions: how do we develop possible futures and whose futures are they? The following section accordingly looks at how the socio-economic system and human actors are represented in current futures studies referring to land use competition.

14.2.3 Social System, Social Change and Actors

Biophysical constraints, as determined by natural conditions and technologies applied, are inseparably linked to human practices and cultures, which constantly interact with and reshape the biophysical world. Therefore, any biophysically feasible future development requires certain human practices and cultures that realize it.

In the case of the agro-food system, there are many futures that are biophysically feasible. A productivist future, in which all diets converge to Western standards with a high share of animal products, while agricultural productivities are constantly maximized and cropland is expanded, as well as a future, in which all humans converge to vegetarian or vegan diets, while agriculture is extensified and cropland remains at current levels (Erb et al. 2016). Both futures presume that certain agricultural technologies, such as crop cultivars or nutrient management techniques, are available or developed and applied. The productivist future, for example, is difficult to imagine without a global industrialization of agriculture, including above other things the application of high yielding crop cultivars, synthetic fertilizers and pesticides.

Is it possible to predict—or at least restrict—which of those many biophysically feasible future pathways is taken by society? Most scientists would tend to agree that such a prediction is not possible with any degree of certainty. That future scenarios are severely restricted nonetheless—often into distinct pathways—has mainly to do with factors other than biophysical boundary conditions.

The integrated assessment model IMAGE, developed by the Netherlands' Environmental Assessment Agency and its applications can serve as a concrete example, how biophysically feasible future pathways are further restricted due to certain reasonings. IMAGE has been applied in many politically highly influential, international assessment reports, such as the Millennium Ecosystem Assessment, the OECD Environmental Outlook, the UNEP Global Environmental Outlooks and the IPCC assessment reports, where it has been used to explore energy and land use futures (Stehfest et al. 2014: 27–29). IMAGE was also instrumental in the support of EU policies related to agriculture, energy and the environment and it played a crucial role in many large research projects on EU and national level.

The construction of the IMAGE and many other models usually leads to a very restricted set of conceivable future scenarios of the agro-food system, which might be described as modernist: while further developed regions continue their development along historical trends, less developed regions gradually catch up, more or less mirroring the path demonstrated by rich countries in the past. With respect to the agro-food system, this particularly involves a general trend towards Western diets, rich in calories and animal products, as well as, more recently, the assumption that bioenergy is increasingly required to fuel an industrial economy set on growing further. Fulfilling this demand requires a (further) industrialization of agriculture, with its known social and ecological implications. IMAGE and the scenarios derived from it are a particularly influential example but no exception in this respect. However, it is questionable whether socio-economic relations derived from data of the past will remain valid for the coming decades. It might be argued that a sustainable, post-fossil society will be at least as different from the industrial society as the current industrial society is from pre-industrial agrarian societies (Haberl et al. 2011). If this presumption is taken serious, it becomes clear that it is not possible to transfer "socio-economic laws" derived from data of the last decades to a future post-fossil society.

In fact, the economic models used by these futures studies represent a very particular model of development, which continues to expand throughout the world, starting from Great Britain in the nineteenth century and closely linked to a fossil fuel-based energy system (Haberl et al. 2011; Sieferle 2001). The accumulation of capital, as represented on a monetary level by a growth of the GDP, is at its very heart. By the assertion of objective economic processes and economic laws, this development model—represented by according models of the agro-food system—constrains the imagination of possible futures.

Although most studies still stick to this very restricted set of futures, an opening towards alternative future can be observed during the past years. For example, the IMAGE model has been applied to show the climate impact of different diets, including a completely vegan diet (Stehfest et al. 2009). On the production side, an increasing number of studies discuss options to switch to an agricultural system relying stronger on agro-ecological methods (Badgley et al. 2007; Badgley and Perfecto 2007; Leu 2004; Schader et al. 2016). And Erb et al. (2016) show the biophysical feasibility of many alternative futures, including some who rely on organic agriculture and a switch to vegetarian diets.

14.3 Imagining Place-Related Land Use Futures

To address the limitations of land use-related future studies in the context of complex local to global dynamics, research increasingly approaches global food system futures and related transition pathways from a more place-based and participatory perspective (Sonnino 2013). Also the FAO recognizes the emergence of a "paradigm for eco-system based, territorial food system planning [that] seeks… not to replace the global supply chains that contribute to food security for many countries, but to improve the local management of food systems that are both local and global" (FAO 2011, cited in Sonnino 2013). Against this background, we argue that:

(a) Facing a globalized food system, which is governed by difficult to regulate market dynamics rather than by national or international top-down policies (Friedman 2005), policies aiming at a sustainability transition of the food system need to develop a stronger focus on grassroots initiatives. Supporting such a transformation through research requires a better understanding of land use patterns and food-supply chains from a place-related perspective in order to be able to picture and understand local patterns of land use and food supply that emerge from the interplay of local and global dynamics.
(b) Research should facilitate the creation of future scenarios of place-based transition practices, governance strategies and the actors (e.g. local governments or civil society) involved.

And, even more, it should create space for innovative imaginations of future developments and action. This calls for participatory research strategies in which various actors contribute their experiences, experiments and imaginaries. In line with the described assumptions of a more place-based research agenda within futures studies on land use and food systems, we neither advocate to abandon any considerations of global interactions, nor to ignore biophysical relations and the constraints they imply. However, we suggest that a global view must be complemented by a much stronger focus on place-based perspectives, which are underrepresented within the discourse revolving around the notion of land use competition.

The cases discussed in this section serve as examples for potentially enriching place-based perspectives and methodologies, which might be applied in a more place-based perspective on feasible and desirable futures. In the following, we briefly discuss these perspectives and methodologies along the ontological dimensions of (a) relations of space and place, (b) biophysical constraints and (c) social systems, social change and actors.

14.3.1 Relations Between Local Place and Global Space

Considerations of global interactions and constraints are important. However, it is important to recognize the difference between a global perspective, in which different places are necessarily represented as a limited set of parameters, and a place-based perspective, which starts from a particular place and a description of its complex socio-ecological system and then asks how this particular place interacts with other places, regions and the world. While there exists a wealth of futures studies on the agro-food system from a global perspective, studies taking a place-based perspective while considering the global embedding of places are virtually absent.

Contributions to this section provide methodological concepts that might be fruitful for discussing global relations from a place-based perspective are Foodshed Analysis, Life Cycle Assessment (LCA) or local in-depth studies.

The concept of "Foodshed" is often being used to explore the potential of a particular region to feed itself (Horst and Gaolach 2014). In his case study, De Ruiter uses Foodshed Analysis for a historical analysis. He illustrates the changes in the United Kingdom's fruit and vegetables supply over the last 25 years, from 1986 to 2009, showing that the United Kingdom has become increasingly dependent on food supply from regions abroad. Foodshed Analysis thus helps to illustrate a nation's or potentially any region's food-related interplay with other parts of the world. Furthermore, it allows to link global dynamics in food-supply patterns to national or local health and sustainability issues. This can shed light on complex dynamics that should be taken into account when conceptualizing future pathways of agro-food systems and of related governance strategies.

The second case study as presented by Theurl applies the methodology of LCA. Her analysis looks at the whole life cycle of food products in defining seven life cycle phases: an agricultural production phase including upstream inputs to the farm; the processing of foods; packaging; storage; emissions from imports; emissions from food distribution (consumer shopping, supermarket logistics, box delivery scheme); and heating in glasshouses. She sheds light on the multitude and complexity of parameters that have to be taken into account when assessing, for example, triggers for CO_2 emissions of local and global food supply chains. It becomes clear that LCA helps to critically assess assumptions related to greenhouse gas emissions of local and global food supply. Furthermore, Theurl shows, on the one hand, how LCA helps to depict crucial, societal interventions points and, on the other hand, the need to imagine, discuss and assess potential futures that might seem unlikely but that are—in principle—possible (i.e. re-establishing local infrastructure for food processing).

In her in-depth case study on urban agriculture, Walthall refers to the debate on global urbanization and its relation to agriculture and local food supply. She argues that agricultural land close to cities is highly in demand for other than agricultural uses and is thus highly contested. At the same time, especially in growing urban areas, farmers are increasingly facing the need or chance to "produce" more than food, which helps them to gain additional income. Against this background, Waltham introduces the concept of multifunctional agriculture (MFA), showing that agricultural activities may—apart from food and fibre—also provide places of recreation for urban people as well as ecosystem services. With her case study of community-supported agriculture close to Berlin, she gives an example of multifunctional, urban agriculture, showing in detail the development, organization, benefits and challenges of such projects.

In the fourth case study, Mayer, while not presenting a specific place-related method, observes and analyses the interaction of two relatively distinct and competing agrarian models in Brazil: on the one hand, a model oriented towards exporting cash crops for the world market practiced by large farms. And on the other hand, a model oriented towards production for the domestic market and subsistence largely practiced by the many smallholders of the country. His case study underlines the importance of differentiating between those two modes also with respect to their relation to local place and global space: while smallholders largely remain embedded in domestic markets and continue to use part of their production for their own subsistence, large estates mainly depend on national and global markets. Obviously, this corresponds to very different modes of interaction between local place and global space, as materialized in global markets.

14.3.2 Biophysical Constraints

It would be a mistake to ignore that every future development of the agro-food system is constrained by a combination of fundamental natural laws, available technologies applied to produce and distribute food as well as the amount of biomass consumption. Yet again, studies from global or large-scale regional perspectives on the future of land use and the agro-food system need to be complemented by place-specific perspectives taking biophysical constraints serious, concerning place-specific relations as well as biophysical interconnections between particular places and the planet as a whole.

Both chapters by de Ruiter and Theurl not only exemplify how global interconnections could be considered in a place-based perspective, they are also examples for possible frameworks, in which biophysical interactions are explicitly considered. They show how globalized food-supply patterns affect (and are affected by) agricultural practices, nutritional habits and people's health. A different perspective on biophysical relations is introduced by Walthall's case study. With the concept of multifunctional agriculture, the case study shows how urban agriculture as a specific way of using land allows for ecosystem services and the biophysical retreat of humans. Furthermore, with the concept of community-supported agriculture, the case study presents a way of organizing the biophysical food-related reproduction of urban communities.

14.3.3 Social Systems, Social Change and Actors

One aim of this section is to introduce new perspectives on societal dynamics in relation to land use science. We believe that the case studies presented here do provide such new perspectives and methodological tools that empower practitioners as well as scientists to perceive the future of land use patterns and its societal organization and dynamics from new and diverse angles.

De Ruiter, for example, by means of Foodshed Analysis throws light on complex global–local dynamics that might be taken into account when conceptualizing future pathways of national agro-food systems and the related governance strategies. Theurl shows how LCA helps to depict crucial societal intervention points. In so doing, her study guides our focus (and that of scientific communities) to the need to imagine, discuss and assess potential futures that might seem unlikely but that are—in principle—possible (i.e. re-establishing local infrastructure for food processing). And Mayer shows the importance of differentiating between modes of production in order to better understand the overall dynamics of land use.

The importance and power of imagining (when facing dominant (global) socio-economics as wells as biophysical dynamics) unlikely but possible futures by means of (local) case studies and local innovations is emphasized by Walthall's case study. She presents and discusses "real-world" agricultural experiments and innovations in urban areas. She shows how the collaboration of different local actors allows for establishing new patterns of local, urban food supply. With this, a set of less established choices of action is added to the dominant pathways. Whether and how these might lead to more sustainable patterns of food supply and land use in different places and on different scales remains a disputed question.

14.4 Conclusion

Many studies on the future of land use competition are characterized by a remote and rather planetary perspective on land use and explore future pathways based on biophysical and economic accounting models. Although such models strongly reduce the wealth of complex socio-ecological realities on the ground, they are an important contribution to explore the safe operating space (Rockström et al. 2009) or biophysical option space (Erb et al. 2016) of large regions or the planet, and they are an important contribution to the debate about global environmental justice. At the same time, it is important to be aware of the particularities and limitations of each perspective and the insights it provides. The global operating space to which a sustainable society has to adhere, however, still leaves open many feasible future pathways. Deciding among feasible pathways and putting them into practice requires the complementation of global studies with place- and context-specific knowledge. This section discusses a number of methodological approaches to strengthening the nuanced analysis of local contexts without losing sight of the evident cross-scalar connections.

References

Badgley, C., Moghtader, J., Quintero, E., Zakem, E., Chappell, M. J., Avilés-Vázquez, K., et al. (2007). Organic agriculture and the global food supply. *Renewable Agriculture and Food Systems, 22*, 86–108.

Badgley, C., & Perfecto, I. (2007). Can organic agriculture feed the world? *Renewable Agriculture and Food Systems, 22*, 80–86. doi:10.1017/S1742170507001986.

Berndes, G., Hoogwijk, M., & van den Broek, R. (2003). The contribution of biomass in the future global energy supply: A review of 17 studies. *Biomass and Bioenergy, 25*, 1–28.

Chen, X., Huang, H., Khanna, M., & Önal, H. (2011). Meeting the mandate for biofuels: Implications for land use, food, and fuel prices. In J. S. G. Zivin & J. M. Perloff (Eds.), *The intended and unintended effects of US agricultural and biotechnology policies* (pp. 223–267). Chicago, MI: University of Chicago Press.

Cohen, J. E. (1996). *How many people can the earth support?*. New York/London: W.W. Norton & Company.

Connor, D. J. (2008). Organic agriculture cannot feed the world. *Field Crops Research, 106*, 187–190.

Ehrlich, P. R. (1971). The population crisis: Where we stand. In N. Hinrichs (Ed.), *Population, Environment and People*. New York: McGraw-Hill.

Erb, K.-H., Lauk, C., Kastner, T., Mayer, A., Theurl, M. & Haberl, H. (2016). *Exploring the biophysical option space for feeding the world without deforestation. Nature Communications, 7,* 11382. doi:10.1038/ncomms11382.

FAO. (2011). *Food, Agriculture and cities. Challenges of food and nutrition security, agriculture and ecosystem management in an urbanizing world.* FAO food for the cities multi-disciplinary initiative position paper, Rome: Food and Agricultural Organization of the United Nations (FAO).

Field, C. B., Campbell, J. E., & Lobell, D. B. (2008). Biomass energy: The scale of the potential resource. *Trends in Ecology & Evolution, 23*, 65–71.

Friedman, H. (2005). From colonialism to green capitalism: Social movements and emergence of food regimes. *Research in Rural Sociology and Development, 11*, 227.

Golub, A., Hertel, T., Lee, H.-L., Rose, S., & Sohngen, B. (2009). The opportunity cost of land use and the global potential for greenhouse gas mitigation in agriculture and forestry. *Resource and Energy Economics, 31*, 299–319. doi:10.1016/j.reseneeco.2009.04.007.

Goodman, D. (1999). Agro-food studies in the "Age of Ecology": Nature, corporeality, bio-politics. *Sociologia Ruralis, 39*, 17–38. doi:10.1111/1467-9523.00091.

Haberl, H. (2015). Competition for land: A sociometabolic perspective. *Ecological Economics, 119*, 424–431. doi:10.1016/j.ecolecon.2014.10.002.

Haberl, H., Fischer-Kowalski, M., Krausmann, F., Martinez-Alier, J., & Winiwarter, V. (2011). A socio-metabolic transition towards sustainability? Challenges for another great transformation. *Sustainable Development, 19*, 1–14. doi:10.1002/sd.410.

Harvey, M., & Pilgrim, S. (2011). The new competition for land: Food, energy, and climate change. *Food Policy, 36*, S40–S51. doi:10.1016/j.foodpol.2010.11.009.

Havlik, P., Schneider, U. A., Schmid, E., Böttcher, H., Fritz, S., Skalskä, R., et al. (2011). Global land-use implications of first and second generation biofuel targets. *Energy Policy, 39*, 5690–5702. doi:10.1016/j.enpol.2010.03.030.

Hoogwijk, M., Faaij, A., van den Broek, R., Berndes, G., Gielen, D., & Turkenburg, W. (2003). Exploration of the ranges of the global potential of biomass for energy. *Biomass and Bioenergy, 25*, 119–133. doi:10.1016/S0961-9534(02)00191-5.

Horst, M., & Gaolach, B. (2014). The potential of local food systems in North America: A review of foodshed analyses. *Renewable Agriculture and Food Systems, 30*, 399–407.

Johansson, D. J. A., & Azar, C. (2007). A scenario based analysis of land competition between food and bioenergy production in the US. *Climatic Change, 82*, 267–291. doi:10.1007/s10584-006-9208-1.

Lapola, D. M., Schaldach, R., Alcamo, J., Bondeau, A., Koch, J., Koelking, C., & Priess, J. A. (2010). Indirect land-use changes can overcome carbon savings from biofuels in Brazil. *Proceedings of the National Academy of Sciences, 107*, 3388.

Leu, A. (2004). *Organic agriculture can feed the world*. Acres—A voice for eco-agriculture 34.

Marchetti, C. (1979). 1012: A check on the earth-carrying capacity for man. *Energy, 4*, 1107–1117.

McMichael, P. (2012). The land grab and corporate food regime restructuring. *The Journal of Peasant Studies, 39*, 681–701. doi:10.1080/03066150.2012.661369.

Müller, D., & Munroe, D. K. (2014). Current and future challenges in land-use science. *Journal of Land Use Science, 9*, 133–142. doi:10.1080/1747423X.2014.883731.

Murphy, R., Woods, J., Black, M., & McManus, M. (2011). Global developments in the competition for land from biofuels. *Food Policy, 36*, S52–S61. doi:10.1016/j.foodpol.2010.11.014.

Popp, A., Dietrich, J. P., Lotze-Campen, H., Klein, D., Bauer, N., Krause, M., et al. (2011). The economic potential of bioenergy for climate change mitigation with special attention given to implications for the land system. *Environmental Research Letters, 6*, 034017. doi:10.1088/1748-9326/6/3/034017.

Rathmann, R., Szklo, A., & Schaeffer, R. (2010). Land use competition for production of food and liquid biofuels: An analysis of the arguments in the current debate. *Renewable Energy, 35*, 14–22. doi:10.1016/j.renene.2009.02.025.

Reilly, J., & Paltsev, S. (2009). Biomass energy and competition for land. In T. W. Hertel, S. K. Rose, & R. S. J. Tol (Eds.), *Economic analysis of land use in global climate change policy* (p. 182). London: Routledge.

Rockström, J., Steffen, W., Noone, K., Persson, A., Chapin, F. S., Lambin, E. F., et al. (2009). A safe operating space for humanity. *Nature, 461*, 472–475. doi:10.1038/461472a.

Rounsevell, M. D. A., Pedroli, B., Erb, K.-H., Gramberger, M., Busck, A. G., Haberl, H., et al. (2012). Challenges for land system science. *Land Use Policy, 29*, 899–910. doi:10.1016/j.landusepol.2012.01.007.

Sachs, W. (1992). *The development dictionary: A guide to knowledge as power*. London: Zed Books.

Sachs, W. (1993). *Global ecology: a new arena of political conflict*. London: Zed Books.

Schader, C., Muller, A., Scialabba, N., Hecht, J., Isensee, A., Erb, K.-H., Smith, P., Makkar, H., Klocke, P., Leibe, F., Schwegler, P., Stolze, M., Niggli, U., ET AL. (2016). Impacts of feeding less food-competing feedstuffs to livestock on global food system sustainability. *Journal of The Royal Society Interface*, In press.

Sieferle, R. P. (2001). *The subterranean forest: Energy systems and the industrial revolution*. Isle of Harris, UK: White Horse Press.

Smith, P., Gregory, P. J., van Vuuren, D., Obersteiner, M., Havlík, P., Rounsevell, M., et al. (2010). Competition for land. *Philosophical Transactions of the Royal Society B-Biological Sciences, 365*, 2941–2957. doi:10.1098/rstb.2010.0127.

Sonnino, R. (2013). Local foodscapes: Place and power in the agri-food system. *Acta Agriculturae Scandinavica, Section B—Soil & Plant Science*, 63, 2–7. doi:10.1080/09064710.2013.800130.

Stehfest, E., Bouwman, L., Vuuren, D. P., Elzen, M. G. J., Eickhout, B., & Kabat, P. (2009). Climate benefits of changing diet. *Climatic Change, 95*, 83–102. doi:10.1007/s10584-008-9534-6.

Stehfest, E., van Vuuren, D., Kram, T. & Bouwman, L. (Eds.). (2014). *Integrated assessment of global environmental change with IMAGE 3.0: Description and policy applications*. Den Hague: PBL Netherlands Environmental Assessment Agency.

Turner, B. L., Lambin, E. F., & Reenberg, A. (2007). The emergence of land change science for global environmental change and sustainability. *Proceedings of the National Academy of Sciences, 104*, 20666–20671. doi:10.1073/pnas.0704119104.

Turner, B. L., & Robbins, P. (2008). Land-change science and political ecology: Similarities, differences, and implications for sustainability science. *Annual Review of Environment and Resources*, 33, 295–316. doi:10.1146/annurev.environ.33.022207.104943.

Verburg, P. H., Crossman, N., Ellis, E. C., Heinimann, A., Hostert, P., Mertz, O., Nagendra, H., Sikor, T., Erb, K.-H., Golubiewski, N., Grau, R., Grove, M., Konaté, S., Meyfroidt, P., Parker, D. C., Chowdhury, R. R., Shibata, H., Thomson, A., Zhen, L., et al. (2015). Land system science and sustainable development of the earth system: A global land project perspective. *Anthropocene,* In press. doi:10.1016/j.ancene.2015.09.004.

Wirsenius, S., Azar, C., & Berndes, G. (2010). How much land is needed for global food production under scenarios of dietary changes and livestock productivity increases in 2030? *Agricultural Systems, 103*, 621–638. doi:10.1016/j.agsy.2010.07.005.

Chapter 15
Exploring a 'Healthy Foodshed': Land Use Associated with UK Fruit and Vegetables Supply

Henri De Ruiter, Jennie I. Macdiarmid, Robin B. Matthews and Pete Smith

Abstract With an agricultural system that is approaching its natural limits and a global rise in obesity and associated diseases, it is vital to consider human health as a primary driver for future food production. Foodshed analysis is used to analyze the origins of food for a particular region and to assess the implications for environmental sustainability. This case study uses the concept of a foodshed to analyze the land area needed to supply the UK with fruit and vegetables, critical components of a healthy diet, over the period 1986–2009, and incites a critical reflection on how to achieve healthy and environmentally sustainable food. The results show that the 'Fruit and Veg' foodshed of the UK has increased over the studied period and that particularly vegetables are increasingly sourced from abroad, suggesting that the UK is increasingly reliant on other countries to satisfy its recommended nutritional needs. Most important 'external' cropland suppliers are Spain, China, and Italy, together contributing over 30 % of total land area for fruit and vegetables abroad. To better understand trade-offs and synergies between land use, health, and food consumption, it is imperative to include land use as indicator in the context of sustainable diets. A major challenge will be how to achieve a shift in consumption toward less land-intensive patterns, without neglecting socioeconomic issues such as social justice. The alignment of nutritional and agricultural policies is urgently needed as it has the potential of tackling several global challenges simultaneously.

Keywords Food security · Trade-offs · Consumption · Social justice · Governance

H. De Ruiter (✉) · R.B. Matthews
Information and Computing Sciences, James Hutton Institute,
Aberdeen, Scotland, UK
e-mail: henri.deruiter@hutton.ac.uk

H. De Ruiter · P. Smith
Institute of Biological & Environmental Sciences, University of Aberdeen,
Aberdeen, Scotland, UK

J.I. Macdiarmid
Public Health Nutrition Research Group, Rowett Institute of Nutrition and Health,
University of Aberdeen, Aberdeen, Scotland, UK

J. Niewöhner et al. (eds.), *Land Use Competition*,
Human-Environment Interactions, DOI 10.1007/978-3-319-33628-2_15

15.1 Introduction

Global diets provide an important link between human health and environmental sustainability (Tilman and Clark 2014), yet only recently has the scientific community begun to analyze the two holistically rather than in isolation, as it is increasingly recognized that dietary changes are needed to solve both environmental and human health problems. Over the last couple of decades, the world has experienced a major nutrition transition, characterized by a shift away from a traditional diet that is dominated by starchy staples—cereals, legumes, and vegetables—toward a diet with a much higher consumption of processed foods and animal products. In particular, there has been an increased consumption of vegetable oils and sugars, an increase in the consumption of animal products, and a shift away from fruits, vegetables, and whole grains consumption (Popkin 2006). This nutrition transition is associated with numerous drivers, including rising global incomes, increasing globalization, trade liberalization, and patterns of urbanization.

The changes in global dietary patterns have important implications for human health and environmental sustainability. On the one hand, a higher per capita caloric availability and a higher consumption of unhealthy food items have resulted in a global obesity epidemic. The number of overweight and obese individuals increased from 857 million in 1980 to 2.1 billion in 2013, and as of yet, there is no country in the world where obesity levels are declining (Ng et al. 2014). The consequences of this global obesity epidemic for human health are huge as obesity has been associated with the steep rise in non-communicable diseases such as diabetes type II, cancer, and cardiovascular diseases.

On the other hand, the global nutrition transition has also major environmental consequences. In particular, the ongoing increase in the consumption of animal products has a large environmental impact, as most of the production of animal products is inherently inefficient and requires vast amounts of resources compared to the production of the same amount of calories and proteins from plant origin. The ongoing nutrition transition happens at a time where the current agricultural system is already approaching or even exceeding its natural limits as it is responsible for almost a quarter of all greenhouse gas emissions and occupies about 40 % of the Earth's land resources (Foley et al. 2005). Moreover, it is projected that the pressure on the Earth's limited land resources will increase considerably in the near future, to satisfy the demand for more land-intensive food items, associated with a range of drivers, including, for instance, the rising use of biofuels as a substitute for petroleum, an ongoing process of urbanization, the need for climate mitigation strategies such as conservation of forests and carbon sequestration projects, and last but certainly not least, the consumption of food (Lambin and Meyfroidt 2011). Historical increases in crop production have been a result of the expansion of agricultural land and in later stages resulted from the intensification of agriculture. However, many scholars now argue that expansion of agricultural land is not a viable option, as almost all potential cropland lies beneath tropical forests. Bringing this land into cultivation will have very negative consequences, such as biodiversity loss and a significant increase in greenhouse gas emissions through deforestation. At the same time, the intensification of agriculture is currently unsustainable because of, for instance, its excessive nitrogen use. Therefore, to limit agricultural land use and its associated negative impacts, new strategies are required, and dietary change is vital to consider as a potential strategy, especially since it has the potential to improve human health simultaneously.

Against this backdrop, it is imperative to understand the complex and multi-faceted relation between food consumption, human health, and land use. A possible tool to explore the relation between food consumption, land use, and socioeconomic dimensions is foodshed analysis. The concept of a 'foodshed,' analogous to a 'watershed,' has been introduced to analyze the origins of food and the sustainability of local food and could serve as an analytical tool to examine the relation between food consumption and land use. This case study uses foodshed analysis and expands it by explicitly considering foods that are particularly important for human health: fruit and vegetables. A low consumption of these food types is associated with many non-communicable diseases and is an important risk factor for the global burden of disease (GBD 2013 Mortality and Causes of Death Collaborators 2015).

15.2 Case Study: The 'Fruit and Veg' Shed of the UK

It is believed that the origin of the term 'foodshed' dates back to 1929, where it was introduced by Walter Hedden in his book 'How Great Cities Are Fed' (Peters et al. 2009). Foodsheds can be used in three different ways: as a spatial concept, as an analytical tool, and as a basis for action (Horst and Gaolach 2015). The current case study uses the concept of a foodshed as an analytical tool to examine changes in the 'Fruit and Veg' foodshed of the UK over the last decades, hereby inciting a broader discussion on the interconnectedness of agriculture and health and analyzing whether foodshed analysis can contribute to this debate. Foodshed analyses are often being used to ultimately answer the question: 'What is the potential of a particular region to feed itself?' (Horst and Gaolach 2015). This case study differs in that it is based on a historical analysis rather than on creating a scenario for future self-sufficiency of the UK. In doing so, it highlights the trends over the last decades and the implications for health and sustainability. In this case study, the concept of a foodshed is understood as the geographical area that is needed to supply food, specifically fruit and vegetables, to the UK. A larger 'Fruit and Veg' foodshed means that more area outside of the UK is needed to supply the fruit and vegetables. Generally, foodshed analyses are conducted on a smaller scale and often assess the origins of food for a region within a country, but here, the national scale is chosen because the most comprehensive supply data is available on this scale.

There is no universal definition of what comprises a healthy diet, but almost all recommendations for a healthy diet include a minimum consumption of fruit and vegetables, which is set at 400 g day-1 in the UK (NHS Choices 2013). In the current case study, the 'Fruit and Veg' foodshed of the UK is used to examine changes in fruit and vegetables supply over the last 25 years, from 1986 to 2009. It must be recognized that this is at best a crude proxy for a healthy diet; however, it can be used to illustrate the broader connection of health and sustainability. Using FAO statistics on trade and food production, the total land area used to produce the total fruit and vegetables supply for the UK is assessed. Trade poses a difficulty for the proper analysis of a foodshed because most trade flows are only recorded from the last step in the global trade chain. For instance, about 5 % of all bananas in the UK are imported from the Netherlands—a country that does not produce bananas—according to FAO statistics. Therefore, the Netherlands is not part of the foodshed of the UK, as it is only a transit country for bananas. To account for this and to incorporate trade flows in a methodologically sound way, we use data obtained from (Kastner et al. 2014). Using these data, it is possible to calculate how much land is required for the UK food supply and to determine the countries where the food was originally produced.

The fruit and vegetables supply data used here refer to the total available, or supplied, fruit and vegetables to the UK. It represents the total quantity of fruit and vegetables produced in the UK added to the total quantity imported and minus all exports. As such, the total 'Fruit and Veg' foodshed of the UK is assessed, without considering food waste, actual intake of fruit and vegetables, or differences in regional and individual consumption. The data used in this case study contain information about all major fruit and vegetables and include a total of 31 different fruit crops and 22 different vegetables. Because of the methodology followed by Kastner et al. (2014), all supply data are obtained in primary crop equivalents. Thus, all fruit and vegetables used in processed food, such as in tomato sauce, are converted back to their primary crop. Grapes are for that reason excluded from the current analysis, as it is difficult to disentangle 'real' grape consumption from wine consumption. Nevertheless, as an aside, it is interesting to note that about 33 % of the total imported area related to fruits is actually related to the import of grapes (both in the form of wine and in the form of normal grapes). In addition, all fruit and vegetables fed to animals are also included in the primary crop equivalents. We assume that this amount is negligible.

15.3 'Fruit and Veg' Foodshed Is Increasing

The data described above show that the total fruit and vegetables supply in the UK increased from 8.8 million tons in 1987 (three-year mean) to 11.8 million tons in 2008, or an increase of 35 %. Total fruit supply increased from 3 million tons in 1987 to 4.5 million tons in 2008 (+48 %), while the total vegetable supply increased from 5.8 million tons in 1987 to 7.4 million tons in 2008 (+27 %). The actual area required to supply all these fruits and vegetables, or the so-called Fruit and Veg foodshed, for the UK equaled 494 k ha in 1987 and increased to 571 k ha in 2008 (+16 %), thus increasing at a slower rate than the increase in supply would suggest. Despite the overall increase in global land area associated with this supply, the domestic land area used for the supply of these fruits and vegetables decreased with 34 % from 201 k ha in 1987 to 133 k ha in 2008 (Fig. 15.1), which means that the total land area overseas increased from 293 k ha to 438 k ha (+49 %).

The most important fruits for the UK, both in terms of supply and land area, are apples, bananas, and oranges. In 1987, apples contributed about 25 % to the total fruit supply (vs. 21 % in 2008), oranges about 15 % (vs. 14 % in 2008), and bananas about 12 % (vs. 20 % in 2008). For vegetables supply, the most important categories are the ill-defined 'vegetables, nes (not elsewhere specified—a term used by the FAO when the type of vegetables is not detailed)' (18 % of total vegetables supply in 1987; 28 % in 2008), followed by tomatoes (21 % and 23 %, respectively) and 'carrots and turnips' (12 % and 10 %, respectively). However, for land area, because of the relatively high yields of tomatoes and 'carrots and turnips,' the top-3 for vegetables' land area consists of 'vegetables, nes,' 'cauliflowers and broccoli,' and green peas.

Changes in overall land area can be caused by four factors: the size of the population, the total consumption of fruit and vegetables per person, the type of fruit and vegetables consumed, and the yields of fruit and vegetables. The population of the UK has increased over the 25-year period considered here: from 57.0 million in 1987 to 61.6 million people in 2008 (FAO, both numbers are three-year averages), leading to an increase in total fruit and vegetables supply and associated land area. Total fruit and vegetables supply per person increased from 423 g person-1 day-1 to 526 g person-1 day-1, suggesting that the total available amount of fruit and vegetables per person has increased in the UK. It is important to keep in mind that this does not necessarily imply an increase in actual intake of fruit and vegetables, since FAO supply data measure total food availability on a country level, and not the food that is actually consumed by consumers. Hence, an increase in fruit and

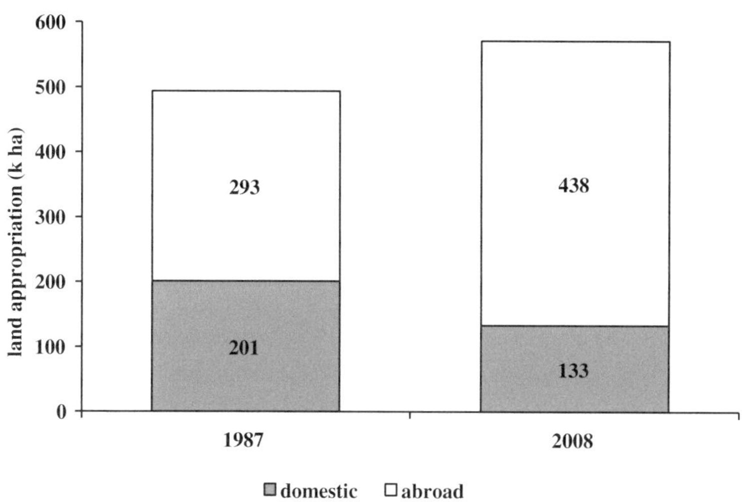

Fig. 15.1 Land appropriation (in k ha) associated with UK fruit and vegetables supply in 1987 (*left*) and 2008 (*right*). *Dark areas* represent domestic land appropriation, while *grey areas* represent land appropriation abroad. Values are three-year averages around the respective years

vegetables supply could also mean that more food is wasted in households or along the food chain, without resulting in an increased intake of fruit and vegetables. During the studied period, the average yield of the supplied fruit and vegetables (as measured by total supplied fruit and vegetables divided by the area for production) has increased from an average yield of 17.8 ton ha-1 to 20.7 ton ha-1, leading to an average decrease in land area per unit of fruit and vegetables production. Despite this reduction in land area per unit of fruit and vegetables, the 'Fruit and Veg' foodshed of the UK has still increased substantially over the studied period.

15.4 Where Does the UK's 'Fruit and Veg' Come from?

When considering the origin of production, a distinction can be made between exotic fruit and vegetables (exotic is here defined as fruit and vegetables that cannot be grown due to the UK's climate conditions, e.g., bananas or pineapples), foreign fruit and vegetables (defined as fruit and vegetables produced abroad and imported to the UK, while the type of fruit or vegetable is also grown in the UK—e.g., apples from France), and domestic fruit and vegetables (grown in the UK—e.g., apples that are domestically produced). In 1987, the weight proportions between the three categories were the following: 19 % of the fruit and vegetables were exotic; 38 % were foreign; and 43 % were domestic. These percentages changed to 24 % exotic; 53 % foreign; and 23 % domestic in 2008 (Table 15.1). This suggests that fruit and vegetables that could have been grown in the UK are increasingly imported from abroad. If we consider the difference between fruit and vegetables in more detail, the bulk of the exotic category consists of fruits, the category foreign is relatively evenly split between fruit and vegetables, and the category domestic consists mainly of vegetables. This suggests that the main shift from domestically produced to imported from abroad is caused by vegetables that were previously grown within the UK, but are now imported from abroad, such as tomatoes. However, the main increase for foreign fruit and vegetables can be seen in the category 'vegetables, nes,' making it difficult to assess which exact type of vegetables is increasingly imported from abroad, since this the type of vegetables is not further detailed by the FAO.

Table 15.1 Relative contribution of the type of food to total supply. Domestic: domestically produced; foreign: foreign produced foods that could have been produced in the UK; exotic: not grown in the UK

Type of food	Share in total supply of fruit & vegetables	
	1987 (%)	2008 (%)
Domestic	43	23
Foreign	38	53
Exotic	19	24

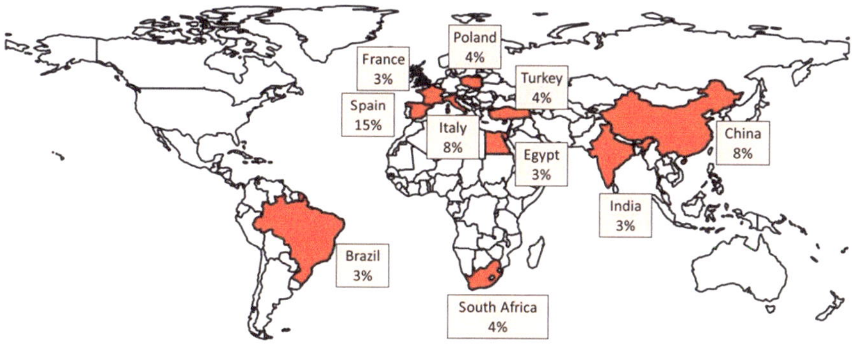

Fig. 15.2 Relative contribution of the ten most important countries to total land appropriation abroad related to the UK fruit and vegetables supply in 2008 (three-year average)

So it can be concluded that fruit and vegetables are increasingly imported from other countries, hereby increasing the 'Fruit and Veg' foodshed of the UK. If we take the land appropriation for fruit and vegetables together for the year 2008, it can be seen that the major 'external land' supplier is Spain, with a share of 15 % in the total land appropriation abroad (this excludes domestic land area) (Fig. 15.2). Other important countries include China and Italy, both contributing about 8 % to the total land appropriation abroad related to fruit and vegetables. Spain and Italy were also the major contributors in 1987; the newcomers in the top-10 for 2008 include China, Poland, Brazil, India, and Egypt. Major increases in land appropriation between 1987 and 2008 related to the UK fruit supply are in Spain (particularly from the import of oranges), Colombia (bananas), and Costa Rica (bananas and pineapples), while major increases in land area for vegetables can be seen in China (vegetables, nes), Spain (cauliflower and broccoli), and India (vegetables, nes). The current case emphasizes the direction of the geographical location of the 'Fruit and Veg' foodshed: Domestic land area for fruit and vegetables has gone down, while land appropriation outside the UK's territory has increased. Table 15.2 shows the UK is increasingly importing fruits and vegetables from outside Europe, thereby extending transport distances and geographical area. It seems that there has been especially a shift from domestic vegetables production to a truly more global sourcing of vegetables.

Table 15.2 Relative contributions of three different regions to the land area associated with the UK fruit and vegetables supply for the years 1987 and 2008 (three-year means)

Region	Percentage of total land area for fruit		Percentage of total land area for vegetables	
	1987 (%)	2008 (%)	1987 (%)	2008 (%)
Domestic	18	10	63	35
EU 15+	21	24	25	28
Rest of the world	60	66	12	37

15.5 Human Health and Land Use

From a health perspective, one could argue that we need to increase the area dedicated to the production of fruit and vegetables rather than decrease. However, an increase in fruit and vegetables consumption will not necessarily be more environmentally sustainable, even if the consumption of other food categories is decreased. Studies have shown that the environmental impact (that is, in this case, greenhouse gas emissions) of fruit and vegetables per calorie can be as high as some of the animal products (Vieux et al. 2012), especially when airfreighted or produced in glasshouses. Therefore, since a diet requires a certain amount of calories, replacing animal products with fruit or vegetables may not always lower the environmental impact of a diet. On the other hand, because of the low caloric content of, for instance, vegetables, replacing a portion of meat requires many more portions of vegetables, which is probably not realistic in practice. Again, this emphasizes the fact that nutrition must be assessed holistically.

Despite consistent and persistent health messages to increase the consumption of fruits and vegetables to at least '5 portions a day,' the UK population on average falls short on this recommendation. Recent statistics show that only 30 % of the adults consume the amount of fruit and vegetables as currently recommended (Bates et al. 2014), despite the fact that our data suggest that the total supply of fruit and vegetables is sufficient (526 g person-1 day-1). This could imply that although national fruit and vegetables supply in the UK has gone up, actual intake could have decreased if more food is wasted. This emphasizes another strategy for improving our diets in term of health and sustainability: the reduction of food waste. Statistics from the UK show that fresh produce—vegetables, salads, and fruits—contributed over a third of the total food and drink waste generated in 2007. This amounts to a total waste of 3 million tons in 2007 (WRAP 2009), which would represent about 25 % of total fruit and vegetables supply as indicated by our data. It must be noted here that the WRAP data represent the amount wasted by households, and hence, this percentage of 25 % is a low estimate for the total fruit and vegetables wasted, since fruit and vegetables are also wasted in other parts of the food supply chain. The WRAP report distinguishes between avoidable (waste that could have been prevented), potentially avoidable (e.g., potato skins—some people eat them, some people would not consider eating them), and unavoidable waste (such as nutshells). Of the total fruit and vegetables waste, about 45 % is avoidable, 43 % is potentially avoidable, and only 13 % is unavoidable. This highlights the considerable potential of reducing food waste as a strategy to create a healthier and more sustainable food system.

15.6 Link Between Nutrition and Agriculture

A healthy diet is, according to many experts, primarily centered on nutrient intakes from the whole diet, and a proper assessment of nutrient availability from the whole diet would thus take this case study a step forward. Some studies are doing this by utilizing linear programming to determine the self-sufficiency of a particular region or city (Ward et al. 2014). Linear programming is a mathematical method which can be used to optimize diets for meeting a certain objective (e.g., a health outcome such as minimizing dietary energy, or environmental outcomes such as minimizing greenhouse gas emissions) while satisfying a number of nutritional constraints (e.g., a minimum intake of certain micronutrients). Using linear programming, it has been shown that a more sustainable diet in terms of greenhouse gas emissions is possible while simultaneously meeting dietary requirements for health (Macdiarmid et al. 2012) and that urban agriculture can make a significant contribution to food security in an Australian context (Ward et al. 2014). These innovative methods try to solve the puzzle that is created by the fact that

> A healthy diet is principally about the nutrient intakes, which can be achieved from many different combinations of foods. In contrast, GHG [emissions] are associated with food items for which the nutrient content can vary (Macdiarmid 2013).

For the current case study, the ultimate aim would be to increase the intake of the right amounts of nutrients, while simultaneously limiting the amount of land needed to produce these food items. Similarly, one could assess where nutrients come from and how this influences the foodshed of the UK. However, improved models are needed for this since nutrient contents vary by crop, by variety, by the time of harvesting, by the type of preparation, and so on.

At the same time, however, it can be argued that these linear optimization models are too reductionist and will never be able to capture the complexity associated with healthy and sustainable food. First of all, these models only focus on a limited set of environmental and socioeconomic indicators. More importantly though, they cannot ultimately deal with a complex food system that may have unexpected and nonlinear feedback loops. This has, for instance, been discussed in the context of a healthy diet: Should nutritional advice be centered on nutrients or on food items? Whereas some nutritionists argue that nutritional recommendations should primarily focus on nutrient intakes, others would argue that this is too simplistic and that a healthy diet is primarily concerned with actual food intake and the interactions between the different components of the diet (Katz and Meller 2014). The same can be argued for environmental sustainability: reducing the concept to, for instance, land use, or aggregated environmental impact per unit of food, will yield important insights, but may miss the broader picture and have unintended consequences.

Although a dichotomy between a reductionist approach and a more holistic view is arguably stylized, it can serve as a useful conceptualization to explore consequences of both approaches for nutrition and sustainable land use. The main difference between the two approaches is that the reductionist approach tries to reduce complex problems to its component parts and consequently tries to optimize these parameters, e.g., the smallest foodshed that is possible, whereas the holistic paradigm stresses that certain problems are too complex and cannot be reduced to their parts and that, for instance, focusing on the smallest foodshed possible will lead to other problems. Agricultural and dietary diversity play an important role in this paradigm as diversity can increase resilience of communities and contribute to a healthier diet. The following section moves beyond the current case study and outlines these two different paradigms and their consequences for accomplishing an agricultural system that uses less land and delivers a healthier diet. Meanwhile, it is emphasized that the two paradigms presented here are not absolute and that both have their part to play in the realization of a more sustainable food system.

15.7 Reductionist Paradigm

The basis of a more reductionist approach to agriculture and nutrition lies at the focus on yield in the twentieth century and is sometimes framed as a productionist paradigm (Lang and Barling 2012). In this paradigm, the primary objectives of agriculture should be to produce as much food as possible for the least cost. As such, the underlying drivers are productivity and efficiency (Lyson 2002). Strategies to include nutrition in the agricultural system involve, among other strategies, soil fertilization, biofortification, and food fortification (Miller and Welch 2013). Soil fertilization is the process whereby micronutrients, such as trace elements like zinc, are added to fertilize the soil and to improve the micronutrient content of the crops. Biofortification refers to conventional plant breeding for enhanced micronutrient content or to genetic engineering to develop crops enriched in vitamins, with 'Golden Rice,' a rice crop enriched in vitamin A as most well-known example. Biofortification is seen as a good way for addressing population-level micronutrient deficiencies in a sustainable way (Miller and Welch 2013). Food fortification—the addition of micronutrients to food items—takes place mostly in developed countries, since the high prevalence of processed foods makes it easier to deploy this strategy. These fortifications are considered a hyperefficient mechanism for supplying essential micronutrients (Dixon et al. 2009). There is evidence that the contribution of processed foods to nutritional recommendations is indeed very substantial. Research from the USA suggests that processed foods contribute to, for instance, 65 % of folate and 46 % of vitamin B12 requirements, though the definition of processed food used in this study is probably broader than commonly associated with the term (Weaver et al. 2014). Nevertheless, it shows that processed foods are an important and maybe a key part of the current Western diet. Consequently, to improve human health while limiting land use and the foodshed, strategies such biofortification and food fortification may contribute to a more efficient agricultural system. If additional micronutrients can be supplied without having to grow additional crops, this results in a lower land use while simultaneously benefiting human health.

From a demand-side perspective, it can be said that the agricultural sector is highly inefficient since only 55 % of total calories produced globally are directly being used for human food, with the rest going to animal feed and biofuels (Cassidy et al. 2013). Because of these 'inefficiencies' in the food system, the authors of this study argue for a new agricultural metric: *people nourished per hectare*, linking agricultural output more directly with human needs. As the authors indicate themselves, their focus on calories is clearly a simplification and factoring in the need for a diverse set of nutrients will complicate the analysis. Moreover, to examine environmental impact more comprehensively, single indicators such as land use are not sufficient and more complex multi-indicator life cycle analyses are therefore being conducted to include more complexity.

The reductionist paradigm has been heavily criticized because it is overly simplistic and does not address the underlying issues of conventional, intensive agriculture, such as issues of farm household, community welfare, and rural development (Lyson 2002). Furthermore, concern is voiced over the possibility that large multinational companies are able to turn genetically engineered products into commodities by intellectual property regimes and are mainly profitable for the companies themselves. Consequently, some say that reductionism has led to a so-called hyperspecialized technoscience (Fardet and Rock 2014), hereby neglecting dietary and agricultural diversity.

15.8 Diversity Paradigm

Therefore, a contrasting view is that the agricultural system should be more diverse, which would in turn create a more resilient system. The underlying assumption is that agricultural biodiversity contributes to dietary diversity. This dietary diversity has been shown to correlate strongly with nutrition status (Frison et al. 2006). However, research has shown that, over the last decades, the global food supplies have followed an opposite trend and have become more homogenous rather than diverse, raising concern about the genetic diversity of the global food supply and overall food security (Khoury et al. 2014). Therefore, more attention to nutritious food is needed to create a more diverse agricultural system. One of the measures being used to asses this agricultural diversity is the nutritional functional diversity (FD) metric, a metric based on plant species composition at the farm (or landscape scale) and the nutritional composition of these plants for a range of nutrients (Remans et al. 2011). Adding a crop variety with a unique combination of nutrients will increase the nutritional FD and removing such a variety will decrease it.

On the demand side, limitations in the reductionist approach focused on nutrients have led to studies that consider the effect of the overall diet and complete dietary patterns (Fardet and Rock 2014). This development is also important for research on the environmental impacts of food. Early studies considered the environmental impact of single food items, but more recent studies consider the impact of a meal (one day) or complete diets (often consisting of a weekly diet, i.e., 7 meals). This rising attention to sustainable diets is significant, as the possibility of convergence around non-communicable diseases, nutrition, and agricultural policies might be largest around the provision of sustainable diets (Candeias and Lachat 2013). The Food and Agricultural Organization of the United Nations defines sustainable diets as:

> [...] those diets with low environmental impacts which contribute to food and nutrition security and to healthy life for present and future generations. Sustainable diets are protective and respectful of biodiversity and ecosystems, culturally acceptable, accessible, economically fair and affordable; nutritionally adequate, safe and healthy; while optimizing natural and human resources (Burlingame and Dernini 2012).

Dietary diversity is a key measure of sustainable diets (Luckett et al. 2015). Nonetheless, the concept of sustainable diets is still ill-defined, and many scholars argue that a better understanding of what comprises a sustainable diet is urgently needed (Johnston et al. 2014). A recent review on the sustainable diets literature shows that most studies consider the impact of eating patterns on the climate by measuring greenhouse gas emissions and to a lesser extent land requirements for food (Auestad and Fulgoni 2015).

A critique on the concept of sustainable diets is that the shift toward a diet low in greenhouse gas emissions or land-intensive products is unprecedented and that the

> *politics* of such profound cultural and behavioral change (…) are completely ignored and "black-boxed." [emphasis in the original] (Goodman et al. 2012).

That is to say, the approach toward sustainable diets has so far mainly been to optimize certain parameters. While this is an obvious first step, it has been argued that questions of food justice, social practice, and associated political questions will have to be more explicitly addressed in the context of sustainable diets. Will the increasing attention to sustainable diets be sufficient to nudge people toward a healthier and more environmentally sustainable diet? Or do we need, as some people argue, sustainable dietary guidelines rather than nutrition dietary guidelines (Lang and Barling 2013), and will this be enough to change consumption patterns? These questions clearly show that more work is needed to understand how a healthy and environmentally sustainable diet might be best achieved. It is interesting to note, however, that in food activist circles, the place of local food and place of production are considered especially important, often more important than the greenhouse gas emissions associated with food production. Inclusion of land requirements for food within the concept of sustainable diets and more specifically the place of origin of food, i.e., the foodshed of a particular region, may function as a useful tool to make the environmental impact of our food more comprehensible and tangible, as opposed to more abstract greenhouse gas emissions. This has been argued by others before, emphasizing the potential of food becoming a problem-solving tool for local communities, with emerging links between socioeconomic development, health, and environmental sustainability (Hamm 2008). Therefore, the concept of a foodshed, which highlights the interdependencies of our current food system, could potentially incite a much more holistic discussion on sustainable diets and contribute to the change that is desired.

15.9 Conclusion

This case study has shown that the so-called Fruit and Veg foodshed of the UK has increased over the last decades in general and the land area abroad related to vegetables supply in particular. In doing so, it concludes that nutrition should be considered in the context of land use competition as a more explicit goal of the agricultural system. Addressing environmental sustainability and human health at the same time could lead to two different, although not mutually exclusive, pathways. The first is to produce micronutrients as efficient as possible. This pathway will favor fortification strategies, the processing of food, and will likely involve genetically optimized crops. The other pathway will focus on sustainable diets and agricultural diversity and could involve food-based dietary recommendations. Challenges in the near future will be the inclusion of land use in the context of sustainable diets and the shift in consumption patterns toward less land-intensive patterns, without neglecting socioeconomic issues such as social justice. Therefore, the alignment of nutritional and agricultural policies is urgently needed and has a lot of potential in tackling several global challenges simultaneously.

References

Auestad, N., & Fulgoni, V. L. (2015). What current literature tells us about sustainable diets: Emerging research linking dietary patterns, environmental sustainability, and economics. *Advances in Nutrition: An International Review Journal, 6*(1), 19–36. doi:10.3945/an.114.005694.

Bates, B., Lennox, A., Prentice, A. et al. (2014). *National diet and nutrition survey*. Results from years 1–4 (combined) of the rolling programme (2008/2009–2011/12).

Burlingame, B. & Dernini, S. (2012). *Sustainable diets and biodiversity. Directions and solutions for policy, research and action.*

Candeias, V. & Lachat, C. (2013). Health, nutrition and agriculture: Can the prevention of NCDs be an entry point for policy alignment in low- and middle-income countries? *SCN News: Changing food systems for better nutrition,* 40.

Cassidy, E. S., West, P. C., Gerber, J. S., et al. (2013). Redefining agricultural yields: From tonnes to people nourished per hectare. *Environmental Research Letters, 8*(3), 034015. doi:10.1088/1748-9326/8/3/034015.

Dixon, J. M., Donati, K. J., Pike, L. L., et al. (2009). Functional foods and urban agriculture: two responses to climate change-related food insecurity. *New South Wales Public Health Bulletin, 20*(2), 14–18.

Fardet, A., & Rock, E. (2014). Toward a new philosophy of preventive nutrition: From a reductionist to a holistic paradigm to improve nutritional recommendations. *Advances in Nutrition: An International Review Journal, 5*(4), 430–446. doi:10.3945/an.114.006122.

Foley, J. A., DeFries, R., Asner, G. P., et al. (2005). Global consequences of land use. *Science, 309* (5734), 570–574. doi:10.1126/science.1111772.

Frison, E. A., Smith, I. F., Johns, T., et al. (2006). Agricultural biodiversity, nutrition, and health: Making a difference to hunger and nutrition in the developing world. *Food and Nutrition Bulletin, 27*(2), 167–179.

GBD 2013 Mortality and Causes of Death Collaborators (2015). Global, regional, and national age-sex specific all-cause and cause-specific mortality for 240 causes of death, 1990–2013: A

systematic analysis for the global burden of disease study 2013. *Lancet*, 385(9963), 117–171. doi:10.1016/S0140-6736(14)61682-2.
Goodman, D., DuPuis, E. M. & Goodman, M. K. (2012). Sustainable diets and governance. In D. Goodman, E. M. DuPuis & M. K. Goodman (Eds.), *Alternative food networks. Knowledge, practice and politics* (1st ed. p. 117). Routledge, Oxon.
Hamm, M. W. (2008). Linking sustainable agriculture and public health: Opportunities for realizing multiple goals. *Journal of Hunger & Environmental Nutrition, 3*(2–3), 169–185. doi:10.1080/19320240802243241.
Horst, M. & Gaolach, B. (2015). The potential of local food systems in North America: A review of foodshed analyses. *Renewable Agriculture and Food Systems,* FirstView: 1–9. doi:10.1017/S1742170514000271.
Johnston, J. L., Fanzo, J. C., & Cogill, B. (2014). Understanding sustainable diets: A descriptive analysis of the determinants and processes that influence diets and their impact on health, food security, and environmental sustainability. *Advances in Nutrition: An International Review Journal, 5*(4), 418–429. doi:10.3945/an.113.005553.
Kastner, T., Erb, K. H., & Haberl, H. (2014). Rapid growth in agricultural trade: Effects on global area efficiency and the role of management. *Environmental Research Letters, 9*(3), 034015. doi:10.1088/1748-9326/9/3/034015.
Katz, D. L., & Meller, S. (2014). Can we say what diet is best for health? *Annual Review of Public Health, 35*(1), 83–103. doi:10.1146/annurev-publhealth-032013-182351.
Khoury, C. K., Bjorkman, A. D., Dempewolf, H., et al. (2014). Increasing homogeneity in global food supplies and the implications for food security. *Proceedings of the National Academy of Sciences, 111*(11), 4001–4006. doi:10.1073/pnas.1313490111.
Lambin, E. F., & Meyfroidt, P. (2011). Global land use change, economic globalization, and the looming land scarcity. *Proceedings of the National Academy of Sciences, 108*(9), 3465–3472. doi:10.1073/pnas.1100480108.
Lang, T., & Barling, D. (2012). Food security and food sustainability: Reformulating the debate. *The Geographical Journal, 178*(4), 313–326. doi:10.1111/j.1475-4959.2012.00480.x.
Lang, T., & Barling, D. (2013). Nutrition and sustainability: An emerging food policy discourse. *Proceedings of the Nutrition Society, 72*(01), 1–12. doi:10.1017/S002966511200290X.
Luckett, B. G., DeClerck, F. A., Fanzo, J. et al. (2015). Application of the nutrition functional diversity indicator to assess food system contributions to dietary diversity and sustainable diets of Malawian households. *Public Health Nutrition, FirstView*: 1–9. doi:10.1017/S136898001500169X.
Lyson, T. A. (2002). Advanced agricultural biotechnologies and sustainable agriculture. *Trends in Bio-technology, 20*(5), 193–196. doi:10.1016/S0167-7799(02)01934-0.
Macdiarmid, J. I. (2013). Is a healthy diet an environmentally sustainable diet? *Proceedings of the Nutrition Society, 72*(1), 13–20. doi:10.1017/S0029665112002893.
Macdiarmid, J. I., Kyle, J., Horgan, G. W., et al. (2012). Sustainable diets for the future: Can we contribute to reducing greenhouse gas emissions by eating a healthy diet? *The American Journal of Clinical Nutrition, 96*(3), 632–639. doi:10.3945/ajcn.112.038729.
Miller, B. D. D., & Welch, R. M. (2013). Food system strategies for preventing micronutrient malnutrition. *Food Policy, 42*, 115–128. doi:10.1016/j.foodpol.2013.06.008.
Ng, M., Fleming, T., Robinson, M., et al. (2014). Global, regional, and national prevalence of over-weight and obesity in children and adults during 1980–2013: A systematic analysis for the global burden of disease study 2013. *The Lancet, 384*(9945), 766–781. doi:10.1016/S0140-6736(14)60460-8.
NHS Choices (2013). *The eatwell plate*. http://www.nhs.uk/Livewell/Goodfood/Pages/eatwell-plate.aspx. Accessed February 10, 2015.
Peters, C. J., Bills, N. L., Wilkins, J. L., et al. (2009). Foodshed analysis and its relevance to sustainability. *Renewable Agriculture and Food Systems, 24*(01), 1–7. doi:10.1017/S1742170508002433.

Popkin, B. M. (2006). Global nutrition dynamics: The world is shifting rapidly toward a diet linked with noncommunicable diseases. *The American Journal of Clinical Nutrition, 84*(2), 289–298.

Remans, R., Flynn, D. F. B., DeClerck, F., et al. (2011). Assessing nutritional diversity of cropping systems in African Villages. *PLoS ONE, 6*(6), e21235. doi:10.1371/journal.pone.0021235.

Tilman, D., & Clark, M. (2014). Global diets link environmental sustainability and human health. *Nature, 515*(7528), 518–522. doi:10.1038/nature13959.

Vieux, F., Darmon, N., Touazi, D., et al. (2012). Greenhouse gas emissions of self-selected individual diets in France: Changing the diet structure or consuming less? *Ecological Economics, 75*, 91–101. doi:10.1016/j.ecolecon.2012.01.003.

Ward, J., Ward, P., Mantzioris, E., et al. (2014). Optimising diet decisions and urban agriculture using linear programming. *Food Security, 6*(5), 701–718. doi:10.1007/s12571-014-0374-0.

Weaver, C. M., Dwyer, J., Fulgoni, V. L., et al. (2014). Processed foods: Contributions to nutrition. *The American Journal of Clinical Nutrition, 99*(6), 1525–1542. doi:10.3945/ajcn.114.089284.

WRAP (2009). Household food and drink waste in the UK.

Chapter 16
Strengthening City Region Food Systems: Synergies Between Multifunctional Peri-Urban Agriculture and Short Food Supply Chains: A Local Case Study in Berlin, Germany

Beatrice Walthall

Abstract Cities and agriculture are fundamentally linked, yet often coevolve in a contradicting manner. On the one hand, many scholars in science and urban planning argue in favor of satisfying urban food demands through local and regional agricultural production. On the other hand, as the process of urbanization occurs, competition between agricultural and non-agricultural land use is intensifying, more often than not to the disadvantage of agriculture in urban and peri-urban areas. In order to be part of sustainable land use in an urbanizing society, studies suggest that agriculture needs to become increasingly multifunctional. However, the interplay of multifunctional agriculture (MFA), food supply systems, and urban areas is not fully understood and requires more attention. Against this background, this chapter explores the potential of MFA within short food supply chains in peri-urban areas. In particular, MFA is seen as a resource for strengthening urban agriculture and city region food systems as a sustainable development. Based on a local case study in Berlin (CSA SpeiseGut), this chapter examines innovative practices and strategies at farm level that foster multifunctionality in community-supported agriculture (CSA). The case study illustrates how multiple functions such as producing local food (production goal), delivering amenities for urban lifestyles (consumption goal), and protecting ecosystem benefits (protection goal) emerged and how they contribute to a city region food system. The chapter reveals that peri-urban farming can indeed become an integrative land-use option when developing synergies between MFA and short food supply chains. In particular, MFA can stimulate the creation of new food networks, which strengthen urban agriculture and city region food systems.

Keywords Community supported · Agriculture · Scales · Land use practices · Governance · Food · Networks

---

B. Walthall (✉)
Centre for Metropolitan Studies, Technische Universität Berlin, Berlin, Germany
e-mail: beatrice.walthall@metropolitanstudies.de

J. Niewöhner et al. (eds.), *Land Use Competition*,
Human-Environment Interactions, DOI 10.1007/978-3-319-33628-2_16

16.1 Introduction: Linkages Between City Region Food Systems,[1] Urbanization, and Multifunctional Urban Agriculture

Urban agriculture is hardly a novel phenomenon. Its presence in the urban fabric can be traced back to the earliest urban settlements, being a fundamental part of more self-reliant city region food systems (Koc et al. 1999; Steel 2008). In fact, as Mougeot (2000: 11) points out, "many cities probably became and continue to be testing grounds for, and diffusors of, innovative farming systems." While the direct interaction between cities and agriculture is timeless, it is nowadays not a common field in the planning and management of urban land use. The changing nature of cities including political, economic, sociocultural, and technological transformations has deeply reshaped food supply systems—creating land-use pressures and opportunities for urban agriculture (Brinkley 2012; Heimlich and Anderson 2001).

As the process of urbanization occurs, a growing demand for housing and infrastructure leads to development pressures (Heimlich and Brooks 1989), facilitating land-use competition, where farming has to compete with other non-agricultural land uses (Robinson 2004: 204 f.; Lovell 2010: 2511). Additionally, rising land prices in the urban fringe (Cavailhés and Wavresky 2003, Hussain and Hanisch 2014: 502 f., Zasada 2011: 640, 644), paired with decreasing profits resulting from agricultural activities (BMELV (Federal Ministry of Food, Agriculture and Consumer Protection) 2010: 8 f.), lead to substantial constraints and raise questions about the economic feasibility of (peri-) urban agriculture. Moreover, specific policies regulating agricultural activities within cities, such as zoning constraints and/or water- and land-use restrictions, push agriculture further from cities, removing valuable ecosystems and disconnecting established food supply systems (Brinkley 2012: 1; Cicia et al. 2011: 326).

While competition between agricultural and non-agricultural land uses is most intense in the urban periphery (Robinson 2004: 206), urban agriculture also presents numerous inclusive rather than competitive interactions with metropolitan areas (Bryant and Johnston 1992; Brinkley 2012; van Veenhuizen and Danso 2007). Besides offering recreational and aesthetic values for urban communities nearby (Lovell 2010: 2500), urban agriculture contributes to a sustainable resource management and maintains fertile soils in and around cities (Brinkley 2012: 1). Moreover, it contributes to job creation by utilizing seasonal and part-time labor (Heimlich and Anderson 2001: 39), provides local, fresh, value-added, and specialty products for the increasing demand of urban consumers (Renting and Dubbeling 2013: 29), enhances the interaction between farmers and the community

[1]According to the FAO (2015), the term city region food system "encompasses a complex network of actors, processes and relationships to do with food production, processing, marketing, and consumption that exist in a given geographical region that includes a more or less concentrated urban center and its surrounding peri-urban and rural hinterland; a regional landscape across which flows of people, goods and ecosystem services are managed."

(Cicia et al. 2011: 328), and improves the self-sufficiency of cities by increasing regional food security (Zasada 2012: 16).

Even though there is a considerable consensus on the value of urban agriculture for city region food systems, predictions suggest that available urban and peri-urban farmland will continue to decline in the future (Zasada 2012: 13). Consequently, in order to be able to compete with non-agricultural land uses and to be sustained in an urbanizing society, a number of authors argue that urban agriculture needs to become increasingly multifunctional (Lovell 2010; Renting and Dubbeling 2013: 25; Zasada 2012).

Against this background, policymakers, researchers, farmers, and consumers are expressing growing interest in what is called multifunctional agriculture (MFA) (Boody 2008: 763). The approach of MFA "refers to the fact, that agricultural activity beyond its role of producing food and fiber may also have several other functions such as the management of renewable natural resources, landscape, conservation of biodiversity (protection) and contribution to the socio-economic viability of [urban and] rural areas (consumption)" (Renting et al. 2009: S112). Essential work on transitions toward multifunctionality has focused on adaptation pathways (van der Ploeg et al. 2002; Wilson 2007) or the integration of multiple functions into agriculture activities (Holmes 2006; Zasada 2012). However, the interplay of MFA, food supply systems, and city regions is not fully understood and requires more attention.

Renting and Dubbeling (2013: 14) argue that developing synergies between MFA and short food supply chains appears to be a significant mechanism for strengthening urban agriculture and regionalized urban food systems at the same time. The authors propose that multifunctional short food supply chains might "assist in dealing with conflicts involving differentiating sustainability and land-use claims [and] stimulate the creation of new coalitions that can better deal with the diverse issues to which peri-urban agriculture gives rise" (Renting and Dubbeling 2013: 45). Therefore, short food supply chains that integrate MFA are an important point of attention when thinking about land-use competition within city region food systems.

Accordingly, this chapter provides insights into the emergence and benefits of MFA within short food supply chains in peri-urban areas. Based on a local case study in Berlin (CSA SpeiseGut), this chapter discusses i) how multifunctional activities (production, consumption, and protection) provided by SpeiseGut emerged and ii) how these activities contribute to a city region food system. The chapter reveals that peri-urban farming can indeed become an integrative land-use option when developing synergies between MFA and short food supply chains. In particular, MFA can stimulate the creation of new networks, which strengthens urban agriculture and city region food systems. Though, MFA also requires a call for cautiousness as it is not merely beneficial for small-scale farmers, but also encompasses several disadvantages as discussed in the closing remarks of the chapter.

16.2 The Emergence of Multifunctional Agriculture: Key Drivers and Motives

For studying the emergence of a multifunctional land uses in the global North, the Geographer John Holmes (2006) offers an explanatory approach. Holmes (Holmes 2006: 142 f.) argues that industrialized and monofunctional agriculture land uses lose their legitimacy through a radical re-evaluation of the three basic purposes underlying human land uses, namely production, consumption, and protection. Against this background, he links the three basic purposes of human land uses to key driving forces contributing to the emergence of multifunctional agriculture.

First, (Holmes 2006: 143) demonstrates that agricultural overcapacity promotes postproductive values, altering agricultural practices through "pluriactivity, extensification, disinvestment, and/or conversion to non-farm uses." Second, he suggests that new urban lifestyles and demands such as tourism, recreational, or residential interests are leading to the emergence of market-driven amenity and land consumption practices. And third, growing awareness of sustainability and preservation issues encourage changing societal values surrounding land protection resulting in the integration of ecosystem management into agricultural production systems (Holmes 2006: 144 f.). Accordingly, Holmes characterizes the transition toward multifunctionality as a "shift from the formerly dominant production goals towards a more complex, contested, variable mix of production, consumption and protection goals" (Holmes 2006: 142 f.). Consequently, the legitimacy of multifunctional land use is being altered as a response to the above-mentioned driving forces and in turn allows the emergence of new visions, infrastructures, and farming practices.

Oostindie et al. (2006) provide further insights into underlying forces promoting the emergence of MFA. By studying farmers' motives to opt for MFA in peri-urban areas, the authors found out that societal limitations (such as urban developments or environmental regulations), the need for extra farm income, and (socially) responsible entrepreneurship (including the desire to meet societal demands or closer contact with citizens) represent the most important motives (Oostindie et al. 2006: 57). To improve their farming activities, Oostindie et al. (2006: 47) identified two different innovative strategies toward MFA: first, adding value through food production (e.g., fresh, organic, specialty food) and secondly, diversifying through the deliverance of new services (e.g., education, leisure, tourism). Given the different motives and possible strategies, Oostindie et al. (2006) conclude that the role of farmers in peri-urban areas could shift gradually from users of production space toward managers of production and consumption space. Linked to driving forces toward a variable mix of production, consumption, and protection goals stated by Holmes (2006), this chapter identifies an additional role of farmers: managers of protection space.

Finally, an important question can be raised: If MFA is a desired outcome in order to deal with land-use competition and strengthen short food supply chains, then how can farmers change their role and farming practices toward multifunctional activities? To discuss the stated question, the CSA model provides a suitable entry point as it simultaneously represents a type of short food supply chain and a form of MFA by comprising a variable mix of multiple functions and goals into agricultural activities.

16.3 Community-Supported Agriculture (CSA)

The CSA concept is based on a direct economic and social partnership between a farmer and consumers. In this arrangement, the farmer provides a supply of fresh and nutritious food throughout the growing season, while consumers support the farmer financially by purchasing a "share" of the harvest upfront (Groh and McFadden 1997: 34; Hinrichs 2000: 299).

Although CSAs are not all likely to be the same and tend to be dynamic according to their needs and organization (King 2008: 117), they share the same core ideas: Farmers and consumers bear the risks, costs, and responsibility for the farm and build a community around food and agriculture (Wild 2012: 9). In terms of risks, there is a constant possibility of crop failure caused by floods, droughts, or insect infestations. In the case of crop failure, the benefits of shared costs come into play: While consumers might have to accept a lower food supply, they still receive some products and the farmer does not have to cover the full cost resulting from crop failure. Since CSA farmers often practice agriculture without government subsidies, this financial support provided by consumers is essential for the continuity of farming activities.

The characteristics of shared costs, risks, and responsibilities, the producer–consumer arrangement, and the development of a community are key elements that distinguish the CSA from other types of (direct) agricultural markets (Groh and McFadden 1997; Hinrichs 2000: 299; van de Kop et al. 2008: 32). At the same time, these characteristics enable a wider scope of action for the farmers, precisely because their labor is financed and they face less economic pressure.

Another distinctive characteristic of a CSA is that it emphasizes the desire to meet human needs and sustain ecological resources, besides financial profits. While economic success is essential to maintain market activities, it is not the only goal. Instead, central goals for many CSA farmers and consumers are the direct provision of locally grown and nutritious food (production goal), based on sustainable agriculture (protection goal), while strengthening small farm incomes and revitalizing the local food economy (consumption goal) (Stagl 2002: 146).

Against this background, the CSA model is a good point of departure to develop innovative synergies between short food supply chains and MFA in order to be an integrative land use and to strengthen city region food systems. The next section sets out an empirical case study that illustrates how multiple functions such as producing local food (production goal), delivering amenities for urban lifestyles (consumption goal), and protecting ecosystems (protection goal) emerged in the case of the CSA SpeiseGut and how they contribute to the Berlin region food system.

16.4 Case Study—CSA SpeiseGut in Berlin, Germany

16.4.1 Methodology

The research is based on an exploratory research design that includes qualitative data from a case study of a CSA in Berlin, Germany. To gain deeper insights into the organizational structures and practices of a CSA, qualitative data were collected with mixed methods including participatory on-site research and individual and group interviews between March 2013 and February 2015. Participating in farm and cultural events, as well as harvesting, processing, and retailing activities, enabled an in-depth analysis of daily practices of the CSA as well as its multi-functional character. Moreover, qualitative interviews using open-ended questions with the farmers and CSA members revealed insights into different perspectives of the participants. The aim was less to generate comparable and generalized information, but rather to obtain detailed information on the emergence of alternative farming practices and activities. In order to do so, the interview questions focused on individual motives to start/join SpeiseGut, benefits of participating in SpeiseGut, perceived strengths and weaknesses of SpeiseGut, the project's progress over the last two years, expansion plans, and future prospects. All interviews were recorded and transcribed for an in-depth content analysis. By coding the data with key words and themes, relevant results could be filtered and systematically structured.

16.4.2 CSA SpeiseGut: Meeting Production, Consumption, and Protection Goals

The 3-hectare CSA farm SpeiseGut is located on the edge of a residential area in the district of Berlin Spandau. The proximity of the CSA to an urban area is not exclusive, but a typical pattern of short food supply chains to reconnect people and food spatially, promote shorter distribution channels, and facilitate direct face-to-face interactions between producers and consumers. In turn, peri-urban areas are a center for societal and lifestyle transition and therefore provide ideal opportunities to restructure farming (Zasada 2011: 640). Therefore, this chapter proposes that innovative land management arrangements unfold, where land-use competition is high and the farmer's role encompasses managing production, consumption, and protection space.

16.4.2.1 Producing Food for Urban Consumers

SpeiseGut was founded in 2013 by a farmer with the intention to provide organic produce to urban consumers without government subsidies. However, this is not an easy goal as productivist agricultural policies, including government subsidies, have contributed to the intensification of agriculture, resulting in massive food surpluses and intense competition within the market. At the same time, many small-scale farmers lack financial resources to start and run their own business, as in the case of SpeiseGut. According to the farmer, investment costs for machineries and land are too high in relation to earnings for him as for many small-scale producers. After more than 16 years of experience in agriculture, the farmer decided to start a CSA as a way out of conventional and industrialized food production.

Producing food for urban demands includes understanding urban consumption interests. SpeiseGut recognizes urban demands by integrating the consumers in the selection of crops. Hence, rather than simply choosing the products, the farmer and the CSA members discuss at the beginning of the planting season what is possible, needed, and desired (Walthall 2013: 69). Throughout the season, members are continuously encouraged to give feedback on the received products (e.g., on quantity and quality). The direct interactions between the farmer and the members enable valuable insights into food preferences and demands. Accordingly, the farmer can adjust and respond to local interests of the community involved.

In 2015, SpeiseGut has about 180 household CSA memberships, ranging from single-person households to couples and families. Membership costs 62.50 euros/month minimum, but members are free to pay more and some actually do. The price for a share is settled at the beginning of each season, when the farmer calculates the budget for the business and divides the total costs by the number of members. Then, the minimum price is being presented and discussed with the members and agreed upon with a vote. Due to the agreement procedure and the fact that individual products do not have a fixed price within a CSA, it is difficult to compare the price for a share with store prices or even other CSAs.

A share from SpeiseGut includes vegetables, fruits, juices, herbs, and flowers in the growing season. In the winter months, preserved fruits and vegetables, noodles, honey, and oils compensate the lack of fresh produce. Dairy and meat products are not part of the assortment yet and, if desired, need to be purchased separately at a different farm or shop.

Each week, SpeiseGut delivers the share to eight districts in Berlin and one district in the nearby city of Potsdam, where members receive their products at self-organized pickup sites. Public transportation throughout the city connects the members with the pickup sites and the farm, which contributes to a car-independent commute.

Although the CSA initiative is the core of SpeiseGut's activities and enables some financial relieve for the farmer, earnings resulting from the memberships are insufficient and additional incomes are absolutely necessary to consolidate and expand SpeiseGut's pursuits (Walthall 2013: 71). Consequently, direct marketing activities with local food businesses and initiatives, i.e., local restaurants, hotels, cantinas, stores, and community gardens, were planned from the beginning of the project. These circumstances can be linked to agricultural overcapacity driving the transition toward MFA, as stated earlier.

After a year of strengthening SpeiseGut's existence, the farmer intended to increase the agricultural operations and income, but could not secure additional farmland. By following a flexible and dynamic approach, the farmer expanded instead by cultivating a 2-ha apple orchard mainly for juice production, opening an on-site restaurant (HavelGut), shortly followed by a regional food shop (Regionalladen) closer to the residential area. Both the restaurant and the store make use of SpeiseGut's products by cooking and/or selling them directly to consumers. Taken together, the farm, the restaurant, and the shop are managed in connection with each other, allowing the farmer to respond and adapt to local conditions and therefore firmly establish his business.

The existing land-use limitations and the need for extra farm income confirm Oostindie et al.'s (2006: 57) stated motives of farmers to opt for MFA in peri-urban areas. Hence, the unsuccessful accumulation of land and the integration of a restaurant and a shop into the farm business indicate that the multifunctional character of SpeiseGut emerged as an outcome of land-use competition.

16.4.2.2 Offering Amenities for Urban Lifestyles

In addition to the production of organic and local food, agricultural land can also be used for other functions, such as consumption. According to Holmes (2006: 144), rural space is increasingly being used, or in his terms "consumed," by urban interests attracted by, e.g., leisure or rural lifestyle. Due to the proximity to cities, peri-urban areas have an advantage of being relatively accessible to urban consumers as in the case of SpeiseGut, which can be reached with public transport. One member specifically values SpeiseGut as an opportunity to engage with rural activities, without having to sacrifice the urban lifestyle (Walthall 2013: 70). Besides countryside experiences, urban consumers are particularly interested in the community and educational services provided by the CSA. One member reports that she feels anonymous and isolated in the urban environment and appreciates the sense of community created through SpeiseGut (Walthall 2013: 67, 76). Other members stated that they were seeking hands-on practices to get a better understanding in how to grow, process, and store food (Walthall 2013: 61).

To build a sense of community and to share knowledge, SpeiseGut offers several planting, weeding, harvesting, and food processing events, including common lunch breaks and social gathering after work. Moreover, throughout the year, SpeiseGut hosts several events in their restaurant HavelGut including book presentations and documentary screenings. Such activities provide regular face-to-face interaction, where farmers and consumers get to know each other and build a community. According to the farmer, these interactions generate transparency and trust within the CSA, which is essential to ensure a positive relationship (Walthall 2013: 43). Moreover, the farmer and members can test and improve different farming and marketing practices, which is essential for niche experimentation and the stabilization of SpeiseGut.

While farm activities help to build a community, they also provide knowledge. Hence, the farmer offers insights into agricultural production by allowing the members to stop by the field or by publishing pictures of the farm activities in newsletters and on social media. For all interviewed members, participating in a

CSA meant gaining practical skills, such as seeding, weeding, harvesting, sorting, washing, transporting, processing, and cooking the produce. Some members mentioned that they did not know all the vegetables and that they had to learn how to cook with them, but enjoyed the creative process of learning (Walthall 2013: 81).

In addition to the CSA, SpeiseGut established farm-to-school projects for educational purposes. Partnerships with local schools offer opportunities for children to learn about agriculture, ecology, nature, nutrition, and health at the farm (Walthall 2013: 71, 81).

Since 2014, SpeiseGut participates in the 2000m2 initiative—a civil society-driven campaign to reveal food system challenges resulting from current land-use patterns.[2] On the grounds of SpeiseGut, a separate "global field" of 2000m2 was cultivated for the public to reflect the proportions of available worldwide cropland and the actual land usage for food, feed, and fuel production. By comparing various land uses for food, feed, and fuel, the field illustrates how much (or little) arable land is actually dedicated for human food production, while disproportionate acreage goes toward animal feed and biofuel. Thus, with the 2000m2 project, SpeiseGut supports accessible knowledge for visitors and consumers by visualizing global land-use patterns and competition on a local scale.

By providing educational services, SpeiseGut responds to societal demands, reconnects urban lifestyles with agricultural practices, helps to build a community, fosters learning experiences, and contributes to the creation of healthy lifestyles. Consequently, SpeiseGut activities go beyond the sole production of food and fulfill consumption purposes to create a valuable land-use alternative in the peri-urban area of Berlin.

16.4.2.3 Providing Ecosystem Protection in Urban Areas

Changing societal concerns resulting from growing awareness of sustainability and preservation issues (e.g., sustainable resource management, biodiversity preservation, or landscape protection) facilitated the revaluation of human land use and led to the reappreciation of ecosystem benefits in agricultural production systems (Holmes 2006: 144 f.). Similar driving forces assisted in popularizing the CSA model. The CSA concept originated in the 1960s in Switzerland and Japan, but was popularized during the 1980s in the USA due to increased environmental awareness and concerns around food safety and health issues (van de Kop et al. 2008: 32, McFadden 2004). More recently, the CSA concept experienced growth not only in the USA, but also in European countries, including Belgium, France, Italy, and Germany.

SpeiseGut cultivates 30 different cultures and 70 varieties including a number of heirloom crops, keeps several bee colonies (Walthall 2013: 64), and manages an old apple orchard. Taken together, these agricultural activities do not only expand the

[2]For further information, see http://www.2000m2.eu/.

available food for urban consumers, but also opens possibilities for the farmer to enhance biodiversity, promote soil fertility, and protect the landscape in a peri-urban area. Seeing it from this perspective, the farmer of SpeiseGut clearly incorporates a role to manage protection space besides production and consumption space. Moreover, consumers can support the farmer in protecting parts of the peri-urban ecosystem by sponsoring the bees and/or apple trees. For that, SpeiseGut offers the adoption of bees starting at 75 euros/year or a tree for 85 euros/year. In turn, the sponsor could receive between 3 and 12 kg of locally produced honey and/or between 30 and 100 L of locally produced apple juice, depending on the sponsorship.

Besides protecting parts of the ecosystem in the peri-urban area of Berlin, SpeiseGut enhances sustainable resource management by harvesting all crops (regardless of size, color, or shape), promoting less transportation, using fewer cooled storage, and having less packaging material (Walthall 2013: 67, 76).

Within two years of operation, SpeiseGut proved to be an innovative and successful project, with 180 CSA memberships, regular customers at the restaurant and shop, as well as a number of bee and tree sponsors. The case study demonstrates that some, though not all, urban consumers are willing to pay the higher price for local and organic food, which in turn promotes biodynamic farming methods, resource management, and landscape protection provided by SpeiseGut.

While urban consumers express growing interests in the CSA, SpeiseGut continuously sought to expand the farmland in order to increase the production for ideally 500 memberships. In 2015, SpeiseGut received access to additional 6 ha of municipal land, managed by the landscape preservation association (Landschaftspflegeverband). When asking the farmer how he gained access to the land, he stated that the city recognized the increasing demand for a local food supply and SpeiseGut's contribution to the urban food system (personal conversation 2015). Since the farmer's land management supports vital ecosystem benefits and meets the requirements by the landscape preservation association, the farmer was eligible to farm on the municipal land.

SpeiseGut's farmland grew from 3 ha to 11 ha, which is a remarkable achievement, given the increasingly difficult circumstances of receiving access to farmland for small-scale farmers. The successful acquisition of additional land, as in the case of SpeiseGut, demonstrates that urban farming can become a competitive land-use option when incorporating multiple functions such as the supply of local food, offering amenities, and landscape protection.

Taken together, the various farm activities illustrate the emergence of new management arrangements, where the farmer uses land for production, consumption, and protection goals simultaneously. Key findings from the case study are summarized in the following (Fig. 16.1) to visualize the broad spectrum of strategies to strengthen the urban agriculture and city region food systems.

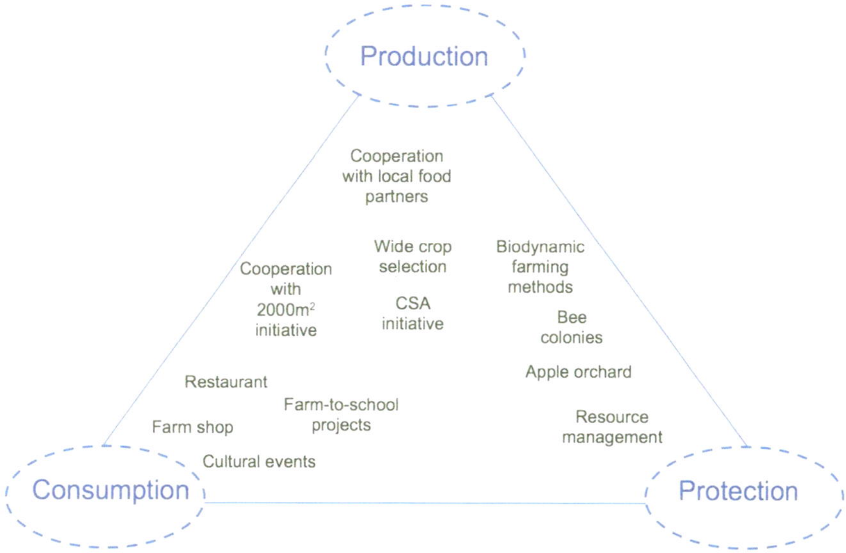

Fig. 16.1 SpeiseGut's initiatives and strategies, positioned according to the relative weights given to production, consumption, and protection goals (own illustration inspired by Holmes 2006)

16.5 Lessons Learned from SpeiseGut

The case study of SpeiseGut is used to illustrate the potential of MFA in short food supply chains to strengthen the urban agriculture and city region food systems. The most relevant findings will be discussed in the following sections.

16.5.1 Enhancing Synergies Between MFA and Short Food Supply Chains

To establish SpeiseGut in the city region food system of Berlin, the farmer used the CSA model as a starting point for his agricultural activities. As the CSA model favors a multifunctional character, synergies between short food supply chains and MFA could be exploited, improved, and expanded by integrating additional activities.

SpeiseGut enhances synergies between MFA and short food supply chains by producing value-added food for urban consumers, while contributing to a regionalized urban food system by reconnecting people and food spatially and promoting shorter distribution channels as well as generating local employment and income. The linkages are enabled through various projects such as the CSA initiative, the apple orchard, the bee colonies, the restaurant, and the farm shop and through cooperation with other local food partners.

Moreover, SpeiseGut creates mutually beneficial linkages between MFA and short food supply chains by offering amenities for urban lifestyles (e.g., sense of community, leisure, cultural events) and providing educational services (e.g., hands-on practices through CSA, farm-to-school projects, 2000m2 initiative) while facilitating direct face-to-face interactions between the producer and consumers and promoting consumer participation in agricultural activities.

Finally, SpeiseGut developed synergies by providing ecosystem protection (e.g., wide crop selection, biodynamic farming methods, beekeeping, landscape protection, resource management) while encouraging consumers to engage and support sustainable local food systems. Each individual project from SpeiseGut might come with its own risks and challenges, but taken together they compensate each other's risks and provide the necessary flexibility for experimentation surrounding agricultural activities.

After developing and improving synergies between MFA and short food supply chains within SpeiseGut, the farmer made evident that peri-urban agriculture is a valuable land-use alternative as it provides a wide spectrum of social, economic, ecological, and spatial benefits for urban areas, supplementary to its contribution to the local food supply. Further, the case study reveals that peri-urban agriculture in Berlin is increasingly being valued not only by farmers or urban consumers, but also by interest groups, local food partners, and city actors.

16.5.2 Creating New Networks

Networking is a key factor in fostering pioneering work, stimulating the development of innovative practices within short food chains, and facilitating transitions toward more self-reliant city region food systems.

The multifunctional character of SpeiseGut enabled the farmer to create a new network between the producers (farmer), consumers (CSA members and other consumers), interest groups (Landschaftspflegeverband, community gardens), land owner (city of Berlin), and other actors in the food chain (restaurants, cantinas, hotels, grocery stores). Additionally, SpeiseGut joins global networks such as Slow Food, Terra Madre, or the 2000m2 initiative. Such networks are essential to build the kind of support structures, which facilitate and maintain new forms of sustainable agriculture, as pointed out by Ostrom (1997: 87). This means the larger and more heterogeneous the network is, the more likely are new imperatives, paradigms, and practices being stabilized and established.

In summary, the case study illustrates that the strengthening of urban agriculture and short food supply systems is based on the creation of networks and linkages between various actors and different elements of the food system. Here, multifunctionality provides an important juncture for multiple projects and stakeholder arrangements. Consequently, MFA proves to be a viable approach to create the required networks and linkages to (re)create a more self-reliant city region food system.

16.6 Closing Remarks

It should be stated that not every CSA achieves such viability as presented in the case study. In contrast, many CSAs are economically quite vulnerable, and farmers often work long hours to earn a living wage (Stagl 2002: 157) and might fail to compete with non-agricultural land uses in the urban fringe. Hence, this article presented a promising approach by making use of synergies between MFA and short food supply chains combined in the CSA model. The integration of various projects into SpeiseGut's activities improved the economic viability while providing a wide spectrum of social, economic, ecological, and spatial benefits for urban areas, which in turn increased the value of the land usage.

Yet, another critique often proposed toward the CSA model in previous literature is the exclusive character of a CSA. While the CSA community offers several social benefits for members, it is also rather fashionable for a certain group of people who are already aware of unsustainable agriculture and food systems, have a compatible lifestyle to participate, and/or have the financial resources to afford a membership (Walthall 2013: 79).

Regardless of its exclusivity, the CSA model may help to reduce the ecological footprint of a city like Berlin, which is heavily dependent on a global food system. For instance, shorter food miles, less packaging, and reduced food waste contribute to lower greenhouse gases. Although it is arguably unrealistic and unfeasible to feed a city of 3.5 million inhabitants merely based on a local food production, approximately 160,000–200,000 people could benefit from a local food supply when considering that Berlin has about 2.000 ha of arable land (Statistische Ämter des Bundes und der Länder 2011: 46).[3]

Yet, some food system scholars have discussed the contradictions of radicalized local food movements and criticized the assumption that the local food scale is inherently good, more socially just and ecologically sound (Born and Purcell 2006). Referring to the misleading assumption as a local trap, the authors (2006: 195) suggest that "there is nothing inherent about any scale" and that local food systems may be "equally likely to be just or unjust, sustainable or unsustainable, secure, or insecure." After all, the argument is not about an either/or declaration, but that shorter supply chains provide a counterbalance to the global food system and in turn can contribute to a more resilient city region food system.

---

[3]Equation is based on SpeiseGut's capacity to supply 80 to 100 people/ha with fruit and vegetables.

Based on the results from case study (CSA SpeiseGut in Berlin, Germany), this chapter shed light on the feasible options for peri-urban land use and possible pathways to strengthen urban agriculture and city region food systems. Arguably, MFA combined with short food supply chains is clearly not a one-fit-for-all solution. Thus, it (re)links productive activities to the consumers in urban areas, reduces the vulnerability of global food supply systems, contributes to the urban socioeconomic and ecological system, and provides a list of choices addressing patterns of land-use competition.

References

Boody, G. (2008). Multifunctional agriculture: more than bread alone. *BioScience, 58*(8), 763–765.

Born, B., & Purcell, M. (2006). Avoiding the local trap. scale and food systems in planning research. *Journal of Planning Education and Research, 26*, 195–207.

Brinkley, C. (2012). Evaluating the benefits of peri-urban agriculture. *Journal of Planning Literature, 00*, 1–11.

Bryant, C. R., & Johnston, T. R. R. (1992). *Agriculture in the city's countryside*. London: Belhaven Press.

Cavailhés, J., & Wavresky, P. (2003). Urban influences on periurban farmland prices. *European Review of Agricultural Economics, 30*(3), 333–357.

Cicia, G., Colantuoni, F., Del Giudice, T., & Pascucci, S. (2011). Community supported agriculture in the urban fringe: empirical evidence for project feasibility in the metropolitan area of naples (Italy). *International Journal on Food System Dynamics, 2*(3), 326–339.

BMELV (Federal Ministry of Food, Agriculture and Consumer Protection) (2010). *German agriculture: Facts and figures*. https://www.bmel.de/SharedDocs/Downloads/EN/Publications/GermanAgriculture.pdf?__blob=publicationFile.

Food and Agriculture Organization of the United Nations (FAO) 2015 *Food for the Cities*. http://www.fao.org/fcit/fcit-home/en/.

Groh, T., & McFadden, S. (1997). *Farms of tomorrow revisited: community supported farms—farm supported communities*. Kimberton (PA): Biodynamic Farming and Gardening Association.

McFadden, S. (2004). The history of community supported agriculture, part i. community farms in the 21st century: Poised for another wave of growth? http://www.newfarm.org/features/0104/csa-history/part1.shtml.

Heimlich, R. E. & Brooks, D. H. (1989). Metropolitan growth and agriculture: Farming in the city's shadow. *Agricultural Economic Report* No. 619, Economic Research Service, U.S. Department of Agriculture, Washington, DC.

Heimlich, R. E. & Anderson, W. D. (2001). Development at the urban fringe and beyond: Impacts on agriculture and rural land. *Agricultural Economic Report* No. 803, Economic Research Service, U.S. Department of Agriculture, Washington, DC.

Hinrichs, C. C. (2000). Embeddedness and local food systems: Notes on two types of direct agricultural market. *Journal of Rural Studies, 16*, 295–303.

Holmes, J. (2006). Impulses towards a multifunctional transition in rural australia: Gaps in the research agenda. *Journal of Rural Studies, 22*, 142–160.

King, C. A. (2008). Community resilience and contemporary agri-ecological systems: reconnecting people and food, and people with people. *Systems Research and Behavioral Science, 25*, 111–124.

Koc, M., MacRae, R. & Mougeot, L. J. A. (1999). For hunger-proof cities. sustainable urban food systems. International Development Research Centre, Ottawa (ON).

Lovell, S. T. (2010). Multifunctional urban agriculture for sustainable land use planning in the united states. *Sustainability, 2*, 2499–2522.
Mougeot, L. J. A. (2000). *Urban agriculture: definitions, presence, potentials and risks, and policy challenges*. Report No: International Development Research Center. 31.
Oostindie, H., Roep, D., & Renting, H. (2006). Definitions, references and interpretations of the concept of multifunctionality in the Netherlands. *European Series on Multifunctionality, 10*, 41–81.
Ostrom, M. (1997). Community farm coalition. In T. Groh & S. McFadden (Eds.), *Farms of Tomorrow Revisited: Community Supported Farms– Farm Supported Communities* (pp. 87–102). Kimberton (PA), USA: Biodynamic Farming and Gardening Association.
Renting, H., Rossing, W. A. H., Groot, J. C. J., Van der Ploeg, J. D., Laurent, C., Perraud, D., et al. (2009). Exploring multifunctional agriculture. a review of conceptual approaches and prospects for an integrative transitional framework. *Journal of Environmental Management, 90*, S112–S123.
Renting, H. & Dubbeling, M. (2013). Synthesis report: Innovative experiences with (peri-) urban agriculture and urban food provisioning—lessons to be learned from the global south. Thematic paper 3, RUAF Foundation.
Robinson, G. M. (2004). *Geographies of agriculture: Globalisation*. Pearson Education Limited, Harlow: Restructuring and Sustainability.
Stagl, S. (2002). Local organic food markets: potentials and limitations for contributing to sustainable development. *Empirica, 29*, 145–162.
Statistische Ämter des Bundes und der Länder. (2011). Agrarstrukturen in Deutschland Einheit in Vielfalt. Regionale Ergebnisse der Landwirtschaftszählung 2010. Stuttgart.
Steel, C. (2008). *Hungry city: How food shaped our lives*. London: Vintage.
Van de Kop, P., Nijhof, K., Kloen, H., & Braun, A. (2008). community supported agriculture: An alternative local food system. *Leisa Magazine, 24*(1), 32–34.
Van der Ploeg, J. D., Long, A., & Banks, J. (2002). *Living countrysides: Rural development processes in europe: The state of the art*. Doetinchem: Elsevier.
Veenhuizen, R. & van Danso, G. (2007). Profitability and sustainability of urban and peri-urban agriculture. agricultural, *Marketing and Finance Occasional Paper* 19, Food and Agriculture Organization of the United Nations, Rome.
Walthall, B. (2013). Combining food and the city: community supported agriculture (CSA) as an impetus for integrating food and urban planning. master's thesis. Humboldt-Universität zu Berlin. Unpublished.
Wild, S. (2012). Sich die Ernte Teilen….Einführung in die Solidarische Landwirtschaft. Printsystem Medienverlag, Heimsheim.
Wilson, G. A. (2007). *Multifunctional agriculture: A transition theory perspective*. Wallingford: CABI International.
Hussain, Z., & Hanisch, M. (2014). Dynamics of peri-urban agricultural development and farmers' adaptive behaviour in the emerging megacity of hyderabad, india. *Journal of Environmental Planning and Management, 57*(4), 495–515.
Zasada, I. (2011). Multifunctional peri-urban agriculture—a review of societal demands and the provision of goods and services by farming. *Land Use Policy, 28*, 639–648.
Zasada, I. (2012) Peri-urban agriculture and multifunctionality: Urban influence, farm adaptation behaviour and development perspectives. Dissertation. Technische Universität München.

Chapter 17
Agribusiness and Family Farming in Brazil: Competing Modes of Agricultural Production

Andreas Mayer

Abstract Brazil has become one of the largest exporting countries of soy, sugarcane, maize, oranges and coffee for the global market. The export-oriented agribusiness model is embedded into a neo-extractivist development model, where state revenues are largely based on exports of primary products from agriculture and mining. While export crops are mainly produced by large-scale agribusiness, traditional staple crops such as cassava, rice or beans, which still provide an important contribution to the diets of Brazilians, continue to be produced by small-scale family farming. The concurrence of both farming modes and the commodification of new land for the production of export crops potentially lead to competition for resources and markets, thus reinforcing the strong inequality in landownership in Brazil. Based on arguments from scientific literature and empirical data, I explore this potentially competitive process, based on land-use expansion patterns for soy and sugarcane (agribusiness), and cassava and beans (family farming). While processes of competition cannot be directly inferred from this analysis, it provides insights into hot spots of dynamic changes in land use and informs local case studies, which are required to deepen the understanding of possible land-use competition.

Keywords Landownership · Governance · Export · Social justice · Commodification

A. Mayer (✉)
Institute of Social Ecology, Alpen-Adria Universitaet Klagenfurt-Wien-Graz,
Vienna, Austria
e-mail: Andreas.Mayer@aau.at

J. Niewöhner et al. (eds.), *Land Use Competition*,
Human-Environment Interactions, DOI 10.1007/978-3-319-33628-2_17

17.1 Introduction—The Increasing Commodification of Land in Brazil

Brazil, the largest country and economic power in Latin America, has exhibited strong economic growth in the first decade of the twenty-first century. On the one hand, public policies and investment schemes supported production for exports to generate revenues, in particular agricultural products and metallic minerals. The abundant state revenues were partly used to finance social welfare programmes and lifted significant parts of the rural population out of extreme poverty. The revenues were also used to better connect remote areas to urban centres in the South-East and South and to advance large infrastructure projects. The economic benefits were undeniable and contributed to the fact that Brazil is now listed among countries with high human development (United Nations Development Programme 2016). On the other hand, large parts of the economy remain oriented towards local markets and subsistence production. These production systems still provide an important share of domestic staples in Brazil and are thus an integral part of local food security and food sovereignty.

Concerning land use and the production of biomass, the export-oriented development model is reflected by large export volumes of soy, sugarcane and derived products. Soy and sugarcane are the two dominant crops that are produced by agribusiness in Brazil (Lapola et al. 2014), and these crops are also among the most important traded crops internationally. Soy exports increased eightfold between 1990 and 2011 (Kastner et al. 2014), soy production grew fourfold (to nearly 75 Megatons in 2011) and sugarcane production tripled during the same period with a total production of 734 Mt in the same year (Food and Agriculture Organization of the United Nations 2014). The revenues from export farming contributed to a significant reduction of the Brazilian trade deficit since the beginning of the twenty-first century, the time when the workers' party (partido dos trabalhadores, abbreviated PT) rose to power. In 2010, the financial surplus from agricultural trade (exports minus imports) accounted for nearly 60 billion US$, around 3 % of the overall GDP. This export-oriented, mechanized large-scale agricultural production is a typical high input–high output system, based on specific crop varieties (some of them genetically modified) and massive inputs of fertilizers and the use of pesticides. Brazil has overtaken the USA as the largest user of agrotoxics in 2008 (Grain et al. 2014). Soy and sugarcane are also important flex-fuel crops, which are—depending on the policy and economic environment—used as animal feed and agrofuels, or are processed to products such as sugar or oils.

Globalization has led to an increasing spatial disconnect between the land where biomass is grown and the places where these agricultural products are consumed (Kastner et al. 2014; Meyfroidt et al. 2010). Export crops have also dominated the expansion of agricultural lands. Kastner et al. (2014) show that at the global scale, the land used to produce agricultural exports grew around 50 % between 1987 and 2008, at a much higher rate than croplands for domestic use which remained virtually constant. This development is even more pronounced in Latin America, where the share of land devoted to exports doubled, and in 2008 accounted for nearly half of the total agricultural lands (Kastner et al. 2014), with Brazil and Argentina as the main agricultural exporters (Schaffartzik et al. 2014).

While the majority of newly commodified land is used for exports, staple crops remain important for local markets. Family farming was responsible for more than a third of the total value generated by agriculture in 2010. The importance of smallholder farming in Brazil has also been confirmed by the last agricultural census from 2006. Family farming still provides a large part of food for Brazilians, with 87 % of the domestic production of cassava, 70 % of (e.g. black) beans, 46 % of corn, 34 % of rice or 58 % of milk (IBGE 2012). It also generates more jobs than industrial farming: while small-scale farming using mixed cropping systems generates one job per 8 hectares cultivated, large-scale mechanized monocultures generate only one job per 67 ha (Altieri and Koohafkan 2008). Currently, 12.3 million workers or 74.4 % of the total workforce in rural areas are employed in family farming. Thus, it is clear that a shift towards large-scale agribusiness will reduce workforce demand for agriculture tremendously (Li 2011).

These two agricultural development models are even mirrored by the parallel existence of two-state institutions (MAPA and MDA), one concerned with agribusiness and the other one concerned with smallholders. The ministry of agriculture, livestock and supply (Ministério da Agricultura, Pecuária e Abastecimento, MAPA) is the governmental institution that lays out the framework for agribusiness in Brazil. MAPA is responsible for public policies to stimulate the development of agriculture and the promotion of agribusiness, as well as the regulation and standardization of services related to the industry. Agribusiness includes small, medium and large farmers, and the whole life product chain from agricultural production, processing and manufacturing to the distribution to the end consumers (MAPA 2016). The main goal of the production of soy and sugarcane is to provide flexible raw materials for a variety of different uses. The original primary goal of agricultural production—providing food security—has become a lesser concern. Critical scientists and social movements critique that these crops are important drivers for social exclusion and urban migration (Wolford 2010).

The Ministry of Agrarian Development (Ministério do Desenvolvimento Agrário, MDA) has been established in 1999 and is responsible for small-scale, family and subsistence farming (MDA 2016). The first pillar of MDA is to oversee the land reform in Brazil and redistribute and demarcate land and land titles. Land tenure in Brazil is highly unequal, with agribusiness owning only 10 % of all farm land titles (i.e. official records of landownership), but 75 % of all agricultural lands, while (family run) smallholders own only 25 % of the agricultural land (IBGE 2012; Abbey et al. 2006). Access to microcredits is the second pillar to strengthen family farming. The Brazilian government distributed around 7.6 billion US$ in 2008/2009 towards family farms through the programa nacional de fortalecimento da agricultura familiar (PRONAF) to strengthen food production of poor rural producers. Various other programmes exist that stimulate demand for family farming products through, for example, public procurement policies.

17.2 A Development Model in Crisis

At the time of writing this manuscript, two iron mining dams in Minas Gerais burst, the tailings entered the Rio Doce and ended up in the Atlantic Ocean. While this accident is not caused by agricultural production, it is related to a development model that is based on the exploitation and commodification of natural assets. Be it mining or agriculture, both are increasingly conducted as highly mechanized, large-scale businesses, where economic survival is based on high world market prices and foreign demand. While prices for iron, soy, sugar and other important Brazilian export goods were relatively high during the first decade of the twenty-first century, they have been falling since 2012, which may be one reason behind the massive economic recession that keeps the Brazilian economy under siege. GDP growth rates nearly came to a standstill, from 3.9 % in 2011 to only 0.1 % in 2014 (The World Bank 2015). Prices for soy meal, sugar or oranges—important agricultural export goods in Brazil—decreased constantly since 2012, reaching values in 2015 that were 40 % lower than 3–4 years before (Index Mundi 2015). Tens of thousands of Brazilians are gathering in the streets to show their anger against the political system that has increasingly failed to provide adequate social security, public health and education (Douglas 2015). Accordingly, this model of development, which has been praised as a role model for countries in the global South, has come under increasing pressure, and the impending signs of failure are difficult to ignore.

Such development models, popular throughout Latin America, have been critically analysed as extractivism or, in its particular current form, as neo-extractivism (Bunker 1985; Bunker and Ciccantell 2005; Burchardt and Dietz 2014; Gudynas 2010; Svampa 2012; Yates and Bakker 2013). (Neo-)extractivism is understood as a development model where states make use of their sovereign rights to exploit their own natural resources, export primary commodities, often associated with a commodification of the last remaining natural assets in territories that were considered as "unproductive". At the same time, governments increase their control over these businesses and foster a more inclusive development by using the revenues for social programmes. Thus, neo-extractivist development models depend on the revenues from resource extraction activities and redistribute parts of these revenues through social welfare programmes (i.e. bolsa familía or fome zero. See discussions on the compensatory state from, e.g. Webber 2014). It is argued that the wider goal was to stimulate economic growth and create employment in other, non-extractive sectors. The main difference of neo-extractivist policies in comparison with the extractivism of the past decades is then to use the generated revenues to overthrow this very same development model and ultimately overcome extractivism.

However, the way Brazil and other countries have been implementing this political path has linked the exploitation of natural resources to the destruction of the environment and the livelihoods of local communities. This has likely contributed significantly to a widening of the gap between political and economic elites and increasingly marginalized communities. The establishment of resource enclaves and the expulsion of local communities, as well as the centralization of political power by an elite who easily controls the abundant natural assets, are negative impacts of such policies. Local populations cannot use the land dedicated to export production, and their access to land and water resources is hampered. Fertilizers, pesticides or other substances used through the extraction and processing of natural resources often pollute water resources. Furthermore, extractive industries often only create job opportunities, which either require highly qualified specialists or are poorly paid. Ultimately, the dependency on world market prices exposes countries to factors beyond the limits of national jurisdiction and may lead to what has been described as development of underdevelopment (Brand and Dietz 2014; Bunker 1985).

Critical scientists and civil rights movements have raised two further debates that underline that the export-oriented development model in Latin America is under pressure: food sovereignty and land grabbing (Borras et al. 2012b; Chaifetz and Jagger 2014; De Schutter 2011; Laroche Dupraz and Postolle 2013). The term food sovereignty was coined by members of the social movement of peasants La Via Campesina in 1996 and describes the right of people to choose their food system. While this definition does not exclude large-scale export-oriented production per se, most proponents of food sovereignty put the peasant communities and populations who produce, distribute and consume food at the core of food systems and policies, rather than the corporations and market institutions that dominate the global food system. Brazil's Landless Workers Movement (movimento dos trabalhadores rurais sem terra, MST) explicitly lists food sovereignty as an overarching framework for agrarian reform and rural development and uses this concept to critique industry led high input and export-oriented production (MST 2006). Land grabbing in Latin America refers to the fact that "recent land investments have consolidated the earlier trend away from (re)distributive land policies in most countries in the region, and are likely to result in widespread reconcentration of land and capital" (Borras et al. 2012a). This trend contradicts scientific findings that smallholders are the key driving force for poverty reduction and rural development (Pokorny et al. 2013; Wiggins 2009).

17.3 Competing Development Models

The Brazilian government has supported the export-oriented agribusiness model as well as family farming through the creation of MAPA and MDA. While agribusiness mainly generates monetary revenues on global markets and provides cheap food for the cities, smallholders mainly generate food for local markets and for subsistence. However, competition between these two modes of agricultural production occurs at different levels: smallholders and agribusiness are both competing for fertile lands suitable for agriculture, are receiving governmental credits and participate in local, regional and global markets, where they compete for demand. While the latter two are not central for this chapter, I here focus on the expansion and commodification of agricultural lands in Brazilian municipalities.

I try to shed some light on this question by drawing on arguments from reports and scientific literature and by exploring how areas used by agribusiness and smallholders have developed in different regions in Brazil between 1990 and 2010. These two contrasting modes of agricultural production are a matter of increasing debate in Brazil (Abbey et al. 2006; Lapola et al. 2014; Martinelli et al. 2010b), and the mounting criticism against an export-oriented development might be a sign that in fact these models are at least partly competing over the same resources.

Olivier De Schutter, the United Nations special rapporteur on the right to food, describes problematic mechanisms linked to competition between smallholders and agribusiness in the context of land grabbing (De Schutter 2011). From a certain perspective, the development of large agroindustrial estates can be seen as an opportunity to modernize agricultural production, to produce large volumes through mono-cropping and to increase competitiveness on regional and international markets. This mode of agricultural production, De Schutter refers to it as the Brazilian farm, is problematic once "agriculture is placed in the broader context of rural development and poverty alleviation" (2011). Allocating marginal lands to investors for the development of large-scale plantations instead of distributing it among local farming households may impact negatively on local food security and food sovereignty. This competition between smallholders and agribusiness for suitable lands can be avoided by integrating smallholders into global commodity markets, e.g. through contract farming or farming cooperatives. However, smallholder products are often more expensive than the produce from large-scale plantations who can produce large volumes at relatively low costs, thanks to the economies of scale. This is the main reason why larger production units retain competitive advantages over smallholders (De Schutter 2011). Commentators taking on a governance perspective, however, have argued that countries (or subnational units) with high levels of social instability and weak public sector institutions are unlikely to benefit from world market orientation (Kosacoff et al. 2008). This raises two closely related questions: is the coexistence of these two farming types on the same territory possible? If so, how can competition be regulated such that small-scale farming is not structurally disadvantaged leading in every case to the rural economy being dominated by export-oriented, large-scale farming (De Schutter 2011).

The differentiation between these two farming modes is typically made by the size of the respective properties. Godar et al. (2014), for example, group farms from smallholders (farm size <100 ha) to very large landholders (farm size >2.500 ha). However, these data are only available in digitalized form for every Brazilian municipality for the year 2006, which precludes the analysis of changes over time. To allow for such an intertemporal analysis, I used data for crops closely linked to a particular mode of production. In Brazil, soy and sugarcane are mostly produced by export-oriented agribusiness, while beans and cassava are mostly produced by smallholders. Soy and sugarcane and their derived products (such as raw or refined sugar or soy cake) are the most important agricultural exports of Brazil in terms of their weight. Despite recent claims that soy can be produced profitably on smaller-sized farms, the majority of soy and sugarcane are grown on larger farms

(Galeano 2012; Greenpeace 2006; Martinelli et al. 2010a; Weinhold et al. 2013; Wilkinson et al. 2012). In contrast, beans and cassava are important staple crops in Brazil and mainly produced by smallholders (Lapola et al. 2014; Martinelli et al. 2010a). As these crops are also linked to particular uses, i.e. food, feed or fuel, this framework allows to explore competition between these uses for a given time period as well as changes over time.

Table 17.1 provides an overview of the four selected crops in terms of the main mode of production in which they are produced (i.e. smallholder farming or agribusiness), and how they are used, i.e. as food for direct human consumption, as livestock feed or for other purposes, such as the production of agrofuels and other energy carriers. Soy and sugarcane are mostly exported, and significant amounts are used for other purposes than food. In contrast, beans and cassava are both mostly used as food in Brazil. In the following analysis, I look at the development of harvested areas of the selected crops as a share of municipal areas over the last two decades, focusing on the decadal time steps 1991, 2001 and 2011. In order to eliminate annual variations in harvested area, 3-year means around the respective years (e.g. 1990, 1991 and 1992 for 1991) are used, as given by agricultural census data (for sources, see Table 17.1).

Figures 17.1a–f show six maps with the share of either export crops (17.1a–c) or domestic staples (17.1d–f) for every Brazilian municipality (n = 5554), where at least at one point in time (1991, 2001, or 2011) an area larger than 1 % of the total municipal area was used to grow soy and sugarcane or cassava and beans. Overall, export crops were less scattered over Brazil than domestic staples during the whole period under observation. While soy and sugarcane occupied more than 1 % of the total municipal area in nearly half of all Brazilian municipalities, cassava and beans took up more than 1 % of the total area in almost two-thirds of all municipalities. Export crops usually are grown in larger slots and require large infrastructure to transport them from the agricultural areas where they are produced to facilities where they are further processed. They also often have a close proximity to ports to ship them overseas. The latter was of less importance in the expansion over the last decade, as remote areas in Mato Grosso became the hot spots of soy expansion (See Fig. 17.1c).

Table 17.1 Main classification and characteristics of the four selected crops

	Soy	Sugarcane	Cassava	Beans
Agribusiness share (US$)	84 %	High	16 %	33 %
Smallholder share (US$)	16 %	Negligible	84 %	67 %
Area harvested (1000 ha, 2010)	23.327	9.076	1.789	3.423
Total production (1000 t, 2010)	68.756	717.463	24.967	3.158
Export share	50 % of production	High	Negligible	Negligible
Food use	Partly	50 %	Mostly	Mostly
Fuel use	<10 %	50 %	Negligible	None
Feed use	Mostly	Negligible	Partly	Negligible

Sources (Banco do Nordeste 2015; Food and Agriculture Organization of the United Nations 2015; Price Waterhouse Coopers 2013; Scholtes 2009; Soybean and Corn adviser 2015)

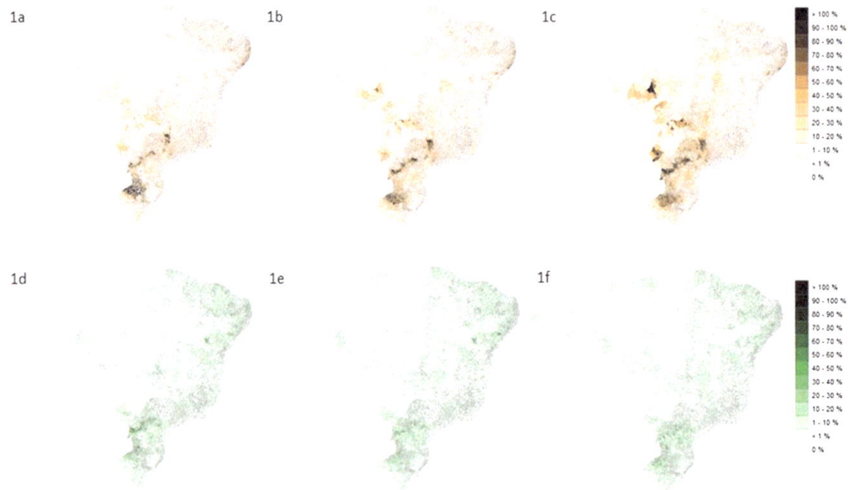

Fig. 17.1 Share of harvested area of export crops (17.1a–c) and domestic staples (17.1d–f) in total agricultural area per municipality in 1991, 2001, and 2011. *Light colours* indicate low shares; *dark colours* high shares. *Source* IBGE, maps by Susan Haertig (www.susanhaertig.com)

Domestic staples, mostly produced in smaller lots and destined for (rural) local markets, do not need such an extensive infrastructure. Shorter distribution networks are an important difference between export crops and domestic staples.

In 1991, the first year of the analysis, export crops were concentrated in the South, the region with the longest record of commercial agricultural production in Brazil. Santa Catarina, Paraná and São Paulo provided by far the largest share of export goods, while there was only minor production in the Brazilian North-East at that time. The Amazon and significant parts of the Cerrado (except Mato Grosso, which already showed first signs of soy expansion) were mostly untouched by the production of soy and sugarcane. The map for 2001 shows that export crops have already reached the fringes of the Cerrado and the Amazon, a trend which continued in the following decade. Until 2011, Mato Grosso and Mato Grosso do Sul became the main soy-producing regions, while São Paulo consolidated its role as the major producer of sugarcane in Brazil. Mato Grosso, especially, is also a hot spot of uneven land distribution. In 2010, 84 % of its soy, which is the main agricultural product in this state, was produced by large-scale plantations. The patterning forms an arch of export crop expansion in the West of Brazil, starting from Paraná and heading upwards to Mato Grosso and the Amazon in Rondônia, along the Pantanal and along the border to Paraguay. Unclear land tenure and high commodity prices incentivize the expansion of agribusiness, often at the cost of smallholders or areas that belong to indigenous groups, such as the Guaraní-Kaiowá people (Frayssinet 2012). Increasing deforestation rates in these areas since 2013 are likely to be linked to the expansion of the agricultural frontier, as the Brazilian environment minister Izabella Teixeira stated: "[…] in Mato Grosso […] pressure for more logging is again strong and coming from agriculture and livestock activities" (Watts 2015). These results are supported by Godar et al. (2014), who analysed the spatial distribution of dominant farm sizes in the Amazon biome.

In 1991, the production of beans and cassava, both important domestic staple crops, was spread across all populated regions in Brazil, with the obvious exception of urban centres. The most significant areas were situated in the consolidated agricultural regions in the South, along the borders to Argentina and Paraguay. Yet, the North-East also exhibits a remarkable density of beans and cassava, except in Alagoas, where sugarcane was by far the most important agricultural crop produced. In 2001, a different pattern begins to emerge. Firstly, a strong contraction of staple crop production can be observed in the South. A comparison with Fig. 17.1b shows that the contraction happened in the very same regions where export crops expanded, thus indicating a trajectory from the production of domestic staples towards export crops. Secondly, a slight contraction is observable in the North-East. However, as there was no remarkable expansion of export crops in the same region, other factors such as cultural and lifestyle changes are likely to have driven this development.

In the South, this trend of contracting areas for staple crops continued until 2011, with the exception of some municipalities where an area increase for staple crops could be observed. This expansion might indicate a growing spatial concentration of the production of beans and especially cassava, the latter being increasingly commodified to be sold as flour or as animal feed to large swine and poultry businesses in the region. In Tocantins, a state at the fringes of the Amazon, export crops expanded in the last decade, as well as along the coastline in the North-East. Staple crops did not show such a massive expansion, as the land occupied remained relatively constant in Tocantins, and the areas only grew slightly in the coastal states in the North-East. In the latter, however, a shift of staple crops away from the coast is emerging. This might indicate, together with the slight increase of export crops in the last decade until 2011 in the same region, competition for fertile lands between smallholders and agribusiness. Nevertheless, qualitative and local case studies are necessary to deepen the understanding of these processes of competition.

The areas for beans and cassava have only expanded in one-fifth of all municipalities between 1991 and 2011 (n = 1178), while in the majority of municipalities the area remained virtually constant. Export crops, however, clearly exhibited a different pattern. In more than half of all municipalities where soy and/or sugarcane were grown in 1991, the area expanded until 2011 (1250 out of 2355 municipalities), and in nearly one third of municipalities the area remained unchanged. Table 17.2 lists the number of municipalities where areas for export crops expanded, and staple crops showed the following pattern: an expansion of domestic staples (I), constant areas for domestic staples (II), less area in 2011 (III) and municipalities where no domestic staples were grown (IV). The latter three categories make up nearly 90 %, while data indicate that only in exceptional cases does export crop expansion also drive a similar expansion of domestic staples in the same municipalities. Table 17.2 also clearly underlines the expansive pattern of export crops in the South of Brazil (together with little expansion of domestic

Table 17.2 Number of municipalities with export crop types occupying larger area per municipality in 2011 than in 1991 and domestic staples occupying larger (pattern I), the same (pattern II), less (pattern III) or no area at all (pattern V) per federal state

Federal State	total municipalities	Pattern I Export and Staple +	Pattern II Export + Staple o	Pattern III Export + Staple -	Pattern IV Export + no Staples
RO	52	0	2	2	2
AC	22	1	0	0	0
AM	62	0	0	0	0
RR	15	0	0	0	0
PA	142	1	0	2	1
AP	16	0	0	0	0
TO	139	0	1	0	7
MA	217	6	0	3	7
PI	224	1	1	0	1
CE	184	1	0	2	0
RN	167	1	1	4	0
PB	222	3	2	8	0
PE	184	0	10	12	1
AL	102	2	2	4	2
SE	75	0	2	5	0
BA	417	1	2	1	0
MG	853	16	22	19	24
ES	77	0	2	1	0
RJ	89	0	1	1	2
SP	639	25	72	84	142
PR	399	81	105	109	38
SC	293	7	43	27	6
RS	497	22	46	44	60
MS	78	3	4	6	10
MT	141	2	1	2	28
GO	246	12	13	22	38
DF	1	1	0	0	0

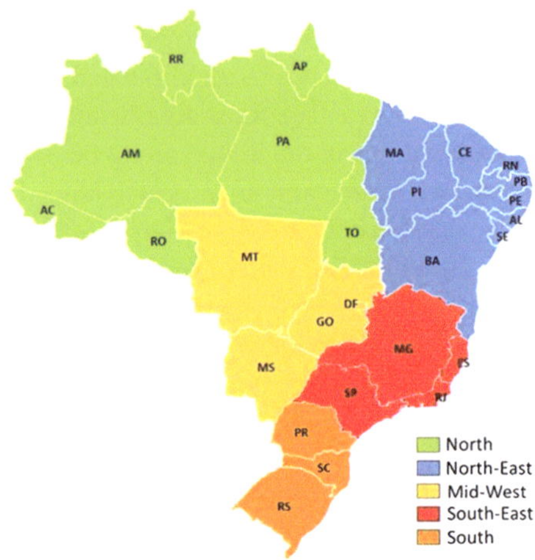

Colours indicate main regions (Grande Região) in Brazil, with *green* North, *blue* North-East, *yellow* Mid-West, *red* South-East, *orange* South. Region codes: *RO* Rondônia; *AC* Acre; *AM* Amazonas; *RR* Roraima; *PA* Pará; *AP* Amapá; *TO* Tocantins; *MA* Maranhão; *PI* Piauí; *CE* Ceará; *RN* Rio Grande do Norte; *PB* Paraíba; *PE* Pernambuco; *AL* Alagoas; *SE* Sergipe; *BA* Bahia; *MG* Minas Gerais; *ES* Espírito Santo; *RJ* Rio de Janeiro; *SP* São Paulo; *PR* Paraná; *SC* Santa Catarina; *RS* Rio Grande do Sul; *MS* Mato Grosso do Sul; *MT* Mato Grosso; *GO* Goiás; *DF* Distrito Federal. Data *Sources* see text, map: "Brazil Labelled Map" by João Felipe C.S., licensed under CC Attribution-Share Alike 3.0 Unported (accessed 01/2016)

staples in the same municipalities), and that the newly commodified land in the Mid-West (together with Tocantins) is nearly exclusively used to grow export crops (pattern IV). Area expansion of domestic staples with no growth of export crop areas was observable in only 2 % of all municipalities in the latter region. In the North-East, however, the area for domestic staples increased in 26 % of all municipalities. This result supports the findings that there might have been a shift of domestic staples away from the coast. A better understanding of the expansion of export crops and simultaneous displacements of domestic staples require further investigations of the social, political and economic drivers of these patterns.

17.4 The Continued Restructuring of Brazilian Agriculture

Large-scale, export-oriented agriculture started its conquest in the South of Brazil, expanded to the Mid-West and reached the fringes of the Amazon biome, where export production was identified as a main driver for deforestation (DeFries et al.

2010). Next to the Amazon, the biodiversity-rich Cerrado was also a hot spot of land conversion. In 2005, half of the approximately 2 mio km^2 have been converted into agricultural lands for livestock and soy plantations (Klink and Machado 2005). Next to deforestation and biodiversity loss, this expansion might also be associated with rural migration and urban sprawl (Barretto et al. 2013).

In this chapter, I have discussed the two main modes of agricultural production in Brazil and explored tentative evidence of competition between these two modes. The production of soy and sugarcane, two crops that are closely linked to export-oriented agribusiness in Brazil, has expanded in terms of number of municipalities where they are grown (+50 %), as well as in terms of the share of the agricultural area per municipality (in 53 % of municipalities this area increased). Staple crops such as beans and cassava have shown a distinct and different pattern. While in the first decade after 1991, the number of municipalities where these crops were grown remained constant, it contracted by nearly 20 % until 2011. Furthermore, the area cropped by beans and cassava increased in only 22 % of all municipalities. In the South of Brazil, the area used to grow export crops increased significantly, while areas for the production of beans and cassava remained constant. In the Mid-West, a hot spot of agricultural expansion, soy area increased massively, mostly into lands hitherto untouched by export-oriented agriculture, raising various social and environmental problems. While this does not imply a direct competition with smallholder farming, indirect effects may occur, as well as conflicts with indigenous people into whose lands the expansion of agribusiness has taken place. Although expulsion effects through the expansion of export-oriented agribusiness of Brazil cannot be inferred from the data presented in this chapter, scientific literature, as well as non-scientific media, and NGO reports suggest that the expansive pattern of export crops and the contracting pattern of domestic staples are linked (Lapola et al. 2014; Scholtes 2009; Wolford 2010).

I have used beans and cassava as examples of typical domestic staple crops mostly produced by smallholders or family farming, and soy and sugarcane as examples of typical export crops mostly produced by large-scale and highly mechanized plantations. However, agricultural practices for the same crops differ widely and range from small-scale, labour-intensive low input–low output systems to capital intensive, large-scale mono-cropping systems. Typical export crops also contribute an important share to domestic diets. For example, while Brazil is the most important exporter of sugar worldwide and has massively increased exports of sugarcane ethanol to Europe, South Korea, and the USA, the largest share of sugarcane is still used for domestic consumption, either as ethanol or as sugar. Additionally, sugarcane plantations in São Paulo and Alagoas are quite different in terms of the size of agricultural plots or yields. Soy is also a considerable feedstock for the Brazilian livestock sector (whose products are also important export goods). Finally, cassava is increasingly produced in large monoculture plantations in some regions, especially in the industrialized South, where it is used to produce flour and animal fodder. Nevertheless, soy and sugarcane as well as beans and cassava have shown distinct expansion patterns, are grown for different purposes, and are responsible for different environmental and social impacts. These factors, as well as the fact that they are both closely linked to different farming modes, support the distinction made in this chapter.

This quantitative analysis of biomass flows delivered first and tentative insights into processes that can be linked to land use competition between export crops and domestic staples in Brazil. Spatial patterns of expansion and contraction of both farming types that are associated with soy, sugarcane, beans and cassava can only be revealed by such a quantitative and comparative approach. These patterns show hot spots with highly dynamic developments, where qualitative, place-specific case studies are needed to deepen the understanding of likely competition between distinct farming systems.

The long-lasting discussion of agricultural production in Brazil, mostly related to climate change and deforestation, is bound to continue. Since the 1990s, large-scale agricultural production for exports has been a primary goal of policy makers and businesses and this has been an economic success story. However, in terms of land distribution and equality, Brazil ranks among those countries with the worst performance (Lapola et al. 2014). Despite the fact that the largest share of domestic food is produced by family farming and the majority of rural settlers directly depend upon access to land, only little progress towards a more equal distribution of land was made during the Lula government, despite the land reform having been announced as a major topic during its legislative period (Welch 2011). The slowly progressing land reform and looming threats to domestic food security were forestalled with compensatory payments through social welfare programmes such as the bolsa familía. These programmes, however, were set up in times of a prospering economy. Now, in times of low economic growth, the problems of such policies are coming to the surface. It is open whether the Brazilian government will (be able to) abandon the neo-extractivist development model and overcome this deep economic recession. In this process, the potential of smallholders plays a crucial role in reducing rural poverty and making sustainable use of the natural resources, e.g. through agroecological production. For this potential to be realized productively, smallholders need formal rights over land and resources, which makes the land reform pivotal for every sustainable development policy in Brazil.

References

Abbey, L. A., Baer, W., & Filizzola, M. (2006). Growth, efficiency, and equity: The impact of agribusiness and land reform in Brazil. *Latin America Business Review, 7*, 93–115.

Altieri, M. A. & Koohafkan, P. (2008). *Enduring farms: Climate change, smallholders and traditional farming communities*. Third World Network (TWN), Penang.

Banco do Nordeste (2015). *Agricultura familiar* (WWW Document). http://www.bnb.gov.br/agricultura-familiar2. Accessed December 17, 15.

Barretto, A. G. O. P., Berndes, G., Sparovek, G., & Wirsenius, S. (2013). Agricultural intensification in Brazil and its effects on land-use patterns: An analysis of the 1975–2006 period. *Global Change Biology, 19*, 1804–1815. doi:10.1111/gcb.12174.

Borras, S. M., Franco, J. C., Gómez, S., Kay, C., & Spoor, M. (2012a). Land grabbing in Latin America and the Caribbean. *Journal Peasant Studies, 39*, 845–872. doi:10.1080/03066150.2012.679931.

Borras, S. M., Kay, C., Gómez, S., & Wilkinson, J. (2012b). Land grabbing and global capitalist accumulation: Key features in Latin America. *Canadian Journal of Development Studies Canadienne Détudes Dév, 33*, 402–416. doi:10.1080/02255189.2012.745394.

Brand, U. & Dietz, K. (2014). Neo-)Extraktivismus als Entwicklungsoption? Zu den aktuellen Dynamiken und Widersprüchen rohstoffbasierter Entwicklung in Lateinamerika. *Polit. Vierteljahresschr.* 88–125.

Bunker, S. G. (1985). *Underdeveloping the Amazon: Extraction, Unequal Exchange, and the Failure of the Modern State*. First Edition ed. University of Illinois Press, Urbana.

Bunker, S. G., & Ciccantell, P. S. (2005). *Globalization and the race for resources, themes in global social change*. Baltimore: Johns Hopkins University Press.

Burchardt, H.-J., & Dietz, K. (2014). (Neo-)extractivism—A new challenge for development theory from Latin America. *Third World, Quarterly, 35*, 468–486. doi:10.1080/01436597.2014.893488.

Chaifetz, A., & Jagger, P. (2014). 40 Years of dialogue on food sovereignty: A review and a look ahead. *Global Food Security*. doi:10.1016/j.gfs.2014.04.002

De Schutter, O. (2011). How not to think of land-grabbing: Three critiques of large-scale investments in farmland. *Journal Peasant Studies, 38*, 249–279. doi:10.1080/03066150.2011.559008.

DeFries, R. S., Rudel, T., Uriarte, M., & Hansen, M. (2010). Deforestation driven by urban population growth and agricultural trade in the twenty-first century. *Nature Geoscience, 3*, 178–181. doi:10.1038/ngeo756.

Douglas, B. (2015). Brazilian president under fire as tens of thousands protest in 200 cities. *The Guardian*.

Food and Agriculture Organization of the United Nations. (2014). FAOSTAT trade database (WWW Document). http://faostat.fao.org/desktopdefault.aspx?pageid=342&lang=en&country=238. Accessed December 17, 15.

Food and Agriculture Organization of the United Nations. (2015). FAOSTAT database (WWW Document). http://faostat3.fao.org/download/Q/QC/E. Accessed January 13, 16.

Frayssinet, F. (2012). *Soy and sugar cane fuel native land conflicts in Brazil*. Inter Press Service. Inter Press Serv.

Galeano, L. A. (2012). Paraguay and the expansion of Brazilian and Argentinian agribusiness frontiers. *Canadian Journal of Development Studies/Revue canadienne d'études du développement, 33*, 458–470. doi:10.1080/02255189.2012.744301.

Greenpeace. (2006). *Eating up the Amazon*. London, UK: Greenpeace.

Gudynas, E. (2010). *The new extractivism of the 21st century: Ten urgent theses about extractivism in relation to current South American progressivism*. America Program.

IBGE. (2012). *Censo agropecuário: 2006: Brasil, grandes regiões e unidades da federação: segunda apuração*. Rio de Janeiro, Brazil: IBGE.

Grain, Martinez-Alier, J., Temper, L., Munguti, S., Matiku, P., Ferreira, H., Soares, W., et al. (2014). *The many faces of land grabbing. Cases from Africa and Latin America* (No. 10), EJOLT report.

Index Mundi. (2015). Soybean meal—monthly price—commodity prices—price charts, data, and news—IndexMundi (WWW Document). http://www.indexmundi.com/commodities/?commodity=soybean-meal&months=60. Accessed December 17, 15.

Kastner, T., Erb, K.-H., & Haberl, H. (2014). Rapid growth in agricultural trade: Effects on global area efficiency and the role of management. *Environment Research Letters, 9*, 034015. doi:10.1088/1748-9326/9/3/034015.

Klink, C. A., & Machado, R. B. (2005). Conservation of the Brazilian Cerrado. *Conservation Biology, 19*, 707–713. doi:10.1111/j.1523-1739.2005.00702.x.

Kosacoff, B., Lopéz, A., Pedrazolli, M., (2008). *Trade, investment and fragmentation of the global market: Is Latin America lagging behind?* United Nations Publications.

Lapola, D. M., Martinelli, L. A., Peres, C. A., Ometto, J. P. H. B., Ferreira, M. E., Nobre, C. A., et al. (2014). Pervasive transition of the Brazilian land-use system. *Nature Climate Change, 4*, 27–35. doi:10.1038/nclimate2056.

Laroche Dupraz, C., & Postolle, A. (2013). Food sovereignty and agricultural trade policy commitments: How much leeway do West African nations have? *Food Policy, 38*, 115–125. doi:10.1016/j.foodpol.2012.11.005.

Li, T. M. (2011). Centering labor in the land grab debate. *Journal Peasant Studies, 38*, 281–298. doi:10.1080/03066150.2011.559009.

MAPA. (2016). Ministério da Agricultura, Pecuária e Abastecimento (WWW Document). http://www.agricultura.gov.br/. Accessed January 22, 16.

Martinelli, L., Joly, C. A., Nobre, C. A., & Sparovek, G. (2010a). The false dichotomy between preservation of the natural vegetation and food production in Brazil. *Biota Neotropica, 10*, 323–330. doi:10.1590/S1676-06032010000400036.

Martinelli, L., Naylor, R., Vitousek, P. M., & Moutinho, P. (2010b). Agriculture in Brazil: Impacts, costs, and opportunities for a sustainable future. *Current Opinion Environment Sustainable, 2*, 431–438. doi:10.1016/j.cosust.2010.09.008.

MDA. (2016). Ministério do Desenvolvimento Agrário (WWW Document). http://www.mda.gov.br/. Accessed January 22, 16.

Meyfroidt, P., Rudel, T. K., & Lambin, E. F. (2010). Forest transitions, trade, and the global displacement of land use. *Proceedings of National Academy of Sciences, 107*, 20917–20922. doi:10.1073/pnas.1014773107.

MST. (2006). #5: Agrarian reform in the context of food sovereignty, the right to food and cultural diversity, friends of the MST. In *Via Campesina Issue Paper*. Presented at the International Conference on Agrarian Reform and Rural Development, Porto Allegre.

Pokorny, B., de Jong, W., Godar, J., Pacheco, P., & Johnson, J. (2013). From large to small: Reorienting rural development policies in response to climate change, food security and poverty. For. Policy Econ. *Forest and conservation policy in a changing climate, 36*, 52–59. doi:10.1016/j.forpol.2013.02.009.

Price Waterhouse Coopers. (2013). *Agribusiness in Brazil: An overview*. Price Waterhouse Coopers.

Schaffartzik, A., Mayer, A., Gingrich, S., Eisenmenger, N., Loy, C., & Krausmann, F. (2014). The global metabolic transition: Regional patterns and trends of global material flows, 1950–2010. *Global Environmental Change, 26*, 87–97. doi:10.1016/j.gloenvcha.2014.03.013.

Scholtes, F. (2009). *Status quo and prospects of smallholders in the Brazilian sugarcane and ethanol sector: Lessons for development and poverty reduction* (No. 43), ZEF Working Papers. Center for Development Research, University of Bonn, Bonn.

Soybean and Corn adviser. (2015). 7 % of Brazil's Soybeans used for biodiesel production. (WWW Document). http://www.soybeansandcorn.com/news/Jun7_10-7-of-Brazils-Soybeans-Used-for-Biodiesel-Production. Accessed December 17, 15.

Svampa, M. (2012). Resource extractivism and alternatives: Latin American perspectives on development. *Journal Für Entwicklungspolitik, 28*, 43–73.

The World Bank. (2015). GDP growth (annual %) | Data|Table (WWW Document). http://data.worldbank.org/indicator/NY.GDP.MKTP.KD.ZG. Accessed December 17, 15.

United Nations Development Programme. (2016). *Human development report 2015*. United Nations, [S.l.].

Watts, J. (2015). Amazon deforestation report is major setback for Brazil ahead of climate talks. *The Guardian*.

Webber, J. R. (2014). 13 Revolution against "Progress": Neo-extractivism, the compensatory state, and the tipnis conflict in Bolivia. In J. R. Webber., S. Spronk (Eds.), *Crisis and contradiction* (pp. 302–333) Brill.

Weinhold, D., Killick, E., & Reis, E. J. (2013). Soybeans, Poverty and Inequality in the Brazilian Amazon. *World Development, 52*, 132–143. doi:10.1016/j.worlddev.2012.11.016.

Welch, C. (2011). Lula and the meaning of Agrarian reform|Friends of the MST (WWW Document). http://www.mstbrazil.org/news/lula-meaning-agrarian-reform. Accessed December 17, 15.

Wiggins, S. (2009). *Can the smallholder model deliver poverty reduction and food security* (No. 08), FAC Working Paper. Future Agricultures Consortium, UK.

Wilkinson, J., Reydon, B., & Di Sabbato, A. (2012). Concentration and foreign ownership of land in Brazil in the context of global land grabbing. *Canadian Journal Development Studies Canadienne Détudes Développement, 33*, 417–438. doi:10.1080/02255189.2012.746651.

Wolford, W. (2010). Participatory democracy by default: Land reform, social movements and the state in Brazil. *Journal Peasant Studies, 37*, 91–109. doi:10.1080/03066150903498770.

Yates, J. S. & Bakker, K. (2013). Debating the "post-neoliberal turn" in Latin America. *Progress in Human Geography,* 0309132513500372. doi:10.1177/0309132513500372

Chapter 18
Local Food Systems and Their Climate Impacts: A Life Cycle Perspective

Michaela C. Theurl

Abstract The creation and maintenance of local food sovereignty and security is a top priority on political agendas. Food systems are characterized by complex dynamic nexuses at different levels and scales, and sustainability assessments necessarily need to approach social, economic, and environmental dimensions. While the provision of food is limited, among other factors, by the quantity of fertile areas, these areas are also attractive for other land uses, e.g., for the cultivation of energy plants or afforestation, which can result in socioeconomic conflicts related to land-use competition. Numerous studies have looked at the global picture, using integrated assessment models, in which biophysical and economic models are coupled to derive potential future pathways and related greenhouse gas (GHG) emissions of feeding and fueling the rising global population. While such models can be useful to show global boundary conditions, their representation of complex local-to-global interdependencies is necessarily limited. Looking at local structures, life cycle assessment (LCA) is a useful tool to analyze a product supply chain and gain detailed insights in order to provide apparently complex data in a simple way for detecting option spaces or informing civic societies. This chapter addresses the local food system of Vienna. It shows what we can learn from LCA about, for example, potential savings of food-related GHG emissions and about gaining efficiencies of local produce in contrast to imports from the international market.

Keywords Life cycle assessment • Food system Vienna • Sustainability • Product supply chain

---

M.C. Theurl (✉)
Institute of Social Ecology, Alpen-Adria Universitaet Klagenfurt-Wien-Graz, Vienna, Austria
e-mail: MichaelaClarissa.Theurl@aau.at

M.C. Theurl
Research Institute of Organic Agriculture, FiBL Austria, Vienna, Austria

J. Niewöhner et al. (eds.), *Land Use Competition*,
Human-Environment Interactions, DOI 10.1007/978-3-319-33628-2_18

18.1 Local Food Systems Within a Global Cycle

The topic of food sovereignty, which has initially been formulated and developed in and for the Global South, has been receiving increasing attention in economically rich countries recently. People increasingly question and rethink their social interactions and their purchase habits of basic necessities such as furniture produced in Sweden or pots from China. Food and eating serve as a tangible example for such reflexivity, showing intensification of public and scientific discourses around "re-connecting" society to nature and of re-establishing consumers' relationships with farmers. In relation to this, local food networks have been established as a counter-movement to the mainstream food system, depending on supermarket chains and international corporations (Milestad et al. 2010). The increasing importance for consumers to eat fresh and healthy, locally produced foodstuffs, the assumed connection of local food networks to higher quality and lower environmental impacts, and the desire to reconnect consumers and producers are drivers for this counter-movement. Furthermore, the world food price crises 2008 triggered a process of rethinking global trade structures and the strong dependency on the dynamics of a global food market (McMichael 2005). Another driver in this development is climate change. Food systems are important drivers of climate change, and local food systems are often believed to be "climate friendlier," that is, to have a smaller carbon footprint than imported goods from other parts of the world.

After the electricity and heat production sector, the agriculture and forestry sector is the second most important contributor to global total anthropogenic greenhouse gas (GHG) emissions (IPCC 2014). The agricultural sector is responsible for nearly 25 % of total global anthropogenic GHG emissions with about 47 % of methane (CH_4, primarily from livestock management, rice production) and 58 % of nitrous oxide (N_2O, primarily nitrogen fertilizer and manure application) emissions (Smith et al. 2014). Accordingly, experts see considerable mitigation activities related to agricultural production systems such as soil management techniques, carbon sequestration, input reductions (supply side), and waste reductions or dietary changes (demand side) within the agricultural production sector.

Against the expectation of a rising and wealthier global population, numerous studies try to understand the complex interrelations of a finite land area, rising demands for food and fiber, and climate change implications (Bajželj et al. 2014; Cardinale et al. 2012; Cassidy et al. 2013; Erb et al. 2009; Foley et al. 2011; Smith et al. 2014; Tilman and Clark 2014, see also Sect. 2). The global models underlying these studies implement GHG emissions from agricultural land-use systems often in a coarse way, focusing on global connections and tendencies (van Vuuren et al. 2011). Studying climate and other environmental impacts of food systems from a more place-specific perspective therefore requires complementary approaches, which are able to grasp the local complexities on a more detailed level, without losing sight of its cross-scalar interactions.

When it comes to assessing impacts of local production structures, the so-called life cycle assessment (LCA) provides a promising approach. Generally speaking,

LCA is an accounting tool to evaluate the environmental impacts of a defined product, which can also be a process, a service, or an organization, from production to use and disposal or from "cradle to grave." Resource inputs such as raw material and energy carrier extraction are considered, as well as wastes and emissions that occur during the different life cycle phases. If recycling plays a role in the overall scope of such an LCA, the evaluation takes place from "cradle to cradle." The application of an LCA follows standardized guidelines that have been internationally harmonized by the International Organization for Standardization (Finkbeiner et al. 2006). Recently, the European Union started a process of developing a product environmental footprint (PEF) and organizational environmental footprint (OEF) standard with the aim to harmonize the existing heterogeneous LCA methods landscape (Finkbeiner 2014) and to support European policies (European Commission 2013).

One main challenge thereby is the definition of the system boundary, which can have a large effect on the results, as it determines the processes that are included in the assessment. It is therefore essential to interpret the results of an LCA study against the background of the chosen system boundaries. For instance, using the example of food items, it has a different validity if environmental impacts are considered from production to farm gate, if they are considered along the whole supply chain from production to use at household level, or if they additionally include consumers' transport and disposal. Since it is not feasible to include all directly and indirectly related processes due to their complexity, the truncation decisions are based on the research question and on practitioners' value judgments (Finkbeiner 2014; Freidberg 2015). Still, LCA can look at processes in a very detailed way and is mostly used to find hot spots and compare different products, processes, or organizations in order to discover potential fields of action to reduce environmental impacts.

The LCA approach is one way to address implications of local food systems for climate change, by using the carbon footprint as one informative LCA-based indicator. The carbon footprint can be calculated for a nation, an organization, an event, an individual (lifestyle), or a product. Carbon footprints are reported in terms of carbon dioxide (CO_2) or carbon dioxide equivalent (CO_2-eq) that is potentially emitted from (in) direct inputs/outputs related to the life cycle of the benchmark or functional unit. In practical terms, the global warming potential (in CO_2-eq) is calculated, which provides information about the relative amount of heat a GHG traps in the atmosphere and how much the functional unit, e.g., one kilogram of carrots, relatively contributes to the greenhouse effect.

In this chapter, inputs and outputs of the following life cycle stages of agricultural food items are investigated: agricultural production (farm gate), packaging, storage, and transportation. The latter stands for distances related to the distribution of the product and in some cases the import to the domestic market. This chapter will examine the question of the climate performance of locally produced food items and under which circumstances food products from local systems have smaller carbon footprints compared to imported items. This chapter will explore local food systems in more detail, asking whether local products from different distribution systems, e.g., supermarket or box scheme, are per se related to lower carbon footprints.

18.2 Conceptualizing Local Food Systems in Vienna

Food systems are complex dynamic systems including processes and components interrelated across different scales and acting in diverse temporal, spatial, and institutional areas (Garnett 2014; Godfray et al. 2010; see, e.g., Friedmann 2005). The definition of what "local" means proves more difficult. As a matter of fact, the scientific literature does not provide one single definition of what a local food system is, and despite a common understanding at European level, member states still follow their individual positions (European Union 2012).

In the 1990s, the food mile concept gained momentum (Paxton 1994; Pretty et al. 2005), defining "local" in terms of total transport miles along the food chain and according to spatial distances between each chain element. Later on, concepts such as Smith and MacKinnon's 100-mile diet approach (2007) took the same lines. Alternatively, Hinrichs (2000) suggests the social relationship between consumer and producer as virtual local boundary, e.g., via direct contact, often defined as social "closeness" (Hinrichs 2000). The definitions of local food systems thus depend on the study site and on the research purpose and can include actor-based analysis of food processing facilities and their technology and volume potential, as well as supply-side analysis by, for example, estimating total production potentials of specific farming systems (e.g., farm type and structure; Carey 2013). Here, local food systems are understood as systems, in which consumers purchase food from farms located within a certain radius via diverse distribution channels such as farmers markets, food hubs (in the case of hospitals or commercial kitchens), local processors, local shops, or even at the farm itself (farm shop). Other examples of rapidly growing local food systems are community-supported agriculture (CSA) and food cooperatives, which are actively shaped by consumers who intend to support local farmers, keep the added value local, and prevent long-distance transportation (see Walthall, Chap. 16).

In this chapter, Vienna's food system serves as study area for the application of LCA methods and the evaluation of local food production. Vienna hosts a well-working and expanding local food network with about 20 self-organized food cooperatives (Food Coops) and three CSA (community-supported agriculture) communities with delivery points all over the city.[1] Besides these initiatives, also box schemes foster the establishment and growth of local food distribution systems. Organic producers play a major role in local food systems and are strategic partners

[1]http://www.ernährungssouveränität.at/wiki/CSA-Betriebe_und_Initiativen_in_%C3%96sterreich#weitergehende_Initiativen http://foodcoops.at/?cat=10

that are open for new ways beyond mainstream food supply chains. Austria has a long tradition in organic production. Currently, 17 % of all farms work according to organic principles and Austrian consumers spend 118€ per capita and year for organically grown food products.[2] At the same time, the consumption of locally produced food is very popular in Austria and political stakeholders have been promoting local products regardless of the prevailing management of the agricultural production system (e.g., conventional and organic management techniques). In particular, the Austrian livestock sector has been producing high-quality meat and milk products and is powerful in promoting their local products. Against this background, the question arises whether and under which circumstances local food production contributes to a climate-friendly food system and whether other aspects of the supply chain than "localness" are more important in terms of environmental impacts such as GHG emissions.

The agricultural surface within Vienna's city boundaries has been decreasing recently but is still relatively high for an urban area, with 16.8 % of the total area in 2010 (without forest use; Statistik Austria 2015). Cereal production is the main agricultural crop in Vienna's north, east, and south, whereas intensive horticulture in protected greenhouses is centered in the eastern part of the city. Vineyards are located in the rolling hills of the west. Vienna's hinterlands are characterized by a diverse agricultural landscape, including intensive cereal and vegetable production with a favorable pannonic climate, including vines and orchards and a little dairy and pig farming.

Historically, there have always been interrelations between urban resource use, transport, and the agricultural production of its urban hinterlands. Already von Thünen (1826) suggested for cities, regarded as an isolated economic unit, that land use would be shaped in concentric circles around the urban center, depending on market price and cost of transport. Since the definition of a local food systems is not straightforward, one might use von Thünen's concept and draw a semidiametric circle of 80 km around a city, which, e.g., corresponds to 50 miles used in Carey (2013), and use this as an outer boundary describing a local system (see also Fig. 18.1).

By drawing a radius of 50 km around the Vienna's city center, we already touch the Austrian–Slovakian border, a radius of 80 km, would establish a transnational local food system, bridging and coupling cultural aspects such as different local crop varieties, language, distribution structures, and consumer preference (Fig. 18.1). The creation of a Vienna–Bratislava, transnational local food system, would be an outstanding innovative flagship project that could be used as a role model for other cities and regions. For the time being, the Austrian public still focuses on domestic local products. Besides transport distance, other elements of the supply chain such as the agricultural management techniques or packaging have a strong influence on the resulting total product carbon footprint (see Result section). Thus, the environmental analysis in the following section is based on a radius of 80 km.

[2]https://www.bmlfuw.gv.at/land/bio-lw/Bioweltmeister.html

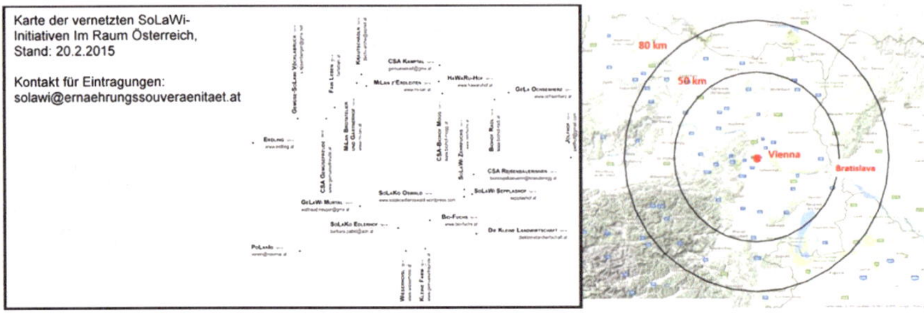

Fig. 18.1 Existing network of CAS in Austria (*left*) and local food system boundary of the case study area of Vienna (*right*)

18.3 Evaluating Food Supply Chains Under Life Cycle Perspectives in Vienna

In the following, the LCA approach is applied to explore local food supply chains in Vienna. Within LCA, the carbon footprint is a promising indicator and addresses the leverage on GHG emissions of the different life cycle stages. The effect of imported foods, local seasonal production, and winter production can be equally addressed as different forms of food distribution and agricultural management strategies. Austria's greatest demand for organically grown food products exists from consumers in Vienna. This study thus investigates the differences between management production practices, using the example of organic and of conventional production systems, and reveals climate change mitigation potentials of organic practices. Moreover, this study explores the impacts of climate change in terms of carbon footprint of different management practices compared to spatially different ways of organizing food distribution (local versus global).

The LCAs of the different food items presented here are based on a consistent methodological framework, based on international guidelines such as IPCC (2006) and standards (ISO 2006a, b), considering Austria-specific characteristics on farm size, soil type, fertilization, crop rotation, and distribution channels (Hörtenhuber et al. 2010, 2011; Lindenthal et al. 2010; Theurl et al. 2014, in preparation-b). GHG emissions are related to 1 kg of product at the point of sale (POS).

The LCA tool is used to analyze the supply chain of food products, which means that seven life cycle phases are investigated: (i) agricultural production phase including upstream inputs to the farm (e.g., fertilizer production, land-use change, pesticides, upstream transport); (ii) processing; (iii) packaging; (iv) storage; (v) emissions from imports (long-distance transport); (vi) emissions from food distribution (consumer shopping, supermarket logistics, box delivery scheme); and (vii) heating in glasshouses. Heating is shown separately, although the energy input is related to the farm stage (red bar in Fig. 18.2).

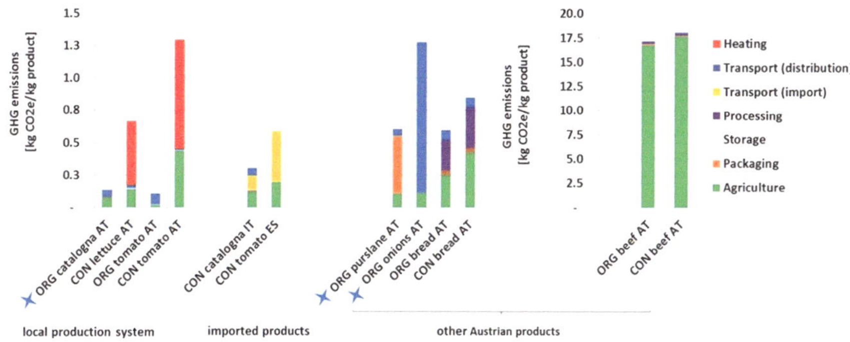

Fig. 18.2 Contribution of different life cycle phases to total carbon footprints of food products from local food systems, imported products, and products produced and consumed in other parts of Austria. *AT* Austria, *IT* Italy, *ES* Spain, blue stars indicate unheated winter harvest systems

The differentiation of diverse life cycle phases allows to additionally discuss the impact of different distribution options of local food systems, namely supermarket logistics with point of sale (POS) in the store, delivery in a box scheme (POS at home), and consumer shopping trips (POS at farm store).

The exploration of different management techniques with different spatial references and specific distribution channels is based on three broad categories: products from (i) local production, (ii) other Austrian production sites, and (iii) the European market (Italy, Spain). Management techniques are understood as conventional and organic cultivation and winter harvest systems and heated winter vegetable production.

As discussed above, the system boundary of the local food system is defined by an 80 km radius around the center of Vienna's first district (see Fig. 18.1). The following local production systems are assessed:

(i) catalogna salad with organic management techniques in open field (box scheme);
(ii) conventional lettuce from heated glasshouses (supermarket logistic);
(iii) organic tomato from unheated plastic tunnels (supermarket logistic); and
(iv) conventional tomatoes from heated glasshouses (supermarket logistic).

The area of the "Seewinkel" is a vegetable production area in the southeast and is of major importance for Vienna's market. Figure 18.1 shows that the area beside Lake Neusiedl with its Pannonia climate lies in a radius between 50 and 80 km. The cultivation area, the climatic conditions, soil types, and precipitation events are important parameters in analyzing food products due to their consequences for yield and environmental impacts such as nitrogen leaching. In this analysis, organic greenhouse tomatoes are produced in this region (ORG tomato AT in Fig. 18.2), while conventional greenhouse tomatoes and traditional greenhouse lettuce are produced within the Viennese city boarders. Catalogna, a popular lettuce variety, serves as an example for produce cultivated in open field typically in lower Austria further east of Vienna (ORG catalogna AT).

To compare impacts of the agricultural management systems (organic vs. conventional) with impacts of the spatial distance between producers and consumers (local vs. imported) on carbon emissions, carbon footprints of vegetables imported from European countries are also presented. Open field catalogna from Italy (distribution via box scheme) and Spanish greenhouse tomatoes from Almeria (supermarket logistic), both under conventional management, are used as examples for imported vegetables (CON tomato ES; CON catalogna IT).

In order to compare these options, other food product LCAs from other Austrian regions are also assessed in this study. While bread and beef are shown as examples of average Austrian products (from different locations, average yields) and average distribution distance in an average Austrian supermarket store. Onions produced in the Alpine region represent the alternative form of direct marketing with the POS in the farm shop, to where the consumer drives with the family car.

For the sake of simplicity, the consumer stage is excluded, e.g., food preparation and household wastes, from all carbon footprint assessments and will be discussed in terms of food products' end-of-life stage qualitatively. One reason is the difficulty to measure individual habits and uses and the waste collection of households in a standardized fashion. Related aspects such as shopping and disposal and their relation to GHG emission reduction will be separately discussed in the following sections.

18.3.1 Are Transport Emissions Lower in Local Food Systems?

It is intuitively plausible that locally produced food requires less transport, as fewer tonne-kilometers seems to be travelled due to shorter supply chains. Since transport is one major driver of climate change (IPCC 2014), products from local food systems therefore are believed to be linked to less GHG emissions than imported products from other European countries or from oversees. However, the results of previous studies show that this does not hold true in general (Sim et al. 2007; Theurl et al. 2014).

The results presented in Fig. 18.2 show that emissions from distribution transport (blue part of each bar) often have less impact on the total carbon footprint of a product than the agricultural production system in which it is produced. It is shown that while transport has a significant share of the total carbon footprint for extensively produced vegetables such as organic open field catalogna or organic tomatoes (ORG catalogna/tomato AT), it is a relatively insignificant factor for heated vegetables, e.g., conventional lettuce and tomatoes and packaged products (CON lettuce/tomato AT). In regard to food produce from other Austrian regions, the results show that emissions from transport are also an insignificant factor for processed products such as bread and for livestock products such as meat (ORG and CON bread AT, ORG and CON beef AT).

Among the studied local food systems, supermarket logistic, as exemplified by conventional lettuce and both variants of conventional and organic tomatoes, can be compared with a box scheme distribution, as exemplified by catalogna (ORG catalogna AT). The results show that although supermarket logistic involves longer traveling distances, emissions related to one kilogram of product are lower compared to the box scheme. The reason for this is that supermarket logistic uses trucks with enormous load capacities (up to 60 tons, e.g., EuroCombis), whereas the box scheme is based on mini-trucks (van, primary data taken from Lughofer 2010). While total emissions per km are higher, emissions per kg of load are usually lower, the larger the vehicle.

The local food supply systems are compared with imported catalogna from Italy (CON catalogna IT) and tomatoes from Spain (CON tomato IT). The results clearly show that when looking at transport emissions only, emissions from long-distance transport (yellow part of the bars) are higher than emissions from unheated local vegetables. On closer examination, however, it becomes clear that for local food systems, transport emissions increase enormously by individual shopping trips to the farm shop by car. The effect of individual shopping trips is demonstrated through the example of onions produced in Austrians alpine regions and sold locally (Fig. 18.2., ORG onions AT). For local and organically grown onions, emissions due to individual 10-km shopping trips to the local farmer are higher than transport emissions for catalogna from Italy or tomatoes from Spain.

The case studies presented here thus seem to suggest that the carbon footprint of vegetables depends on the production system (intensive/extensive) rather than on the spatial distance between producers and consumers. By comparing the same cultivars in terms of organic and conventional management production, it is shown that emissions from the agricultural phase are always lower for organic production (green part in each bar in Fig. 18.2). One can thus conclude that a shift to organic production might mitigate the climate impact considerably, although literature shows that especially for the relation of impacts to one kilogram functional unit, this is not always the case (Meier et al. 2015).

However, it is important to note that the fact that organic cultivation has a lower carbon footprint performance than the conventional counterpart only holds true under specific conditions that underlie the calculation principles of a study. The central assumption here is that the organic and conventional product is produced in the same region and under the same socioeconomic conditions (Theurl et al. in preparation). As a consequence, it is assumed that external factors such as local knowledge, farm size, and the socioeconomic embeddedness remain the same. Also, emissions due to land cover change, potentially caused by the usually lower yields for organic crops, as well as emissions due to changing storage needs, are not considered in these calculations.

18.3.2 *Year-Round Fresh Vegetables from Local Production?*

While local production is appreciated, consumers increasingly demand that most fruits and vegetables are available throughout the year. During winter, the demand for vegetables such as tomatoes or peppers can be covered either by imports from warmer regions such as the south of Spain or North African countries or by vegetables produced locally in heated glasshouses. Both options are linked to high carbon emissions, the former due to emissions from transport and the latter due to emissions from fossil energy used to heat the glasshouses. Although Vienna has high technical standards regarding the energy efficiency of glasshouse production, the vegetables produced locally in heated glasshouses are often linked to more carbon emissions than vegetables imported from warmer regions (Theurl et al. 2014). Spanish tomatoes (CON tomato ES) have even lower emissions than a heated lettuce from local production in Vienna (CON lettuce AT). This shows the importance of local biophysical conditions of production and distribution.

An alternative to heated winter production systems are low-energy and resource winter harvest systems that are gaining special resonance from farmers and consumers in Austria in the last years (Theurl 2015; Theurl et al. in preparation-a). Examples for such unheated winter harvest systems are presented with a blue star in Fig. 18.2. While neither additional energy for heating nor for lightning is used in these systems, cultivation time is longer compared to summer systems and yields are relatively low (Theurl et al. in preparation-a). Considering the evidence that climates are getting warmer and that consumer demands for regionally produced fresh salads also during winter constantly increase, these systems may well represent a valuable option for local producers and especially for small-scale family farms. Moreover, these systems are of high interest for urban gardening communities or individual windowsill cultivation. Carbon footprints of unheated winter harvest systems are in general lower. One point why this does not hold true in general is the fact that the functional unit is one kilogram at POS, and if the product is wrapped, the picture changes dramatically. Particularly small packaging units, e.g., 100 g baby leaf salads, can have a major influence on the overall carbon footprint (see orange part of ORG purslane AT in Fig. 18.2). In the case of winter purslane, packaging makes up 74 % of the carbon footprint, which is nearly as high as heated and unpacked conventional lettuce. One kilogram of wrapped winter purslane is even similar to one kilogram of bread, which is processed (wheat is milled) and baked (ORG and CON bread AT in Fig. 18.2).

Compared to high-yielding and intensive tomato production in Austria, special green winter harvest is suggested as a promising alternative (system with blue star in Fig. 18.2). Besides the reasons already described above, these special winter products open new distribution channels, e.g., for high standard gastronomy, which is also able to pay higher prices for good-quality produce than the individual consumer.

18.3.3 End of Life in Local Food Systems

Roughly one-third of edible calories are lost along the life cycle of a product. Carbon emissions of food products could be reduced significantly by the reduction of food wastes and losses, as the same amount of food can be produced with less inputs, if wastes are reduced (Beretta et al. 2013; Kummu et al. 2012). Although food wastes could be reduced at all stages along the food chain, there is evidence that households are the most important sources of food wastes in industrial countries (Gustavsson et al. 2011). Our case study, the city of Vienna, has a well-functioning recycling management. In the case of biomass, the city collects household food wastes in order to process them in transformation plants. There is a composting plant where so-called A+ compost is produced, suitable for certified organic production, which is also at free disposal for citizens. A biogas plant transforms at least 22,000 tons of kitchen refuse from hospitals, catering companies or restaurant, and household organic waste collection tins into useful methane gas, which is then fed into the local gas grid.[3] Gruber et al. (2015) estimated that roughly 0.5 kg CO_2 was emitted by a combination of composting and waste incineration in an end-of-life scenario for potatoes and rice, which will add to existing carbon footprints (energy use for the household phase not included).

Particularly in vegetable production, a lot of unused material is already produced at the farm stage. In general, biomass from organic production systems in Austria is already composted and reused to a large extent in arable farming systems, for example, by the incorporation of crop residues at the field and in vegetable systems (e.g., composting of plant parts). Yet conventional systems, especially when cultivated in soil-less systems, still have a large unexplored potential to reuse plant residues in biogas facilities. In the case of Vienna, conventional producers already started with delivering their unused plant materials to the biogas plant. Further, unsold milk products, frozen goods, or finished products from retailers have a great potential to be used in fermentation plants. Thus, on the one hand, biogas plants might be considered as an essential option to reuse food wastes and transform them to useful energy. On the other hand, operators of such plants depend on biomass inputs that in the future might compete with increasing human food demand. Therefore, the sustainability pathways might have to dig into systemic optimization at different scales, aiming at reducing food waste and losses. Against this background, there is the need for a better understanding of how local food system stakeholders act and react and how pathways might develop.

[3]https://www.wien.gv.at/english/environment-leisure/vienna-biogas.html

18.4 Implications for Consumer Choice and Policy Incentives

There is no straightforward answer to the question, whether local products are more sustainable, less energy intensive, and healthier. LCA and carbon footprints help analyzing the question from a bottom-up perspective and are able to show the GHG emissions related to specific life cycle stages of a product's life cycle. CO_2 labeling derived from LCA assessments can be an important instrument in influencing consumer choices toward a sustainable food system. At the same time, CO_2 labeling has become a strong marketing instrument. In Austria, one of the biggest retailers is promoting a separate organic brand of Austrian food products and labels the difference to conventional products via carbon and water footprint, biodiversity potential, and the potential higher regional effect to local food webs. However, studies show that environmental awareness alone is not enough to allow for a system change (Schmidt 2009), as price is of crucial importance for purchase decisions besides taste and quality of products (Kearney 2013).

For a more profound transformation of the global food system, agro-food systems such as CSAs and social movements such as the food sovereignty movement forced by la Via Campesina[4] might play an important role, providing new ideas and niches from which the future food system might develop. These niches can be promoted not only by consumers but also by public procurement (Morgan 2008). City councils, for example, today often rely on big retailers and are often unaware how resilient food systems from the city surrounding areas can be integrated into general city planning approaches in a fruitful and sustainable way (Carey 2013).

Yet, no matter which transformative pathway is favored, the aggregated results show that it is important to analyze the climate effects of different food stuffs in an integrated perspective, taking empirical evidence into account. Here, for example, it was shown that contrary to what is often believed, the carbon footprint of local products is not lower in all instances, but depends on the specific conditions concerning the local or interregional production and distribution of food and most of all on the product itself. Livestock products, especially meat products, are related to very high GHG emissions.

By taking such an integrative view on local food systems, taking the whole supply chain into consideration, it was shown that transport distance between producer and consumer is only one factor in determining a low-carbon product. Crucial is the organization of food distribution in terms of distribution systems such as the box scheme, supermarket logistics or individual shopping trips, and the kind of transport vehicle and efficiency. Important is further the management of the production and—as shown in the case of unheated vegetables—whether the product is packed or not. Consumers have become increasingly environmentally minded,

[4]Established in the early 1990s, la Via Campesina is a transnational social movement that fosters social rights and sustainable food production and responsible for coining the concept of "food sovereignty."

demanding fresh, healthy, and locally produced vegetables. However, local shopping trips by the individual family car outweigh the CO_2 saving effect of local, especially organic production and even imported salads from Spain during winter months. One missing element in these results is the (carbon mitigation) potential of closing loop by biomass recycling and waste reduction, which is likely to have favorable consequences. Here, the potential of winter harvest vegetable production without heating has been demonstrated, which currently represent niche products for a growing market. Moreover, the expansion of these low-energy cultivation methods in small-scale urban gardening as well as in vacant glasshouses and the establishment of efficient local infrastructure and distribution logistics together with consumer–farmer cooperations and energy-efficient food-collecting systems are key factors in fostering local food systems.

Acknowledgments Many thanks to my colleagues Christian Lauk, Juliana Lutz, and Stefan Hörtenhuber for their valuable comments and discussions and Thomas Drapela for the help with the graphic representation of the case study area. Funding from ERC-STG2010-263522 LUISE and EU 7th Framework Programme project-308552 DESIRE is gratefully acknowledged.

References

Bajželj, B., Richards, K. S., Allwood, J. M., Smith, P., Dennis, J. S., Curmi, E., & Gilligan, C. A. (2014). Importance of food-demand management for climate mitigation. *Nature Climate Change, 4*, 924–929. doi:10.1038/NCLIMATE2353.

Beretta, C., Stoessel, F., Baier, U., & Hellweg, S. (2013). Quantifying food losses and the potential for reduction in Switzerland. Spec. Themat. *Issue Urban Mining, 33*, 764–773. doi:10.1016/j.wasman.2012.11.007.

Cardinale, B. J., Duffy, J. E., Gonzalez, A. H. D. U., Perrings, C., Venail, P., Narwani, A., et al. (2012). Biodiversity loss and its impact on humanity. *Nature, 486*, 59–67. doi:10.1038/nature11148.

Carey, J. (2013). Urban and Community Food Strategies. The case of Bristol. *International Planning Studies, 18*, 111–128. doi:10.1080/13563475.2013.750938.

Cassidy, E. S., West, P. C., Gerber, J. S., & Foley, J. A. (2013). Redefining agricultural yields: From tonnes to people nourished per hectare. *Environmental Research Letters, 8*, 034015. doi:10.1088/1748-9326/8/3/034015.

Erb, K. H., Haberl, H., Krausmann, F., Lauk, C., Plutzar, C., Steinberger, J. K., Müller, C., Bondeau, A., Waha, K. & Pollack, G. (2009). Eating the planet: Feeding and fuelling the world sustainably, fairly and humanely—A scoping study. Report commissioned by Compassion in World Farming and Friends of the Earth, UK, *Social Ecology Working Paper*. Institute of Social Ecology and PIK Potsdam, Vienna, Potsdam.

European Commission. (2013). *Building the single market for green products. Facilitating better information on the environmental performance of products and organisations.*

European Union. (2012, July). *Rural Review Issue 12—Local food and short food supply chains.* A publication from the European Network of Rural Development.

Finkbeiner, M. (2014). Product environmental footprint—Breakthrough or breakdown for policy implementation of life cycle assessment? *International Journal of Life Cycle Assessment, 19*, 266–271. doi:10.1007/s11367-013-0678-x.

Finkbeiner, M., Inaba, A., Tan, R., Christiansen, K., & Klüppel, H.-J. (2006). The new international standards for life cycle assessment: ISO 14040 and ISO 14044. *International Journal of Life Cycle Assessment, 11*, 80–85. doi:10.1065/lca2006.02.002.

Foley, J. A., Ramankutty, N., Brauman, K. A., Cassidy, E. S., Gerber, J. S., Johnston, M., et al. (2011). Solutions for a cultivated planet. *Nature, 478*, 337–342. doi:10.1038/nature10452.

Friedmann, H. (2005). From Colonialism to Green Capitalism: Social movements and emergence of food regimes. In *New directions in the sociology of global development, research in rural sociology and development* (pp. 227–264). Emerald Group Publishing Limited.

Freidberg, S. (2015). From behind the curtain: Talking about values in LCA. *The International Journal of Life Cycle Assessment,* 1–5. doi:10.1007/s11367-015-0879-6

Garnett, T. (2014). Three perspectives on sustainable food security: Efficiency, demand restraint, food system transformation. What role for life cycle assessment? *Journal of Cleaner Production, 73*, 10–18. doi:10.1016/j.jclepro.2013.07.045.

Godfray, H. C. J., Crute, I. R., Haddad, L., Lawrence, D., Muir, J. F., Nisbett, N., et al. (2010). The future of the global food system. *Philos Transactions R Society London B Biology Science, 365*, 2769–2777. doi:10.1098/rstb.2010.0180.

Gruber, L., Brandstetter, C., Bos, U., Lindner, J. & Albrecht, S. (2015). LCA study of unconsumed food and the influence of consumer behavior. *The International Journal of Life Cycle Assessment,* 1–12. doi:10.1007/s11367-015-0933-4

Gustavsson, J., Cederberg, C., Sonesson, U., van Otterdijk, R., & Meybeck, A. (2011). Global food losses and food waste: extent, causes and prevention. Rome: Food and Agriculture Organisation of the United Nations (FAO).

Hinrichs, C. C. (2000). Embeddedness and local food systems: Notes on two types of direct agricultural market. *Journal Rural Studies, 16*, 295–303. doi:10.1016/S0743-0167(99)00063-7.

Hörtenhuber, S., Lindenthal, T., Amon, B., Markut, T., Kirner, L., & Zollitsch, W. (2010). Greenhouse gas emissions from selected Austrian dairy production systems—model calculations considering the effects of land use change. *Renewable Agriculture Food System, 25*, 316–329. doi:10.1017/S1742170510000025.

Hörtenhuber, S., Lindenthal, T., & Zollitsch, W. (2011). Reduction of greenhouse gas emissions from feed supply chains by utilizing regionally produced protein sources: The case of Austrian dairy production. *Journal of the Science of Food and Agriculture, 9*, 1118–1127. doi:10.1002/jsfa.4293.

IPCC. (2006). IPCC guidelines for national greenhouse gas inventories, Vol. 4, Agriculture, forestry and land use. In S. Eggleston., L. Buendia., K. Miwa., T. Ngara, K. Tanabe (Eds.), *Prepared by the national greenhouse gas inventories programme*. IPCC NGGIP-TSU, IGES, Hayama, Kanagawa, Japan.

IPCC. (2014). *Climate change 2014: Synthesis report. Contribution of Working Groups I, II and III to the Fifth Assessment Report of the Intergovernmental Panel on Climate Change* [Core Writing Team, R. K. Pachauri, L. A. Meyer (Eds.)]. Geneva, Switzerland.

ISO. (2006a). ISO 14040, Environmental management—Life cycle assessment: Principles and framework.

ISO. (2006b). ISO 14044, Environmental management—Life cycle assessment: Requirements and guidelines.

Kearney A. T. (2013). *Lebensmittel: Regional ist gefragter als bio*. Wien.

Kummu, M., de Moel, H., Porkka, M., Siebert, S., Varis, O., & Ward, P. J. (2012). Lost food, wasted resources: Global food supply chain losses and their impacts on freshwater, cropland, and fertiliser use. *Science of the Total Environment, 438*, 477–489. doi:10.1016/j.scitotenv.2012.08.092.

Lindenthal, T., Markut, T., Hörtenhuber, S., Theurl, M. & Rudolph, G. (Eds.). (2010). *Greenhouse gas emissions of organic and conventional foodstuffs in Austria*. Presented at the VII. International conference on life cycle assessment in the agri-food sector (LCA Food), Notarnicola, B.

Lughofer, E. (2010). Treibhausgas-Emissionen bei unterschiedlichen Vermarktungsformen von Gemüse-eine Fallstudie unter Berücksichtigung des Einkaufsverkehrs.

McMichael, P. (2005). Global development and the corporate food regime. In *New directions in the sociology of global development* (pp. 265–299) Research in rural sociology and development. Emerald Group Publishing Limited.

Meier, M. S., Stoessel, F., Jungbluth, N., Juraske, R., Schader, C., & Stolze, M. (2015). Environmental impacts of organic and conventional agricultural products—Are the differences captured by life cycle assessment? *Journal of Environmental Management, 149*, 193–208. doi:10.1016/j.jenvman.2014.10.006.

Milestad, R., Bartel-Kratochvil, R., Leitner, H., & Axmann, P. (2010). Being close: The quality of social relationships in a local organic cereal and bread network in Lower Austria. *Journal Rural Studies, 26*, 228–240. doi:10.1016/j.jrurstud.2010.01.004.

Morgan, K. (2008). Greening the Realm: Sustainable food chains and the public plate. *Regional Studies, 42*, 1237–1250. doi:10.1080/00343400802195154.

Paxton, A. (1994). *The food miles report: The dangers of long distance food transport*. London: Safe Alliance.

Pretty, J. N., Ball, A. S., Lang, T., & Morison, J. I. L. (2005). Farm costs and food miles: An assessment of the full cost of the UK weekly food basket. *Food Policy, 30*, 1–19. doi:10.1016/j.foodpol.2005.02.001.

Schmidt, H.-J. (2009). Carbon footprinting, labelling and life cycle assessment. *International Journal of Life Cycle Assessment, 14*, 6–9. doi:10.1007/s11367-009-0071-y.

Sim, S., Barry, M., Clift, R., & Cowell, S. J. (2007). The relative importance of transport in determining an appropriate sustainability strategy for food sourcing. *International Journal of Life Cycle Assessment, 12*, 422–431. doi:10.1065/lca2006.07.259.

Smith, P., Bustamante, M., Ahammad, H., Clark, H., Dong, H., Elsidding, E. A., et al. (2014). Agriculture, forestry and other land use (AFOLU). In O. Edenhofer., R. Pichs-Madruga., Y. Soukuba (Eds.), *Climate change 2014: Contributions of working group III to the 5th assessment report of the IPCC*. Intergovernmental Panel on Climate Change, Geneva, Switzerland, Cambridge, UK.

Smith, A. D., & MacKinnon, J. B. (2007). *The 100 mile diet: A year of local eating*. Toronto: Vintage Canada.

Theurl, M. C. (2015). CO_2-Fußabdruck. *In: Endbericht Des Projektes Winterernte: Saisonaler, Energie-Extensiver Und Innovativer Gemüseanbau.*

Theurl, M. C., Haberl, H., Erb, K.-H., & Lindenthal, T. (2014). Contrasted greenhouse gas emissions from local versus long-range tomato production. *Agronomy Sustainable Devlopment, 34*, 593–602. doi:10.1007/s13593-013-0171-8.

Theurl, M. C., Hoertenhuber, S. J., Lindenthal, T. & Palme, W. (in preparation-a). Vegetable production quo vadis? Winterharvest in Northern countries.

Theurl, M. C., Lindenthal, T., Hoertenhuber, S. J. & Drapela, T. (in preparation-b). Organic and conventional food. An empirical sustainability analysis from Austrian production systems.

Tilman, D., & Clark, M. (2014). Global diets link environmental sustainability and human health. *Nature*. doi:10.1038/nature13959.

van Vuuren, D., Lowe, J., Stehfest, E., Gohar, L., Hof, A., & Hope, C. (2011). How well do integrated assessment models simulate climate change? *Clim. Change, 104*, 255–285. doi:10.1007/s10584-009-9764-2.

von Thünen, J. H. (1826). *Der isolierte Staat in Beziehung auf Landwirtschaft und Nationalökonomie*. Jena: Fischer.

Part IV
Waterscapes: Competing for Land and Water

Chapter 19
A Water Perspective on Land Competition

Antje Bruns, Tobias Krueger, Bruce Lankford, Fanny Frick, Catherine Grasham and Christina Spitzbart-Glasl

Abstract This chapter reflects on land competition from a water perspective. Conceptual thoughts are enriched with evidence drawn from case studies as well as other published studies about both land and water. At the same time, it lays down an analytical framework for these case studies. Starting with a discussion of the inherent relationship between land and water, we explore recent disconnects in land and water studies that make it difficult to collate empirical evidence and comprehensive understanding of how competition between water and land are inherently linked. For us the term competition refers to gaining access to or control over—either land or water—and thus simultaneously captures social and material dimensions. To address these linkages, we employ the concept of waterscapes. One way of seeing waterscapes is through the lens of the competition that occurs at specific places, in various positions and on/across various scales, thereby capturing a combined view of land and water. The notion of waterscapes is mainly used by scholars from the fields of political ecology and critical geography thinking to explore how power is wielded, and in determining when and where who or what

A. Bruns (✉) · F. Frick
Governance and Sustainability Lab, Trier University, Trier, Germany
e-mail: brunsa@uni-trier.de

F. Frick
e-mail: fanny.frick.1@student.hu-berlin.de

T. Krueger · F. Frick
Department of Geography and IRI THESys, Humboldt-Universität zu Berlin, Berlin, Germany
e-mail: tobias.krueger@hu-berlin.de

B. Lankford · C. Grasham
School of International Development, University of East Anglia, Norwich, UK
e-mail: B.Lankford@uea.ac.uk

C. Grasham
e-mail: C.Grasham@uea.ac.uk

C. Spitzbart-Glasl
Institute of Social Ecology, Alpen-Adria University, Vienna, Austria
e-mail: Christina.Spitzbart@aau.at

J. Niewöhner et al. (eds.), *Land Use Competition*,
Human-Environment Interactions, DOI 10.1007/978-3-319-33628-2_19

gets how much water/land. We briefly review the different notions of competition in disconnected literature concerning land and water in order to instil a further analytical dimension: whilst the term “competition” is increasingly used in land change science to refer to the global rush for land, water scholars refer rather to the various means of water governance.

Keywords Critical geography · Waterscape · Governance · Land–water coupling · Resources · Water as a hybrid

19.1 Introduction

Land competition is about more than land. Competition for land typically takes place where investors see a “high growing potential, which means land with lots of rainfall or land that can be irrigated” (Skinner and Cotula 2011: 1). Consequently, (large) land acquisitions are frequently accompanied by the securing of water rights. The global rush for land is thus—implicitly or explicitly—a rush for the blue gold (Allan et al. 2013). This perspective forms the background for this chapter, in which we explore competition for land as seen through the lens of water.

Current struggles over land are connected to water—and other resources—and as such, land competition is directly linked to how we use and allocate water in terms of quality, quantity and timing (Hodgson 2004). Aspirations concerning food security go hand in hand with an increase in irrigation; urbanisation, itself more than the mere expansion of urban land, but also implying changes in consumption, involves higher water demand for cities; and bioenergy production as a means of securing energy supply and mitigating the effects of climate change is not only a driver of land competition but also of competition over water for various purposes. It is well documented that an increase in water use often leads to overexploitation, pollution or degradation of aquatic ecosystems (Molle and Wester 2009). In addition, societal problems and new inequalities are also emerging as a result.

Solutions for just and ecologically sustainable water management often lie in the realm of land management. However, as argued by Sikor it is the market-driven revalorisation of land generating the competition in the first place (Sikor et al. 2013).

19.2 The Relationship Between Land and Water

Land and water are intimately linked to one another. Water can either enable or constrain land-based activities which, may, in turn, compromise water quantity and quality. We can make a distinction between cases where having too little water would limit land-based activities and cases where too much water would itself be limiting. The first situation puts the focus on water availability, which shall here be defined as the presence of water of appropriate quality for a specific use. Not only then are the quantity and quality of water interrelated; water availability must fundamentally be considered as being relative to its use. We must further consider where and when water is available; the timing of water availability is important for ecosystems and human activities. Water availability is frequently a matter of access through biophysical, technological, cultural, economic or political means: it needs to be pumped, dammed, purchased, negotiated for or otherwise appropriated.

Of the different land-based uses of water, irrigated agriculture accounts for 70 % of water withdrawals at a global level (FAO 2012). The coupling of food and energy production with water and to each other has prominently been termed the water–food–energy nexus (Ringler et al. 2013). This framework includes water for agricultural food, feed and fuel production as well as for hydropower and the cooling function of water in power stations. This may, however, draw attention away from other land-based uses of water: for drinking and sanitation, for water-based livelihoods such as fishing, for terrestrial "unmanaged" ecosystems, settlement and tourism infrastructure as distinct from drinking and sanitation (e.g. in lawn watering), for the agricultural production of fibre, for forestry, landscape value (individual and community identity, aesthetics and recreational value), and in other industries (processing and cooling).

In general, anybody who uses water automatically limits the availability of that same water for somebody else, and yet conversely no water use means that water is fully consumed. Some of the water used will return to where it came from or be passed on to other uses, albeit seldom with the desired quality. Part of the losses to efficiency in irrigation also eventually become return flows, but along the way the water may have picked up chemicals used in agriculture. Water pollution occurs through sediment, nutrients, pesticides, veterinary pharmaceuticals and faecal coliforms amongst other pollutants and is a major externality of agricultural production worldwide. Domestic and industrial water uses partly return as treated or untreated wastewater and are often reused in wastewater irrigation (Qadir et al. 2010). The function that water performs in producing landscape value is strictly non-consuming. A special link between water quantity and quality is in salinisation, where use of freshwater draws in seawater, or groundwater, the salt content of which becomes concentrated in the soil and soil water in semi-arid and arid environments. As an alternative to focussing on pollution, certain ecosystems such as wetlands and forests have been seen in terms of their filtering function for improving water quality.

The reverse side of water scarcity is having too much water. An excess of water due to flooding or high groundwater levels places constraints on land-based activities—which is why we build flood protection measures and land drainage systems in low-lying areas (Bruns 2015). Floods act as vectors of pollution as well as of nutrient input. As with water availability, the temporal dynamics of excess water are crucial. Whether flooding is a frequent or rare event, and hence expected or surprising, will help shape land-based activities, including any mitigation, adaptation or coping strategies (de Vries 2011). Land-based activities, in turn, exert an influence on flooding and groundwater levels. Levees protect settlements but may lead to more or greater floods downstream (Di Baldassarre et al. 2014). Sealed urban surfaces increase surface run-off generation and concentration at the expense of groundwater recharge. Channel straightening for navigation and for drainage increases run-off concentration. Wetlands and "natural" floodplains buffer flood water. Agricultural land use influences partitioning of precipitation into slow and quick flows, i.e. soil matrix and groundwater flows versus run-off and drainage, mediated by soil processes.

When disentangling land–water couplings, we might usefully start with the classic hydrological cycle of precipitation, storage, discharge and evaporation/transpiration that is mediated by climate, land use and soils. Within this concept, land-based activities are connected by the river, lake, reservoir or aquifer they share or because they lie in the same catchment (watershed). It should be noted that surface and subsurface catchments do not usually overlap. Spatially distant land-based activities are connected through atmospheric water transfers as evaporation and transpiration in one area of the globe and then become precipitation elsewhere. Although they are part of the classic hydrological cycle, these atmospheric couplings have only recently come to be seen as a problem, in the concept of the "precipitation-shed" Keys et al. (2012), for example, based on a modelling study demonstrated that deforestation in the Congo and West Africa reduces precipitation in the Sahel via transpiration and atmospheric advection. Further overarching land water feedback exists via climate in which land-based greenhouse gas emissions remodel the water cycle by changing the climatic energy inputs that drive the cycle in the first place (mainly through temperature changes).

What the classic hydrological cycle does not convey are the social dimensions of land–water couplings. Hence, the recent turning of members of the water science community towards the concept of the hydrosocial cycle (Linton and Budds 2014) amongst those social dimensions is virtual water, i.e. the amount of water used in the production of a commodity or service (Allan 2003). The international trade in goods and services thus "shifts" virtual water, opening up another level of supply-demand dynamics that can exacerbate as well as alleviate water availability problems. For example, a country that is "water-poor" could choose to import water-intensive products, thereby alleviating the pressure on its own water resources (Conway et al. 2015). If the exporting countries were "water-rich", all would be well. However, if an exporting country is water-poor then water availability problems would merely be displaced. In reality, water-poor countries frequently export to water-rich countries since they are subject to market forces (Allan 2003).

In an effort to connect the biophysical and social dimensions of land–water couplings, one might say that water flows along gradients, but not in the classic hydrological sense that refers to the differences in hydraulic head. Water does indeed flow according to hydraulic gradients, mainly downward, but also according to power gradients (Swyngedouw 2009). An important means to managing and controlling the flow are water infrastructures that mediate who gets how much water of what quality and when. Since water is a globally finite resource, water flows may be viewed as the outcomes of the competition for water. What we observe as land use is then, at least in part, the outcome of such competition for water. The fact that water is locally renewable and reusable, and excess water is also undesirable, creates ever more subtle nuances in water competition.

19.3 Competition Over Land and Water

Water and land systems are almost always studied independently from each other, as they have traditionally been separated into different disciplines that are characterised by very different approaches (Bakker 2009; Falkenmark 1997). In the previous section, however, we showed that land and water are in fact inseparably related. To achieve a greater integration of landscape and water analysis, integrative research needs to be promoted between "biologists and geoscientists, between those with focus on land use and those focussing on water, between water flow experts and water quality experts, and between natural scientists and political scientists and decision makers" (Falkenmark 1997: 454). Although this sentence was written almost 20 years ago, it still holds true as Krueger et al's recent commentary shows (Krueger et al. 2016).

Humans influence land systems and freshwater flows in numerous ways—mainly for the satisfaction of societal needs arising through population increase, economic growth, industrial development, and demand for energy (Falkenmark 1997; Gheewala et al. 2011). The rising demand for agricultural products (bioenergy, food, feed and fibre) is driving land-use change and ultimately competition for land, and moreover, the need to adapt to climate change and increasing efforts to conserve ecologically valuable areas are turning land into a scarce resource.

Large-scale land acquisitions are both a phenomenon and a driver of land competition and are an increasingly documented global issue (Borras et al. 2012; White et al. 2012). Yet the underlying driving forces for this development are manifold (population growth; dietary changes; political incentives for bioenergy crops), whereas the pace and scale of resource competition has markedly accelerated during the global credit crisis which manifesting itself in 2007 (Allan et al. 2013). Although there is no globally reliable data on the extent of large land deals and the methodologies used to quantify them, "there is a consensus that land grabbing is underway and that is significant" (Borras et al. 2012: 2).

Whilst water definitely is a factor in land deals (Smith 2010), it has long played a secondary role: the production of land-based products implicitly must include water, yet its role has only recently been addressed as a driver or object of land deals in the literature (Franco et al. 2013). Furthermore, as Allan et al. reveal "it is not widely understood that the vast majority of the water consumed by an individual or an economy is embedded in its food consumption" (Allan et al. 2013: 43). Even though population growth is commonly seen as a direct driver for the increasing water demand, it is mainly dietary changes that have led to the increase in consumptive water. Hence, water security is very closely linked to cultural and economic globalisation seen in global food markets and international trade agreements (see Chap. 14).

It is only in recent years that scholars have been placing their focus on water itself and pointing out that whereas land grabbing has long been the focus, "water as both a target and driver of this phenomenon has been largely ignored despite the interconnectedness of water and land" (Mehta et al. 2012: 193). Similarly Rulli et al. argue that "land grabbing is not only a rush for land but also for the freshwater resources available therein" (Rulli et al. 2013: 892).

19.3.1 Resource Characteristics

Despite the inherent coupling of land and water, it might be useful to give a brief examination of their inherent characteristics. This will later help us to understand how competition plays out within the waterscape and shapes it. As a preliminary remark, we find it interesting that the notion of competition that is at the heart of this book is less often used in water studies (except in Lankford et al. 2009; Franco et al. 2013) than in land change science (e.g. Haberl et al. 2014; Haberl 2015). These scientific communities obviously engage differently with interaction processes that can either be framed as competition (suggesting a resource economy view) or as struggles over resource control [inspired by political ecology (Borras et al. 2012; Cotula 2012; White et al. 2012)].

Whilst land is a resource that is bound to a specific place, water is fluid and as such part of a cyclic process of renewal. As a flow, water connects different places (upstream/downstream) which transcends political boundaries of any scale. Due to its fluid character, conflicts over the appropriation or allocation of water may affect a whole river basin (Franco et al. 2013). Water is a classic example of a common good (theoretically, at least), where property rights can be collective, state-owned or private. Water rights are, however, often attached to land rights, as Sjöstedt showed for sub-Saharan Africa, which means that enhancing access to water is highly dependent on land policies and land tenure (Sjöstedt 2011). At this point, we can only refer to the complex nature of land rights—in many countries characterised by legal pluralism.

Only with the surge in the privatisation and commodification of water since the 1990s and the controversial declaration of water as an economic commodity in the Dublin Principles, has water been increasingly redefined as such (Bakker 2002, 2009). Whilst water has to some extent been a contested resource throughout history, this commodification of water amidst narratives of "water scarcity" and "water security" creates a new relevance for the concept of water competition as we are "witnessing the privatisation not just of the service or the accompanying infrastructure but of the resource itself" (Mehta et al. 2012: 198). The process of commodifying water led, however, also to what Walsh calls the "unbundling of land and water rights" (Walsh 2011: 94). He reminds us that water has been usually been bound to the land(owner) and only its marketisation has led to an unbundling into two commodities.

Despite these alarming trends, there is still more evidence that competition for water fosters cooperation rather than conflict (Zeitoun and Mirumachi 2008). How a limited resource can be managed sustainably has been an established field of research for many years and there is accordingly a huge body of work drawing on a theoretical framework of shared pooled resources (Ostrom 1999) with a focus on cooperation and governance. Since the resource characteristics of land were discussed throughout the book, we will not review them in depth but will merely refer to the table (see Table 19.1).

In the light of the specific resource characteristics listed in the table, we can see that water competition brings another level of complexity to an understanding of land competition. Not only is water fluid and renewable in nature, but the mechanisms of water governance are even more complicated due to its multi-level governance that itself is structured through "unequal power relations; fuzziness between legality and illegality and formal and informal rights; unclear administrative boundaries and jurisdictions" (Mehta et al. 2012: 194). Together with opposing conceptions of water as a life-giving resource vs. an economic commodity and various strong cultural belief systems, water competition is fraught with numerous diverging interests and unclear mechanisms that need to be brought to light. A second analytical aspect that appears particularly through a water perspective is that of scale and place: water can link extremely distant places and thus links competition situations across scale.

19.3.2 *The Governance of Land and Water*

A governance perspective that links land and water and goes beyond a rhetorical claim that it overcomes arbitrary separations whilst implementing a more holistic governance approach is much needed (Palmer et al. 2009). Governance can be

Table 19.1 The connects and disconnects of land and water

	Water as a resource	Land as a resource
Resource characteristic	Finite (locally) but renewable resource on a global level	Finite resource
	Flow resource, transcending territories and boundaries	Bound to a specific place/locality
Property regime	Varying: from common pool resource to an increasingly commodified and privatised resource	Varying: from customary land law (often oriented towards communal use) to state land law (often private); also legal pluralism
	Water rights often attached to land rights	
Contemporary discourse	Common good versus economic good, shared responsibility versus commodification and privatisation	Large-scale land acquisitions; Land prices—for housing in cities as much as for farming
	Water as a basic need	Food security
	Right to water and water rights	Land sovereignty

understood as an "integrated system of formal and informal rules, rule-making systems, and actor-networks at all levels of human society" (Biermann et al. 2009: 4) and it shapes the way in which decisions are made and how land and water resources are ultimately governed.

Environmental governance represents a subject of increasing study, especially in the water domain at regional levels (e.g. Pahl-Wostl et al. 2010) but proves to be equally important as regards global challenges that affect land and water at the same time (Mehta et al. 2012). Without improved knowledge of land–water couplings "it is almost impossible to design appropriate management tools or even the adaptive intervention experiments needed to inform policy and management" (Steffen 2009: 1301). Yet, whilst the key role land plays for water governance has been recognised and is even a fundamental part of Integrated Water Resources Management, mechanisms for operationalisation and implementation remain weak. The European Water Framework Directive (WFD) can serve as an interesting example of political willingness to implement integrated land–water management that predominantly builds on river basins as a central coordination unit (Molle 2007). A key concern of river basin management in Europe is high nitrogen and phosphorous input from agriculture. For this, the WFD aims at addressing the long neglected land–water nexus by referring to core ideas of Integrated Water Resources Management as being a holistic approach despite river basin management plans that have been introduced in response to binding legislation, and a generally growing awareness that land and water are still highly disconnected policy spheres (Bandaragoda 2006; Bruns 2015). Over the past years and even decades, we have gathered sufficient evidence that the sectoral strategies embedded in institutional responsibilities inhibit any integrative governance across sectors, actors and scales.

Moreover, land governance has been widely discussed in the light of contemporary debates on food security, land sovereignty, and the increasing land demands made for environmental protection and international climate agreements (such as REDD+) both in science and policy. Yet, similarly to water, a global governance institution with a "comprehensive mandate to govern land" (Sikor et al. 2013: 522) is still missing (see also Lambin and Meyfroidt 2011). Whilst the expansion of agricultural land and resulting competition is influenced by the political economy on the global level, underlying governance changes on local, national and global level continue to accumulate in the now global rush for land and other resources (Lambin and Meyfroidt 2011; Sikor et al. 2013). The increase in global competition for land is enabled by world trade regulations and international markets that are probably more influential than national regulations on land or water management. Sikor et al. therefore come to the conclusion that we are currently experiencing a shift from territorial land-use governance that is based on localised regulations (on land reforms, on land consolidation programs or on definitions of conservation and protected areas) towards a governance of flows (Sikor et al. 2013). With global challenges and trade flows, land governance is becoming increasingly detached from territory. Instead, production standards or certification schemes are the new governance instruments for land. It is exactly this shift that leads to a regulatory gap which fails to address contemporary challenges which range from degradation and pollution to a growing divide between those who have (e.g. access to land and water) and those who have not. Inequalities are intensifying on various levels and land governance will be able to address emerging water challenges less than ever.

The global land matrix is providing initial evidence that there is a severe lack of regulation and governance in the field of land deals which ultimately have effects on water (The Land Matrix Global Observatory 2016). In Mali, more than 160,000 ha of land were acquired between 2000 and 2012, for which water rights were part of the leasing contracts with flat rate water fees (Boroto et al. 2014). Only recently have these dynamics been gaining in visibility, owing to rising prices and more vehement protests by those who are excluded from access to either land or water as well as due to the decision-making processes that led to these trajectories (Rulli et al. 2013; Simpson et al. 2008; Wolford et al. 2013). Some countries, such as Brazil and Argentina, have even introduced policies and laws to restrict foreign land deals and to regulate the global rush for land that threatens their national sovereignty (Oliveira 2013; Perrone 2013) as well as the overall resource base.

19.4 Analytical Framework: Waterscapes

This section argues that the concept of waterscapes presents a useful analytical tool for dealing with both water and land(scapes) simultaneously. It aims at integrating socio-ecological elements and the socio-hydrological cycle into a single analytical framework. Accounting for waterscapes draws attention to the inherent coupling of landscapes and water through processes caused by competition. As Rockström puts it, "land use choices are also water choices, and will always lead to alterations in the flow of freshwater and ecosystem services elsewhere" (Rockström et al. 1999: 5). Nevertheless, water cannot be treated as a purely economic commodity, as it also supports life itself and the resilience of non-human environments. Within natural ecosystems, water plays an important role in providing essential physical, chemical and biological functions (Rockström et al. 1999; Falkenmark 1997; Falkenmark and Folke 2002).

19.4.1 From Water in Landscapes to Waterscapes

Historically, the manifold aspects of the water cycle have been addressed separately and as part of independent fields of study. For a long time, the prevalent paradigm in water management was notably characterised by utilitarian values and the allocation of water for economic development (Cortner and Moote 1994). According to the traditional "hard path" management paradigm, water supply relied heavily on engineering structures and a centralised infrastructure, resulting in increased environmental impact (Gleick 2002; Cortner and Moote 1994). Increasingly the evidence, however, suggests that this traditional approach to water (both water studies and water management) is currently in the process of being transformed. Since the 1970s, management practices have been shifting towards a new "soft path" paradigm, in which alternative practices aim at eliminating the division of water along political boundaries and the fragmentation of ecosystems into economic commodities (Cortner and Moote 1994).

In addition to this ecosystem-based approach, recent studies have expressed a growing interest in how we perceive the relationship between nature and societies. There is a growing scientific community that argues theoretically and study empirically the ways in which nature and society are inherently interwoven and not clearly separable into two distinct spheres. Thus, the notion of waterscape was initially used by scholars who argued that water is more than mere H_2O as a visible material matter, but also entails societal preferences and symbolic dimensions (Swyngedouw 1999; Budds 2009; Loftus 2007). The waterscape framework is therefore closely related to the science of hydrosociology first proposed by Falkenmark (1979) which targets the incorporation of human activities as endogenous parts into the hydrological cycle but since then further revised.

19.4.2 Using Waterscapes as an Analytical Lens to Address Land–Water Couplings Under Competition

The focus on the interconnectedness between humans and the hydrological cycle opens new ways of conceptualising complex processes such as competition within the water sphere is itself constituted through socio-political processes that structure the waterscape at different scales (Molle 2007). We can identify three crosscutting issues as being relevant for our analysis of how competition plays out in waterscapes. Firstly, the initial framing of water as a hybrid is an issue; secondly, the importance of place and scale and thirdly of governance and social power relations are the very socio-political processes that shape decisions.

Water as a hybrid

The waterscape concept is mainly employed to express the mutual causality between human societies and the natural environment, since it is impossible to sustain the physical/human divide when studying water (Bakker 2009). Societies significantly shape water flows in the landscape by means of laws, politics and the construction of hydraulic infrastructure. Society is in turn also shaped and influenced by water (Bakker 2009). The waterscape approach observes these highly dynamic societal relationships with water and sees the socio-hydrological cycle as coupled pair of entities that "continue to make and remake each other" in a discursive socio-ecological process (Budds 2009: 420). The notion of waterscapes has been predominantly shaped by Swyngedouw (1999, 2004) who label water a hybrid thing that is of material, discursive and symbolic relevance. By perceiving water as material, non-material and as social space that is subject to human activities, one can overcome the limitations of traditional perspectives (Zimmer 2011; Budds and Hinojosa-Valencia 2012).

The concept of waterscapes embodies these socio-economic, societal, political and ecological dimensions of the hydrological cycle as coupled and coproduced entities.

Analysing competition in waterscapes thus means that we analyse both the physical geographies of resources and the social processes including discursive strategies.

Reflections on Scale, Position and Place within the Waterscape

Competition plays out at different scales, locations and places, as we will illustrate in our case studies. The concept of scale is employed differently across scientific disciplines (Moss and Newig 2010). Within the field of geography and other natural sciences, for instance, scales have presented an influential concept. Geographical analyses have long been positioned within—apparently given—spatial scales, be they local, regional, national, or global (Gibson et al. 2000; Brenner 1998). Scales have provided established and interwoven measures that have secured "social, economic and political relations" within a hierarchical framework of territorial organisation (Brenner 1998: 478).

Since the 1990s, however, increasingly critical research has resulted in spatial scales no longer being conceived as pre-given domains of social interaction (Brenner 1998). On the contrary, scales are increasingly viewed as socially constructed and politically contested (Agnew 1997; Swyngedouw 1997; Brenner 1998). The idea of scales being constantly restructured in terms of social and political struggles is central to the concept of "(1) scale is socially constructed, (2) scale is both fluid and fixed; and (3) scale is fundamentally a relational concept" (Norman et al. 2012: 54). The concept of scales as socio-political constructs can be employed by individuals or groups to their advantage, which is also significant with regard to social power relations (Molle 2007). The choice of scale can be employed as "a tool of disempowerment, exclusion, and domination" (Brenner 1998: 479) and allows different stakeholders to justify certain strategies (Molle 2007). In the past, the "politics of scale" metaphor has been stretched to cover a lot of different meanings and has been used to illustrate ways in which traditional scalar dimensions are constrained by politics, technologies, institutional designs, and measures (Lebel et al. 2005).

Scales are important for environmental management and play a particularly important role in water management for which the river basin is often seen as a given management unit (Molle 2009). Environmental governance faces several challenges in terms of scaling, as administrative levels do not necessarily correlate with environmentally relevant spatial scales. According to some authors, this lack of conformity causes several problems and may result in serious inefficiencies and spatial externalities (Moss and Newig 2010). In line with this, Trottier argues that a "river basin or any other spatial unit can hardly be considered a 'natural' unit for water management, just as borders can hardly be considered to be 'natural' for a state" (Trottier 2008: 204). The river basin management approach therefore clearly illustrates the social and political construction of scale. Molle even argues that the river basin scale is used "to justify particular interventions, patterns of governance, or modes of water allocation" (Molle 2007: 359). Whilst scalar perspectives are crucial for water governance, they often reify socio-political demarcations and divides (Molle 2009; Norman and Bakker 2009; Cohen and Davidson 2011). Water and land governance needs a scale-sensitive approach and is faced with the remaining question as to what scale is the most relevant to water governance whilst acknowledging that water in the landscape is in a continuous flow and cannot be fixed into pre-given and imposed spatial orders.

Competition plays out in places and locations, or across scales

The drivers and impacts of competition over land and water summarised above require an analysis that goes beyond scalar container thinking that the waterscape idea would suggest. The concept of waterscapes not only characterises the socio-spatial configurations constituted by social and ecological processes and instead is also designed to establish alternative spatial scales (Budds and Hinojosa-Valencia 2012).

Lebel et al. (2005) conceptualised in their research on the governance of water resources in the Mekong region ways in which the politics of scale, place and position play out in its river basin. The authors refer to political processes in river basins. We apply this to competition situations and argue that waterscapes are a more flexible unit of analysis within our process-related perspective. Following Lebel et al. (2005), we use the "politics of scale" representation to refer to situations in which different actors scale information, decisions and policies in different ways. Scalar competition can occur concerning the best levels for policies, strategies and management schemes (as e.g. in the IWRM debate). On an international level, the notion of security is increasingly employed (FAO 2012; GWP 2014; Hanjra and Qureshi 2010). This hints primarily at the scope and significance of information which is considered to be higher on the international/national scale than on the local scale; secondly, it is a means to securing resources. Food, water and energy security aim to legitimise extraordinary policy, although this may contradict developmental innovations from other sectors or at different scales. As the Ethiopian case study (Grasham this volume) illuminates: "competition over water can be driven by universal discourses about food security and the need for more irrigated agriculture that manifests itself in specific places in a river basin. The 'politics of position' refers to politics among locations that depend on their relative physical position, for example, between upstream and downstream water users or those on different banks of a river" (Lebel et al. 2005: 18). This is a very common competition situation in river basins: be it the competition between several urban areas that are located by the same river and depend on the water (such as the cities and municipalities along the Colorado River in the arid US) or newly built dams that reduce the efficiency of downstream utilities (such as the Grand Ethiopian Renaissance Dam in Ethiopia that is currently under construction and heavily contested by the Egyptian state, whose Nile agriculture depends on the water). The "politics of place refers to the unfolding of power relations amongst stakeholders that arise because of the special characteristics of the places interacting above and beyond those arising from levels

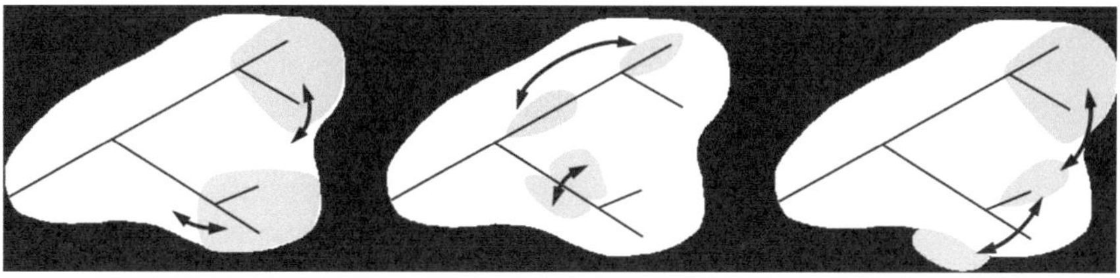

Fig. 19.1 Scale, position and places (*Source* own compilation)

or position" (Lebel et al. 2005: 18). This can be applied to most competition situations. As Franco et al put it "water may be the context of a grab, it may be the object of a grab, or it may be both at the same time" (Franco et al. 2013: 1652). So in all land-based activities where the availability or quality of water is altered and where this has any local impact, it is altering the place (Fig. 19.1).

To add to scale, position and place, in his treatment of the efficiency dimensions of irrigation, Lankford (2013) considers that land and water interconnect via the concepts of nestedness and neighbourliness which in turn depend on the varying soil–water routes available to irrigation water. Nestedness occurs when water losses move beyond the initial use in irrigation to the wider river basin or aquifer for reuse by the wider economy or the environment in general. Neighbourliness is found when water losses are picked up by immediate neighbouring water uses lateral to the location of those losses. In turn many social and environmental relations are dictated by these irrigation losses, their nestedness, neighbourliness, timeliness and degradation in water quality. The amounts of land and water involved in this process are very considerable and explains why and how in river basins and aquifers with irrigation present, water allocation on the basis of efficiency savings is both an imperative and yet highly problematic.

Finally, there is the need to hint at the temporal dimension that is central for competition situations, especially when long-established water uses are competed for by newer water uses, as is the case in many countries where developmental needs require both space (land) and water.

Waterscapes as politicised environments

Water governance and social power relations between different actors play an essential role in shaping the interaction between human activities and the flow of water. Waterscapes therefore present a politicised environment, sometimes labelled "landscape of power", which is formed in response to complex social, cultural and political power relations (Swyngedouw 2004: 29). This approach may be related to the tradition of political ecology that analyses the way in which water and power constantly shape and reshape each other. Through this conceptualisation, the interdependence of water cycle transformation and the differing positions of actors and users is emphasised (Perreault et al. 2012; Swyngedouw 2004, 2009). Power configurations amongst particular socio-ecological arrangements determine access to water, water quality, water rights and the flow of water (Perreault et al. 2012;

Swyngedouw 2009). Analysis of how waterscapes is actually produced, maintained or transformed can therefore illuminate how different actors shape the environment through their daily practices that are located in distinct places and embedded in varying scalar strategies. These practices are—as was discussed earlier—deeply connected with what we do on land. The multi-level governance of land and its revalorisation in the aftermath of the global economic crisis and aspirations to secure land and other resources is shaping contemporary waterscapes.

Reciprocal analysis of land–water couplings and power within the waterscape framework forms a suitable lens through which socio-ecological change can be examined by focusing on human interactions with the water cycle and the ways in which water discloses social perspectives (Budds 2013; Loftus and Lumsden 2008; Swyngedouw 2004, 2009). At this point, however, it becomes clear that linkages between the water cycle and (land-based) human interactions are very broad and cannot only be analysed in terms of governance structures (Bakker 2009). The waterscape framework therefore aims to relate the mutual constitution of water, cultural norms, urban forms and politics of water governance in a holistic way (Molle 2007). It becomes necessary to look "beyond the formal institutions that govern water, to the multiple social relations through which water governance is enacted over space and time" (Budds and Hinojosa-Valencia 2012: 133). By going beyond traditional framings the waterscape idea reflects the hybrid nature of water and embodies the uneven distribution of social power across multiple temporal and spatial scales (Perreault et al. 2012; Budds and Hinojosa-Valencia 2012). Most importantly, power and empowerment can thus be shown to flow "across scales in a direction opposite to water flows" (Molle 2007: 370).

Hence, the study of land competition through the lens of water(scapes) is a way to decentre the analytical focus towards fluid scalar configurations that are neither bound to political administrative boundaries nor to watersheds as given units of analysis. Instead, it would allow us to analyse how economic, social and political factors interact with the physical material resource base in driving the contemporary revalorisation of land and water that ultimately leads to competition for these resources. Furthermore, these perspectives allow a people-oriented approach that is often absent in global land change science but much needed to address competition for resources comprehensively.

19.5 Case Studies on Competition for Land and Water

Competition for either land or water is increasingly important for shaping waterscapes. The underlying dynamics are embedded in processes and contexts that are driving the rush for land and water, but the impact manifests itself in physical space, whether it be in irrigation schemes or in urban infrastructure management that struggling to provide enough water of sufficient quality for all. The following chapters shed light on different cases where competition over land and water is interlinked and the ways in which the process of getting access to and control over either land or water is constantly (re-) forming the waterscape: the case study on the Wien River in Vienna (Wienfluss) comes from the perspective of an environmental history that explains how past competition situations are still visible in the contemporary waterscape. The way everyday practices involved in competition situations are shaping the waterscape to degrees that vary according to localities (upstream/downstream) and places is illustrated in a second case, that of the Densu Delta in Ghana. The third case presents evidence of the way international and national developmental innovations are translated in policies and resource management strategies which trigger competition situations at regional and local levels.

Environmental History Perspective: Wien River

The case study on Wien River focusses on an historical transformation process and looks at both material and immaterial conditions, consequences and legacies of past activities. The physical resource base is discussed on two levels: the transition from an agrarian socio-metabolic regime based on energy from biomass to an industrial regime based on energy from fossil fuels which forms the framework of this case study. With this background, the water in this river loses significance as a resource of energy and materials whilst the space the river demands becomes a desired resource in an expanding city. The complex linkages between material arrangements, users and their practices with its high potential for competition, conflicts and need for negotiation and compromise are vividly pictured in the "Wienflussaufnahme". The case study also touches on how new power relations and authorities have influenced the transformation of the Wien River valley. Scale and place issues are not explicitly discussed, although the concept of socio-natural sites is suited to doing this: this concept is spatially open, the boundaries of a socio-natural site being defined by the analytical gaze of researchers and the questions they ask. A river links upstream and downstream water and land-use into a "longitudinal continuum". The flexible analytical boundaries of socio-natural sites make this concept especially appropriate when studying rivers, since it can correlate practices and material arrangements on the "macro" level (e.g. in a watershed) and "micro" levels (e.g. at the location of a weir) scale The case study on Wien River provides several examples of the way the river interconnects different users and places—often in an undesirable way prone to conflicts, such as when flood protection measures shift flood risk to other places or give rise to water pollution upstream resulting in poor water quality and even disease downstream.

Land–Water Interactions in an urbanising Delta

The journey along the lower Densu River in Ghana exposes multiple land–water interactions that are mediated by contestation over the power to control these resources amongst different actors. For instance, the conversion of agricultural land into housing land by local traditional chiefs has resulted in an overabundance of water in a reservoir managed by the Ghana Water Company which provides the capital city Accra with potable water. As a consequence, the company needs to spill water from the dam increasingly often, exposing downstream settlers to flooding. Actors engaged in these contestations represent varying interests in different places, and at different levels of scale. An example of competing global and local interests is the implementation of wetland conservation as part of the international Ramsar convention and against local chiefs' interest in gaining revenues from land sale, and the interests of migrants in finding a place to settle. Material flows link places as water is channelled from the Weija treatment plant to Accra and waste is brought from Accra to Oblogo, and also because upstream pollution and modification of water streams cause poor water quality and flood risk downstream. In the Densu Delta waterscape, not only is the water an unstable, fluid resource, but the same can be said of land which is lost through erosion along the coast, and reclaimed from the Sakumo lagoon by filling it with gravel and waste.

Competition over Irrigation for Food or Water for Urban Dwellers

The third case is concerned with the land and water resources within the Haramaya watershed in the semi-arid eastern highlands of Ethiopia. This study is of particular interest as there used to be a flourishing lake in the watershed that supplied a large urban population, a local rural population and several smallholder irrigators with water until it dried up completely in 2004. Today, the lake has partially returned, but it is a shadow of its former self. In this case study, land–water coupling is particularly acute since land-use change over time has had an influence on water-use competition and vice versa. The chapter illuminates how the unsustainable abstraction of water resources can be attributed to rapid water resource developments at the local and regional levels which have been spurred by policies and strategies influenced by international and national development visions. It explores the way competition between urban water supply and smallholder irrigation has been exacerbated by these visions. Using the concept of the waterscape, a holistic analysis of the influence of actors involved in governance of the watershed has become possible. The actors exist at different levels of government and water development processes are influenced by a top-down approach to management with limited lateral linkages between actors. This has resulted in a fractured institutional framework governing water resources in Ethiopia, where water supply and irrigation developments are being steered independently of each other.

The case studies illustrate the dynamic coproduction of waterscapes through competition over the power to control water and land as resources for development. Urban development is a key driver of this competition in all three cases. Framing water as a hybrid served to reveal both material and social relations that shape the respective waterscapes. As changes in water and land use interact, places and scales are connected and disconnected in myriad ways—place and scale thus become relative entities. Governance and power relations determine decisions taken on land and water use and thus shaping social, material and scalar relations. This chapter and the three case studies thus introduce waterscape as a frame to analyse water- and land-use competition simultaneously.

Acknowledgments We would like to thank Emily Raab and Viviana Wiegleb for their support. This research is partly funded by BMBFGlobal Change Research under the reference number 01LN1316A.

References

Agnew, J. (1997). The dramaturgy of horizons: Geographical scale in the 'Reconstruction of Italy' by the new Italian political parties, 1992–1995. *Political Geography, 16*(2), 99–121. doi:10.1016/S0962-6298(96)00046-7.

Allan, J. A. (2003). Virtual water—the water, food, and trade nexus. Useful concept or misleading metaphor?: Water international. *Water International, 28*(1), 106–113. doi:10.1080/02508060.2003.9724812.

Allan, J. A., Keulertz, M., Sojamo, S., & Warner, J. (Eds.). (2013). *Handbook of land and water grabs in Africa: Foreign direct investment and food and water security* (1st ed.). London: Routledge international handbooks. Routledge.

Bakker, K. (2002). From state to market?: Water mercantilización in Spain. *Environment and Planning A, 34* (5), 767–790. doi:10.1068/a3425

Bakker, K. (2009). Water. In N, Castree., D. Demeritt., D. Liverman., & B. Rhoads, (Eds.), *A companion to environmental geography* (pp. 515–532). Wiley.

Bandaragoda, D. J. (2006). Water-land linkages: A relatively neglected issue in IWRM. In P. P. Mollinga, A. Dixit, & K. Athukorala (Eds.), *Integrated water resources management: Global theory, emerging practice, and local needs* (pp. 172–185). New Delhi, Thousand Oaks, Calif: Sage Publications.

Barnes, J. (2012). Pumping possibility: Agricultural expansion through desert reclamation in Egypt. *Social Studies of Science, 42*(4), 517–538. doi:10.1177/0306312712438772

Biermann, F., Betsill, M. M., Gupta, J., Kanie, N., Lebel, L., Liverman, D., et al. (2009). Earth system governance. *People, places, and the planet: Science and implementation plan of the earth system governance project 20.* Accessed January 21, 2016.

Boroto, J. R., Chiramba, T., Mafuta, C. & Williams, T. O. (2014). *Analysis of impacts of large-scale investments in agriculture on water resources, ecosystems and livelihoods; and development of policy options for decision makers: Summary of initial findings.* http://www.grida.no/publications/water-grab/

Borras, S. M., Franco, J., Wang, C. (2012). *Competing political tendencies in global governance of land grabbing*. Discussion paper.

Brenner, N. (1998). Between fixity and motion: Accumulation, territorial organization and the historical geography of spatial scales. *Environment and Planning D, 16* 459–482.

Bruns, A. (2015). Wasser- und Bodenschutz in der Landschaftsplanung. In W. Riedel, H. Lange, E. Jedicke, & M. Reinke (Eds.), *Landschaftsplanung* (pp. 1–9). Berlin Heidelberg, Berlin, Heidelberg: Springer.

Budds, J. (2009). Contested H_2O: Science, policy and politics in water resources management in Chile. *Geoforum, 40*(3), 418–430. doi:10.1016/j.geoforum.2008.12.008.

Budds, J. (2013). Water, power, and the production of neoliberalism in Chile, 1973–2005. Environment and Planning D. *Society and Space 31*(2), 301–318. doi:10.1068/d9511

Budds, J., & Hinojosa-Valencia, L. (2012). Restructuring and rescaling water governance in mining contexts: The co-production of waterscapes in Peru. *Water Alternatives, 5*(1), 119–137.

Cohen, A., & Davidson, S. (2011). The watershed approach: Challenges, antecedents, and the transition from technical tool to governance unit. *Water Alternatives, 4*(1), 1–14.

Conway, D., van Garderen, E. A., Deryng, D., Dorling, S., Krueger, T., Landman, W., et al. (2015). Climate and southern Africa's water-energy-food nexus. *Nature Climate Change, 5*(9), 837–846.

Cortner, H., & Moote, M. (1994). Trends and issues in land and water resources management: Setting the agenda for change. *Environmental Management, 18*(2), 167–173.

Cotula, L. (2012). The international political economy of the global land rush: A critical appraisal of trends, scale, geography and drivers. *Journal of Peasant Studies, 39*(3–4), 649–680. doi:10.1080/03066150.2012.674940.

de Vries, D. H. (2011). Temporal vulnerability in hazardscapes: Flood memory-networks and referentiality along the North Carolina Neuse River (USA). *Global Environmental Change, 21* (1), 154–164.

Di Baldassarre, G., Kemerink, J. S., Kooy, M., & Brandimarte, L. (2014). Floods and societies: The spatial distribution of water-related disaster risk and its dynamics. *Wiley Interdisciplinary Reviews: Water, 1*(2), 133–139.

Falkenmark, M. (1979). Main problems of water use and transfer of technology. *GeoJournal, 3*(5), 435–443.

Falkenmark, M. (1997). Society's interaction with the water cycle: A conceptual framework for a more holistic approach. *Hydrological Sciences Journal, 42*(4), 451–466. doi:10.1080/02626669709492046.

Falkenmark, M., & Folke, C. (2002). The ethics of socio-ecohydrological catchment management: Towards hydrosolidarity. *Hydrology and Earth System Sciences Discussions, 6*(1), 1–10.

FAO. (2012). *Coping with water scarcity: An action framework for agriculture and food security*. Rome: FAO Water Reports.

Franco, J., Mehta, L., & Veldwisch, G. J. (2013). The global politics of water grabbing. *Third World Quarterly, 34*(9), 1651–1675. doi:10.1080/01436597.2013.843852.

Gheewala, S., Berndes, G., & Graham, J. (2011). The bioenergy and water nexus. *Biofuels, Bioproducts and Biorefining, 5*(4), 353–360.

Gibson, C. C., Ostrom, E., & Ahn, T. (2000). The concept of scale and the human dimensions of global change: a survey. *Ecological Economics, 32*(2), 217–239. doi:10.1016/S0921-8009(99)00092-0.

Gleick, P. (2002). Water management: Soft water paths. *Nature, 418*(6896), 373.

GWP, Global Water Project. (2014). *Coordinating land and water governance: An essential part of achieving food security*. Perspectives paper.

Haberl, H. (2015). Competition for land: A sociometabolic perspective. *Ecological Economics, 119*, 424–431. doi:10.1016/j.ecolecon.2014.10.002.

Haberl, H., Mbow, C., Deng, X., Irwin, E. G., Kerr, S., Kuemmerle, T., et al. (2014). Finite land resources and competition. In K. C. Seto & A. Reenberg (Eds.), *Rethinking global land use in an urban era* (Vol. 14, pp. 35–69). Cambridge, MA: The MIT Press.

Hanjra, M. A., & Qureshi, M. E. (2010). Global water crisis and future food security in an era of climate change. *Food Policy, 35*(5), 365–377. doi:10.1016/j.foodpol.2010.05.006.

Hodgson, S. (2004). *Land and water—the right interface*. Rome.

Keys, P. W., van der Ent, R. J., Gordon, L. J., Hoff, H., Nikoli, R., & Savenije, H. H. G. (2012). Analyzing precipitation sheds to understand the vulnerability of rainfall dependent regions. *Biogeosciences, 9*(2), 733–746.

Krueger, T., Maynard, C., Carr, G., Bruns, A., Mueller, E. N., & Lane, S. (2016). A transdisciplinary account of water research. *WIREs Water, 3*, 369–389. doi:10.1002/wat2.1132.

Lambin, E., & Meyfroidt, P. (2011). Global land use change, economic globalization, and the looming land scarcity. *Proceedings of the National Academy of Sciences of the United States of America, 108*(9), 3465–3472. doi:10.1073/pnas.1100480108.

Lankford, B. A. (2013). *Resource efficiency complexity and the commons: The paracommons and paradoxes of natural resource losses, Wastes and Wastages*. Abingdon: Earthscan Publications.

Lankford, B. A., Tumbo, S., & Rajabu, K. (2009). Water competition, variability and river basin governance: A critical analysis of the Great Ruaha River, Tanzania. In F. Molle & P. Wester (Eds.), *River basin trajectories: Societies, environments and development* (pp. 171–195). Cambridge, MA: CABI.

Lebel, L., Garden, P., & Imamura, M. (2005). The politics of scale, position, and place in the governance of water resources in the Mekong region. *Ecology and Society, 10*(2), 18.

Linton, J., & Budds, J. (2014). The hydrosocial cycle: Defining and mobilizing a relational-dialectical approach to water. *Geoforum, 57*, 170–180. doi:10.1016/j.geoforum.2013.10.008.

Loftus, A. (2007). Working the socio-natural relations of the urban waterscape in South Africa. *International Journal of Urban and Regional Research, 31*(1), 41–59. doi:10.1111/j.1468-2427.2007.00708.x.

Loftus, A., & Lumsden, F. (2008). Reworking hegemony in the urban waterscape. *Transactions of the Institute of British Geographers, 33*(1), 109–126.

Mehta, L., Veldwisch, G. J. & Franco, J. (2012). Introduction to the special issue: Water grabbing? Focus on the (Re)appropriation of finite water resources. *Water Alternatives, 5*(2), 193–207.

Molle, F. (2007). Scales and power in river basin management: The Chao Phraya river in Thailand1. *The Geographical Journal, 173*(4), 358–373.

Molle, F. (2009). River-basin planning and management: The social life of a concept. *Geoforum, 40*(3), 484–494. doi:10.1016/j.geoforum.2009.03.004.

Molle, F., Wester, P. (Eds.). (2009). *River basin trajectories: Societies, environments and development. Comprehensive assessment of water management in agriculture series* (vol 8). Cambridge, MA: CABI.

Moss, T., & Newig, J. (2010). Multilevel water governance and problems of scale: Setting the stage for a broader debate. *Environmental Management, 46*(1), 1–6. doi:10.1007/s00267-010-9531-1.

Norman, E. S., & Bakker, K. (2009). Transgressing scales: Water governance across the Canada–U.S. Borderland. *Annals of the Association of American Geographers, 99*(1), 99–117. doi:10.1080/00045600802317218.

Norman, E., Bakker, K., & Cook, C. (2012). Introduction to the themed section: Water governance and the politics of scale. *Water Alternatives, 5*(1), 52–61.

Oliveira, G. D. (2013). Land regularization in Brazil and the global land grab. *Development and Change, 44*(2), 261–283. doi:10.1111/dech.12009.

Ostrom, E. (1999). Revisiting the Commons: Local Lessons. *Global Challenges. Science, 284* (5412), 278–282. doi:10.1126/science.284.5412.278.

Pahl-Wostl, C., Holtz, G., Kastens, B., & Knieper, C. (2010). Analyzing complex water governance regimes: The management and transition framework. *Environmental Science & Policy, 13*(7), 571–581. doi:10.1016/j.envsci.2010.08.006.

Palmer, D., Fricska, S. & Wehrmann, B. (2009). *Towards improved land governance*. Land tenure working paper, vol 11. UN-HABITAT.

Perreault, T., Wraight, S., & Perreault, M. (2012). Environmental injustice in the Onondaga Lake waterscape, New York State, USA. *Water Alternatives, 5*(2), 485–506.

Perrone, N. M. (2013). Restrictions to foreign acquisitions of agricultural land in Argentina and Brazil. *Globalizations, 10*(1), 205–209.

Qadir, M., Wichelns, D., Raschid-Sally, L., McCornick, P. G., Drechsel, P., Bahri, A., & Minhas, P. S. (2010). The challenges of wastewater irrigation in developing countries. *Agriculture Water Management, 97*(4), 561–568.

Ringler, C., Bhaduri, A., & Lawford, R. (2013). The nexus across water, energy, land and food (WELF): potential for improved resource use efficiency? Aquatic and marine systems. *Current Opinion in Environmental Sustainability, 5*(6), 617–624. doi:10.1016/j.cosust.2013.11.002.

Rockström, J., Gordon, L., Folke, C., Falkenmark, M., & Engwall, M. (1999). Conservation ecology: Linkages between water vapor flows, food production and terrestrial ecosystem services. *Conservation Ecology, 3*(2), 5.

Rulli, M. C., Saviori, A., & D'Odorico, P. (2013). Global land and water grabbing. *Proceedings of the National Academy of Sciences of the United States of America, 110*(3), 892–897. doi:10.1073/pnas.1213163110.

Sikor, T., Auld, G., Bebbington, A. J., Benjaminsen, T. A., Gentry, B. S., Hunsberger, C., et al. (2013). Global land governance: From territory to flow? *Current Opinion in Environmental Sustainability, 5*(5), 522–527. doi:10.1016/j.cosust.2013.06.006.

Simpson, T. W., Sharpley, A. N., Howarth, R. W., Paerl, H. W., & Mankin, K. R. (2008). The new gold rush: Fueling ethanol production while protecting water quality. *Journal of Environmental Quality, 37*(2), 318–324. doi:10.2134/jeq2007.0599.

Sjöstedt, M. (2011). The impact of secure land tenure on water access levels in sub-Saharan Africa: The case of Botswana and Zambia. *Habitat International, 35*(1), 133–140. doi:10.1016/j.habitatint.2010.06.001.

Skinner, J. & Cotula, L. (2011). *Are land deals driving 'water grabs'? The global land rush*. http://pubs.iied.org/17102IIED

Steffen, W. (2009). Interdisciplinary research for managing ecosystem services. *Proceedings of the National Academy of Sciences of the United States of America, 106*(5), 1301–1302. doi:10.1073/pnas.0812580106.

Swyngedouw, E. (1997). *Neither global nor local: 'Glocalization' and the politics of scale. Spaces of globalization: Reasserting the power of the local*.

Swyngedouw, E. (1999). Modernity and hybridity: Nature, regeneracionismo, and the production of the Spanish waterscape, 1890–1930. *Annals of the Association of American Geographers, 89*(3), 443–465.

Swyngedouw, E. (2004). *Social power and the urbanization of water: Flows of power*. Oxford, Oxford: Oxford University Press.

Swyngedouw, E. (2009). The political economy and political ecology of the hydro-social cycle. *Journal of Contemporary Water Research & Education, 142*(1), 56–60.

The Land Matrix Global Observatory. (2016). Agricultural drivers for land deals. http://www.landmatrix.org/en/get-the-idea/agricultural-drivers/. Accessed January 31, 2016.

Trottier, J. (2008). Water crises: Political construction or physical reality? *Contemporary Politics, 14*(2), 197–214. doi:10.1080/13569770802176929.

Walsh, A. (2011). The commodification of the public service of water: A normative perspective. *Public Reason, 3*(2), 90–106.

White, B., Borras, Jr, Saturnino, M., Hall, R., Scoones, I., & Wolford, W. (2012). The new enclosures: Critical perspectives on corporate land deals. *The Journal of Peasant Studies, 39* (3–4), 619–647. doi:10.1080/03066150.2012.691879.

Wolford, W., Borras, S. M., Hall, R., Scoones, I., & White, B. (2013). Governing global land deals: The role of the state in the rush for land. *Development and Change, 44*(2), 189–210. doi:10.1111/dech.12017.

Zeitoun, M., & Mirumachi, N. (2008). Transboundary water interaction I: Reconsidering conflict and cooperation. *International Environmental Agreements, 8*(4), 297–316. doi:10.1007/s10784-008-9083-5.

Zimmer, A. (2011). *Everyday governance of the waste waterscapes—A Foucauldian analysis in Delhi's informal settlements*. Rheinische Friedrich-Wilhelm Universität Bonn.

Chapter 20
Travelling Through the Densu Delta: Location, Place and Space in the Waterscape

Fanny Frick

Abstract In Greater Accra, Ghana, pressures on formerly undeveloped land increase with rising demand for land as a resource for housing development. Land–water interactions create a waterscape inflicted by risk: ongoing development negatively impacts downstream communities as it reduces the natural capacity of local wetlands to regulate high discharge levels and eventually leads to flooding. The absence of water and sanitation infrastructure exacerbates flood-related risks, and the conversion of land surface changes ecosystem functions, which further contributes to flooding and water pollution. The purpose of this contribution is to explore material and spatial socionatural land–water interactions that are shaping the Densu Delta waterscape. Dynamics in space in the waterscape are explored through the lens of competition over land- and water-use control at different sites along the flow of the lower Densu Delta. In the case study area, urban development is both a cause of flood risk and a driver of risk reduction. As well-off areas gradually become served with basic infrastructure, less affluent communities along the road also experience less flooding and improved access to coping strategies. Competition over access to land thus often becomes a struggle over infrastructure. As a result of these ongoing processes, the nature of competition over land and water (use) is constantly changing in space, time and scale. Both land and water are unstable, fluid resources. The results show that competition between actors over the ownership of land is played out at a local level, between traditional authorities and individuals. Competition over the use of land by contrast is played out between local interest groups interested in further development on the one hand and, on the

F. Frick (✉)
Climate Change and Sustainable Development Lab & Geography Department, Humboldt-Universität zu Berlin, Berlin, Germany
e-mail: fanny.frick.1@student.hu-berlin.de

F. Frick
Integrative Research Institute on Transformations of Human-Environment Systems (IRI THESys), Humboldt-Universität zu Berlin, Berlin, Germany

J. Niewöhner et al. (eds.), *Land Use Competition*,
Human-Environment Interactions, DOI 10.1007/978-3-319-33628-2_20

other hand, governmental authorities at city and national level intending to preserve the land from being developed. International interests including ecosystem conservation, investment in land for development and industrial production mediate the competition over both land and the use of water.

Keywords Ghana · Infrastructure · Flooding · Fluid resources · Competition

20.1 Introduction

Human settlements have developed near rivers and coasts for centuries. As urbanization continues to concentrate on coastal areas across the globe, river deltas, floodplains and wetlands are being altered. In combination with climate change impacts, this development implies a growing exposure of people to flood risk—because of both more people living in flood-prone areas and reduced capacities of coastal ecosystems to level off the impact of storm surges and high run-off levels (Kundzewicz et al. 2013; McGranahan et al. 2007; Newton et al. 2012; Nicholls et al. 1999, 2008). People are shaping and are being shaped by coastal ecosystems and water flows. For instance, poor and vulnerable groups tend to live in the most flood-prone area of an urbanized floodplain, whereas more affluent groups usually live on less exposed sites, or where exposure is reduced by protective infrastructure such as levees. Such infrastructure in turn diverts flood risk to other areas, reproducing socio-economic inequalities and differential vulnerabilities (Boruff et al. 2005). At the same time, water bodies are a source of livelihoods for many coastal and riverine settlers. It is in this sense that floodplains and society have coevolved over time and constitute a socionatural entity (Di Baldassarre et al. 2013).

Less than 20 km west of Accra, the expanding capital of Ghana, lies the Densu Delta (Fig. 20.1). The Densu wetland and the Sakumo Lagoon through which the river discharges into the Atlantic Ocean are protected as a Ramsar Site (convention for the protection of wetlands), to maintain these ecosystems as habitat for migratory bird species (Denutsui et al. 2012). Only a few kilometres upstream, the Densu River has been dammed in the 1950s to provide water for irrigation of agricultural land and to supply Accra with water (Kuma and Ashley 2008). However, as the population has increased in size over the past decades, industrial, agricultural and urban land uses have intensified and are increasingly degrading the ecological functions and natural resources in the Densu River, the Weija reservoir, the Densu wetlands and the Sakumo Lagoon (Fianko et al. 2009; Kusimi 2008a). Pressures on formerly undeveloped land have increased with rising demand for land as a resource for housing development, fuelled by litigation and speculation on land (Barry and Danso 2014; Doan and Oduro 2012). At the same time, the combined processes of sedimentation, pollution and urbanization have led to an increasing number of households being exposed to flooding, shortage of potable water and disturbance of livelihood strategies (Kuma and Ashley 2008; Kusimi 2008a; Osei et al. 2011).

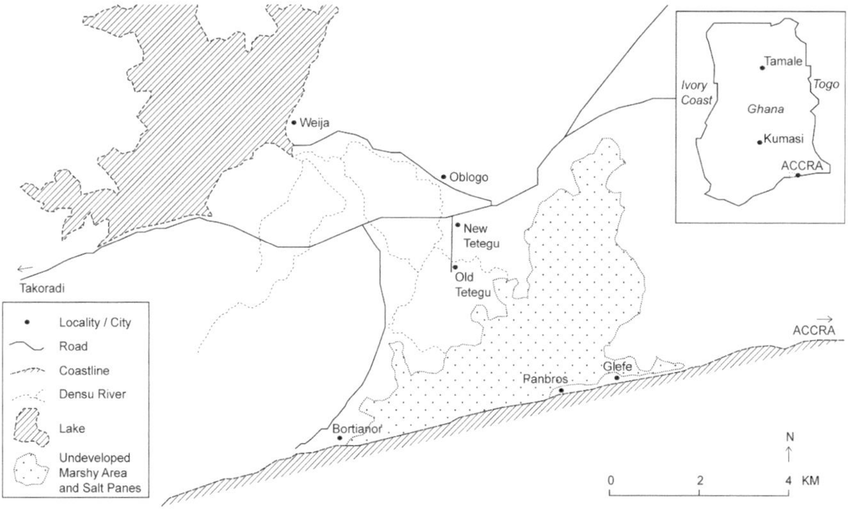

Fig. 20.1 The Densu Delta. *Source* Own design based on Osei et al. (2010) and Google Maps/Google Earth (2016)

In light of the multifaceted character of social and ecological interactions linked to water, the concept of waterscape has been introduced by critical geographers. A waterscape can be understood as 'a socionatural entity, in which water and power are coproduced and expressed' (Budds and Hinojosa 2012, p. 133). It is constantly shaped and altered as coproduction is played out in material forms such as infrastructure and land use, as well as in discourse and social practices around water (Ahlers et al. 2014). While more conventional studies on situations similar to that described on the Densu Delta focus on competition and conflict over resources, research on waterscapes and the hybrid landscapes seeks to elicit the 'subtle contestations' over resources, environments and landscapes that shape people's livelihoods (Goodman et al. 2008). The latter approach emphasizes the action of disputing rather than the end of gaining, winning, defeating or establishing superiority that is entailed in 'competition' (cf. Oxford University Press 2015, see also Chap. 1). A waterscape analysis implies changes in the distribution, and the use of resources is to be understood not as the outcome of a competition over the resources themselves but over the power to control resource use. (Funder and Marani 2015). Subtle contestations as well as competition over water and land control between actors are played out in land- and water-use change on the ground. Contestations of meanings are a means of conscious and unconscious competition over resource control and shape coproduction in a waterscape. Water acts as an agent shaping social relations and discourse, and vice versa, and space and scale are constantly reconfigured (Allan 2004; Budds 2008; Budds and Hinojosa 2012).

The purpose of this contribution is to explore material and spatial socionatural land–water interactions that are shaping the Densu Delta waterscape, more specifically: What are the objects, actors, interactions and scales involved in contestation and competition over land- and water-use control at different sites along the Densu Delta? A journey along the river serves to identify the different forms of competition over land- and water-use control in the Densu Delta waterscape in its spatial dimension. Here, the river is taken as a point of reference for an imagined transect through the waterscape's spatial dimension. Selected sites are analysed for objects and modes of competition over land- and water-use control. The covariation of anthropogenic activities and river flows and the reach of their interactions are discussed subsequently. Methods are a review of literature and grey literature, transect walks and narrative interviews with actors in the case study area. Fieldwork in the Densu Delta was conducted between April and July 2014.

20.2 A Journey Along the Densu Delta

The Densu River, which originates in the Eastern Region in Ghana, was dammed at Weija in 1952, creating the Densu Reservoir for irrigation of close-by agricultural lands and water supply to Accra. After the first dam breached as a result of flooding in 1968, the present dam was erected in 1978 (Kuma and Ashley 2008). The quality of water collected in the reservoir is poor, mainly due to discharge of industrial and agricultural pollutants upstream, but also due to nearby domestic discharges. The pollution of the water contained in and discharged from the reservoir potentially enhances flood-related and fishing-related risks[1] as the water acts as a vector for diseases. Upstream land-use change further contributes to siltation of the Weija reservoir, thus reducing its storage capacity (Kusimi 2008b). Water discharge into the reservoir currently exceeds extraction and necessitates regular spillage during the rainy season, as a manager of the Ghana Water Company (GWCL)explains:

'The dam was built purposely for two things: one for Ghana Water [Company] to extract 30 % of whatever was in the reservoir to treat for the people of Accra, and two for Irrigation Development Authority to also extract 60 % for irrigation purposes. There used to be (…) farmland (…) but for some time now the aspect of irrigation has been suspended because the farmlands have been converted into building lands and people have built on it. So in short what it means is that the 60 % that the Irrigation [Development] Authority was supposed to extract for irrigation purposes is still sitting in the reservoir. However, we have reached our maximum of 30 % (…). If we have to extract more water from the reservoir then we have to sit with the Water Resources Commission, sit with the Irrigation Development Authority and strike an agreement that (…) we maybe take additional 30 % for the purposes of treating water. That is, when we have an expansion [of treatment facilities]' (Interview, GWCL manager, Weija, 2014).

[1] I use the definition by Wisner et al. of risk as 'a compound function of the natural hazard and the number of people, characterized by their varying degrees of vulnerability to that specific hazard, who occupy the space and time of exposure to the hazard event' (Wisner et al. 2004).

For the treatment process to meet WHO standards for drinking water, the company is using chemicals and lime imported from China and Europe. At accelerating inflation rates, this procedure is becoming increasingly unsustainable. Hence, an expansion of the treatment plant currently appears unlikely. Small-scale subsistence fishery is practised in the reservoir, a livelihood strategy among the local population that—according to the GWCL official interviewed—potentially exacerbates pollution of the water and is therefore monitored by GWCL.

The Weija dam has five gates, which are closed during the dry season and partly opened during the rainy season when water levels in the reservoir are rising as a result of increased rainfall in the Eastern Region, to regulate water levels and prevent overspill. Roughly once or twice in two years, incoming water quantities are so large that the full opening of all five gates of the dam is necessary to avoid uncontrolled overspill of the dam. In this case, water levels downstream rise significantly and result in flooding of streets and houses in close-by neighbourhoods, as occurred in June 2014. To prevent such damaging events from happening, water levels and rainfall data are monitored on a daily basis at Nsawam station roughly 30 km upstream and reported to the Weija dam management station to guide decisions on opening the dam gates. Sites close to the dam along the reservoir and the river shore are being converted into plots and developed for housing. GWCL sends out warning letters at the beginning of each rainy season and visits chiefs and other spokespersons of communities located downstream on the day preceding a dam release to warn residents of upcoming floods.

The topography of the landscape allows water from Weija to be channelled to Accra by gravity. The first neighbourhood along the way is Oblogo, situated on the slopes of the Densu valley. In the midst of the settlement only a few kilometres downstream of the Weija dam, a large open dumpsite is situated. 1250 tons of solid waste has been brought here daily from Accra for more than 10 years from 2002 onwards without any environmental health protection measures: pollutants have leached into the groundwater and from there travelled into the Densu River, which serves as a raw water source for many households of the area, and have ended up in the Densu wetland (Osei et al. 2011). Fire outbreaks on the dumpsite have frequently led to the spreading of toxic fumes in recent years. According to a community leader, this is a much more immediate risk faced by the inhabitants than that of flooding, as they can adapt to the latter. The dumpsite is now closed and being covered with gravel for beautification purposes by a local investor—a measure that does not, however, halt groundwater leaching.

South of the Accra–Cape Coast highway, the river opens into the Densu wetland, stretching over approximately 20 km^2 from here to the coast and providing habitat for rare migratory bird species and marine turtles, besides providing protection from flooding. The delta has been declared a Ramsar Site, protected under the international convention through the national Wildlife Division. It is being encroached on from all sides because of limited support from both municipalities and traditional leaders:

'Wetlands (…) absorb more runoff water and therefore they tend to minimize or reduce flooding, but [municipal authorities] don't see it that way. They want us to go and build some physical structures that require huge sums of money; they don't want wetlands to perform its natural functions. (…) The other challenge is the fact that the land belongs to the people, to traditions, to clans, to chiefs (…). We are trying to adopt a management approach that will seek the support of all stakeholders. But because the area is close to the urban sector (…) the economic value of the land keeps increasing and it will be difficult for you to convince a chief who owns the land to leave mangroves in the area as compared to selling of the land to a developer' (Manager at the Wildlife Division, Accra, June 2014).

Regardless of the rapid deterioration of the conservation site, it continues to be visited by eco-tourists.

The first settlement through which the Densu flows into the delta is New Tetegu, a settlement established in the course of relocation of Old Tetegu after the latter had been hit by severe floods in the 1960s. The rapid development of middle- and high-income housing in the area has given the community's landowners association considerable power in negotiation for infrastructure. Anticipating the property taxes to be gained from the community, the municipality has granted building permits to all houses in the community in 2010, which paved the way for infrastructure development by governmental authorities. Gutters, water pipes, electricity and a river wall have been installed, and the road up to the bridge leading to Old Tetegu has been paved. The community has moreover negotiated with Ghana Water Company that dam openings be reduced to a minimum, reducing the chance of flooding in the community.

New Tetegu is delimited by the bifurcation of the Densu River. Old Tetegu comprises an old fishermen's village, Dzelukope, as well as large areas of new development. Dzelukope used to be cut off from other communities by the two streams of the Densu. It is now connected to New Tetegu by a bridge that has presumably been erected in the interest of investors looking to acquire land south of Dzelukope. Here, wetlands are being reclaimed for plot development by the community's chiefs who act as caretakers for the custodian, the Aplaku stool chief from Jamestown in central Accra. Litigation over landownership between different proclaimed chiefs is strong, and the set-up of the chieftaincy system is intransparent. Regardless of Greater Accra planning documents that designate the area as a 'green belt', the caretakers plan to sell plots from the community's current edge up to the seafront. Built into the wetland, all of the houses in Old and New Tetegu are exposed to flooding and inundation, particularly in the rainy season. Residents recall flood events in the past that have destroyed homes or made them uninhabitable for weeks. They have, however, developed a number of strategies to bear the flooding, such as raising houses and navigating flooded streets in canoes, and do not perceive flooding as a major risk. What is a major concern particularly to fishermen in Dzelukope is that 'the Densu River is being contaminated [because] some people at the other side [of the river], they put gutters, their excess water, urine and that kind of thing they just (…) pipe into the river. It worries the fishes and even it affects us because (…) we feed on the fish and at times we use the water raw. At times when they [the GWCL] close the pipe borne water we depend on the Densu River. We use it to bath, clean our utensils; when it is difficult people use it to cook' (Fisherman in Dzelukope, May 2014).

Following the western river arm from Dzelukope, the swampy land is pervaded by scattered construction sites of large concrete houses and plots marked for development. The vast majority of these houses are uninhabited, and most construction sites seem abandoned. Speculation and litigation is high in the two neighbourhoods to which these plots belong, namely Old Tetegu and Bortianor, both controlled by influential stools (Barry and Danso 2014). The stream finally enters the sea at Bojo beach, a resort popular with expats and the upper middle class living in Accra. Here, surrounding mangroves and a wide beach conceal the vast anthropogenic interferences with the ecosystem. The beach offers a view of Faana–Bortianor, an isolated traditional fisher village of wooden huts that can be reached by canoe only.

Along the eastern stream, the Densu water passes basins for salt production managed by Panbros Salt Industry. The company was established in the 1950s by a Greek investor. Southeast of Panbros Salt Industry, sandwiched between the Sakumo Lagoon and the Atlantic Ocean, lies Glefe, an old fishermen village of Ga people[2]. Panbros Salt Industry is a major employer of Glefe residents, but at the same time is seen as the origin of frequent flooding, as the salt basins prevent discharge of excess water: ‘Panbros Salt Industry, they are the number one problem why this place [gets flooded]. [They blocked the lagoon outfall], so whenever it rains [the water] can’t (…) get through here to the sea and whenever the level of the sea is high then the water become[s] stagnant. It cannot flow into the sea again, so it enter[s] into people’s houses, people’s rooms and you see them coming out calling NADMO, AMA to do something’ Community leader (assemblyman of Glefe-Mpoase), May 2014.

Stagnating rainwater in the lagoon is, however, not the only source of flood risk in Glefe. The village is constantly moving inland as a result of ongoing shoreline erosion. On top of that, Glefe has undergone rapid development over past decades, mainly from low-income migrants from across Accra and neighbouring countries who construct their houses on land obtained from local chiefs. As the land between the lagoon and the sea is limited, reclaiming more land from the lagoon by depositing gravel, stones and waste is a common practice. New settlers purchasing land are often unaware of the exposure before experiencing the first rainfall. Sand is mined at the beachside of Glefe as cheap building material for raising structures, creating a tension between those at risk from shoreline erosion and those on reclaimed land in the lagoon. Since part of Glefe’s land area is below sea level, simple gutters are no option for waste water and storm water discharge. A few households have constructed individual drains, which end abruptly on the only street, creating ponds of waste water. Other than that, water is left to drain away into the unpaved ground. In the light of poor sanitary conditions in the community, government officials describe the community as the new ‘Sodom and Gomorrah’—referring to the colloquial name of the allegedly most deprived informal settlement of Accra. Just as in surrounding communities, land litigation and conflict over chieftaincy are common in Glefe. Land speculation has added to the chaotic layout of the settlement that impedes proper infrastructure development. Neither ambulances and fire brigades, nor any other emergency vehicle

[2]Ga are the indigenous people of Accra.

can access the dense settlement. Solid waste is dumped next to the lagoon in the absence of waste collection services and thus adds to the pollution of the lagoon and consequently the Densu wetland.

20.2.1 Competition and the Land-Water Nexus in the Densu Delta Waterscape

In sum, the journey through the Densu Delta illustrates that different objects, actors, modes of competition and scales are involved in contestation over land- and water-use control at different sites.

At the starting point of the journey, around the Weija reservoir, settlers, governmental agencies (here, the Irrigation Development Authority) and a private company (Ghana Water Company Limited, GWCL) compete for controlling water as a resource for fishery, agricultural production and extraction of potable water, respectively. Upstream of the dam, the main object of contestation is the quality of water, whereas it is mainly the quantity of water flows in contestation between GWCL and downstream settlers. Both objects of contestation are not stable, but have historically developed in socioecological interactions (cf. Chap. 22)[3]. In the context of ongoing urbanization, both quality and quantity are becoming increasingly contested. Modes of competition over water quality control are formal regulations enforced by GWCL in combination with more 'informal' control mechanisms among fishermen. Modes of competition over water quantity control include a formal agreement between governmental agencies and GWCL, as well as less formal arrangements between residents' association, assembly man, traditional authorities and GWCL following the petitioning of GWCL by traditional authorities.

In Oblogo, land–water interactions are shaped by the competition over land between actors responsible for managing Accra's waste and individuals protecting their backyards. The location of the Oblogo landfill in a risk-prone informal settlement as well as the absence of waste management in Glefe suggest a power game in Accra that is played out in the competition over control of land as a resource for development. Only a very small fraction of Accra's population, living in high-income areas of the city, is served by household waste collection services. Less affluent areas are served by waste collection from communal containers, which are often inadequate in size and irregularly collected, and thus create unsanitary

[3]Changes in land use from rural to urban have created an imbalance in the management of the reservoir and thus spurred competition over land-and water-use control. The construction and management of the dam has changed local opportunities and necessities of land use, for instance by enabling irrigated agriculture and housing development in areas that had previously been exposed to regular flooding. In combination with rising demand for (and thus economic value of) land for development, this has led to the emergence of control over land as a resource for development as an object of competition among and between traditional and governmental authorities.

conditions such as is the case in Glefe. Waste collected from both households and from containers is brought to open dumpsites that are unsanitary and thus present health risks to adjacent settlers. Regardless of such risks, the selection of sites for dumps is guided by the price of the land and accessibility for trucks, rather than considerations of environmental justice (Boadi and Kuitunen 2003; Oteng-Ababio 2013). As a result, the location of dumpsites serving Accra coincides with that of poor and indigenous areas in the urban fringe and has caused social unrest in Oblogo in the past (Oteng-Ababio 2013).

Competition over control of land as a resource for development is particularly evident in the Densu wetlands. Governmental authorities representing different interests and levels of decision-making, traditional authorities and landowners, Panbros Salt industries, residents and new migrants all have a stake in either developing the land for residential use, or keeping it free from development. Beyond competition over the land use, traditional authorities contest landownership. The unclear situation of landownership in Tetegu, Bortianor, and Glefe is leading to physical fights over plots (Abbey 2013). Guards are being contracted by chiefs to protect land and tear down buildings they consider illegitimate (Ghana News Agency 2013). Development that grossly contradicts planning guidelines and standards is not a particularity of the Densu Delta area, but can be observed across Greater Accra as well as urban Ghana as a whole. Remainders of colonial land management in current urban planning, unaccountable use of expropriated land by the government as well as the co-optation of land market mechanisms by traditional chiefs in the urban fringe contribute to ongoing land disputes (Mensah 2003). In competition over land as a resource for development, the modes of competition are again a mixture of formal regulations (byelaws and land-use planning by AMA), management contracts between Panbros Salt and Wildlife Division, mechanisms not formally codified and often based on violence and speculation and traditional leadership.

The periodical overabundance of water in the wetland that eventually results in flooding creates tensions between the interest of conservationists (namely the Wildlife Division, acting as a national representative of the international Ramsar convention) to restore and protect wetlands and their ecosystem service of flood prevention on the one hand and the desire to protect developed plots in the wetland from floods on the other hand. The competition over land as a resource for development enhances the scattered development of Accra's peri-urban areas, with 'pockets of undeveloped land surrounded by vast stretches of built-up areas' (Doan and Oduro 2012, p. 1320). As a consequence, compacted soils and impermeable surfaces are spreading into the wetland, thus continuously reducing its capacity as a buffer for water retention. The levees built in Tetegu trigger what is known as the 'levee effect' by which the risk of flooding is transferred to settlements upstream and downstream (Di Baldassarre et al. 2013). Accordingly, Glefe was severely affected by the opening of the Weija dam in June 2014. Besides, levees tend to create an intriguing sense of security that attracts people to risk-prone areas (Di Baldassarre et al. 2013), indicated in Tetegu by low perceptions of risk from flooding among most interviewees, even shortly after the June 2014 flood event.

20.2.2 Competition and Contestation in Controlling Unstable Resources

The analysis of actors, objects, modes and scales reveals two characteristics of the Densu Delta waterscape read through the lens of space: firstly, the contestation over land- and water-use control is chiefly driven by the occupation of land in the river's buffer zone and the lagoon. Rising land prices and demands for land for housing development create an incentive for further land reclamation, which in turn alters water flows. Thus, both land and water are fluid resources, there is no stable pool of these resources over which actors compete. In and around Weija, urbanization of agricultural land has altered the demand for irrigation water and reduced the out-take of water for this purpose from the reservoir. As the treatment plant's capacity is limited, this has resulted in regular overabundance of water in the reservoir during the wet season, making dam spillage necessary. Variabilities in the quantity of discharge water across the delta are thus the consequence of both natural rainfall and urbanization patterns. In Tetegu, ongoing development of the wetland implies that land which is part of the river's discharge area is being occupied by residential land use—in other words, the boundaries between land and water are redefined, and contestation over land-use control involves the contestation over the control of water flows. This is even more explicit in Glefe where land is claimed from a lagoon, i.e. a permanent water body. Contestations over land- and water-use control are thus not only strongly interlinked. The stocks of these resources themselves are constantly reconfigured and are thus 'fluid'.

Secondly, the different faces of contestation and competition over land- and water-use control at different points in the Densu Delta are linked by moments of land–water interactions and shape the spatial dimension of the waterscape. Water flows link Weija to Nsawam in the Eastern Region, to Accra and Tetegu and the Sakumo Lagoon at Glefe. Waste flows from Accra to Oblogo and continues to flow into groundwater as leakage on site. In Oblogo, local waste water flows separate the affluent and serviced New Tetegu from the old fisher village by discharging into and thus polluting the Densu that forms the natural boundary between the two parts of the locality. Infrastructure in Weija connects the place to Accra via water pipelines and separates the Weija lake and upstream areas from downstream localities in the form of the dam. The separation is paralleled by governance arrangements that are designed to protect water quality in the reservoir and improve it at the treatment plant for Accra's water supply, whereas downstream of the dam water quality is not monitored or protected by any formal arrangement. Local institutional arrangements compete with national representatives responsible for implementation of the international interest in wetland conservation over the control of land use. Rights to sell and access land and infrastructure have repeatedly been the object of contestation in all localities analysed, which has spurred haphazard urbanization of the delta. And finally, divergent discourses on risk from flooding and water pollution are used by different actors to legitimize and contest wetland land and water use in the Densu Delta.[4] Discursive coalitions can be identified between community representatives of Tetegu and the GWCL official in presenting flood risk in the Densu Delta as 'manageable', provided the infrastructure is in place.

Both the fluid nature of land and water and the moments of their interaction create a waterscape that has a spatial dimension but at the same time shifting boundaries that are difficult to define. The moments of land–water interaction, which have only briefly been addressed here, indicate that the Densu Delta waterscape forms part of urban Accra, particularly in providing infrastructure for water, urban development and waste management. It is at the same time a peri-urban area providing ecosystem services to the city (water supply, land) and for local livelihood strategies that are conventionally considered 'rural', such as fishery. The wetland's value as a habitat for migratory birds moreover constructs it as part of a global network of conservation sites; however, this global dimension of the waterscape might vanish as the site is threatened to lose its Ramsar Site status as a result of ongoing environmental degradation (interview with Wildlife official, May 2014). The analysis therein points at how scales are 'jumped' and 'shifted' in multi-level power dynamics and relations in the contentious geographies of a waterscape (Goodman et al. 2008).

20.3 Conclusion

'Not only does the valley rule the stream, […] but increasingly, human activities rule the valley' (Allan 2004). The cross-scale interactions, legacy effects, covariance of anthropogenic activities and ecological processes and nonlinearities identified by Allan in stream ecosystems are equally present in the Densu Delta. To begin with, the waterscape is shaped by the competition between actors over land, the use of land and the use of water. The competition between actors over the ownership of land is played out at a local level, between traditional authorities and individuals. Competition over the use of land by contrast is played out between local interest groups interested in further development on the one hand and, on the other hand, governmental authorities at city and national level who intend to preserve the land from being developed.[5] International interests including ecosystem conservation, investment in land for development and industrial production mediate the competition over both land and the use of water. Spatial interactions and legacy

[4]Compare, for instance, Tetegu's residents' statement that 'flooding is not a problem' with their ongoing activities in constructing a river wall and with the academic literature presenting the area as highly flood-prone Nyarko (2000).

[5]In general, local interest groups act in the interest of certain individuals, on a short term basis, whereas official government actions represent national and long-term interests. The main interests of local actors are access to affordable and safe land for housing purposes and economic gains from land plot sales, respectively. National- and city-level interests are the protection of residents from environmental hazards and the fostering of socio-economic development.

effects are manifested for instance in Weija, where the dam as large-scale infrastructure is reducing everyday risks from flooding of nearby households while enabling encroachment of the wetlands.

Water and land are linked once more as the development of land for housing is influenced by the abundance of surface waters as a source of livelihoods in fishery, tourism and salt production, for delivering water to the city, and as a source of environmental risks. At the same time, the quality and distribution of water in the case study area is transformed by the modification of land surfaces and land-use change, e.g. through pollution, decrease of infiltration rates and enhanced surface run-off, and infrastructure development.

'Thinking in terms of waterscapes avoids confining analyses to conventional scalar containers, and taking scale choices for granted' (Budds and Hinojosa 2012). The findings from this case study support the argument that rather than looking at watersheds, a flexible take on scale and space is key in understanding and governing a waterscape. Both water and land in the Densu Delta waterscape are non-constant resources, and their availability is fluid: water because it flows, and because of its seasonal variability in quantity and variability in quality in both space and time, and land because new land is being reclaimed at the same time as sites are becoming uninhabitable due to enhanced flood risks. Water, as a result of the land-use changes, has become an unusable resource and a source of environmental risk to health, property and lives. It is undergoing a transformation from a resource for livelihoods to also being an environmental hazard, thereby also transforming the quality of land as a resource for socio-economic development. Land has become a commodity of economic speculation. The quality of land for housing development is further transformed in uncontrollable ways as upstream modifications and development enhance the risk of flooding on plots further downstream. Here, land competition is mediated through space and by the river as the vector that connects places. As the stock of the resource(s) being competed for is unstable, the nature of competition in the Densu Delta waterscape is constantly changing in space, time and scale.

Looking at competition and contestation over land- and water-use control allowed for a specific understanding of the socionatural production of the Densu Delta waterscape as it is played out in its spatial and material dimension. Non-material dimensions of the waterscape such as discourse and rights have only been touched on in this analysis. A closer analysis of these components would allow for a better understanding of the sociopolitical reproduction of risk and vulnerability in the waterscape.

Acknowledgments This research has been (partly) funded by the German Federal Ministry of Education and Research (BMBF) under the (project funding) reference number 01 LN 1316 A and Heinrich-Böll-Stiftung under the funding reference number P105800. The paper is a contribution to THESys Summer School. I would like to thank Esinam Attipoe, Esther Dansu-Wiredu, Martin Oteng-Ababio and Kofi Owusu for their invaluable comments in interpreting my findings and all interview partners in the Densu Delta for sharing their knowledge and experiences with me.

References

Abbey, E. E. (2013). Bortianor youth protest sale of land. *Daily Graphic Ghana.*

Ahlers, R., Cleaver, F., Rusca, M., & Schwartz, K. (2014). Informal space in the urban waterscape: Disaggregation and co-production of water services. *Water Alternatives.*

Allan, J. D. (2004). Landscapes and riverscapes: The influence of land use on stream ecosystems. *Annual Review of Ecology, Evolution, and Systematics, 35*(1), 257–284. doi:10.1146/annurev.ecolsys.35.120202.110122

Barry, M., & Danso, E. K. (2014). Tenure security, land registration and customary tenure in a peri-urban Accra community. *Land Use Policy, 39*, 358–365. doi:10.1016/j.landusepol.2014.01.017

Boadi, K., & Kuitunen, M. (2003). Municipal solid waste management in the Accra metropolitan area, Ghana. *The Environmentalist, 23*, 211–218.

Boruff, B. J., Emrich, C., & Cutter, S. L. (2005). Erosion hazard vulnerability of US coastal counties. *Journal of Coastal Research, 215*, 932–942. doi:10.2112/04-0172.1

Budds, J. (2008). Whose scarcity? The hydrosocial cycle and the changing waterscape of La Ligua River Basin, Chile. In M. K. Goodman, M. T. Boykoff & K. T. Evered (Eds.), *Contentious geographies: Environmental knowledge, meaning, scale* (pp. 59–79). Ashgate Studies in Environmental Policy and Practice

Budds, J., & Hinojosa, L. (2012). Restructuring and rescaling water governance in mining contexts: The co-production of waterscapes in Peru. *Water Alternatives, 5*(1), 119–137.

Denutsui, D., Akiti, T. T., Osae, S., Blankson-Arthur, S., Tutu, A., Buah-Kwofi, A., et al. (2012). Investigating sea water influence and water quality assessment for different purposes in Densu Delta wetland, Accra, Ghana. *Elixir Agriculture, 42*, 6069–6073.

Di Baldassarre, G., Kooy, M., Kemerink, J. S., & Brandimarte, L. (2013). Towards understanding the dynamic behaviour of floodplains as human-water systems. *Hydrology and Earth System Sciences, 17*(8), 3235–3244. doi:10.5194/hess-17-3235-2013

Doan, P., & Oduro, C. Y. (2012). Patterns of population growth in peri-urban Accra, Ghana. *International Journal of Urban and Regional Research, 36*(6):1306–1325. doi:10.1111/j.1468-2427.2011.01075.x

Fianko, J. R., Osae, S., & Achel, D. (2009). Impact of anthropogenic activities on the Densu River in Ghana. *Water and Environment Journal, 23*(3), 229–234. doi:10.1111/j.1747-6593.2008.00137.x

Funder, M., & Marani, M. (2015). Local bureaucrats as bricoleurs. The everyday implementation practices of county environment officers in rural Kenya. *International Journal of the Commons, 9*(1), 87–106.

Ghana News Agency. (2013, December 19). Land guard terrorises land owners at Bortianor.

Goodman, M. K., Boykoff, M. T., & Evered, K. T. (2008). Contentious geographies: Environmental knowledge, meaning, scale. In M. K. Goodman, M. T. Boykoff & K. T. Evered (Eds.), *Contentious geographies: Environmental knowledge, meaning, scale* (pp. 1–23), Ashgate Studies in Environmental Policy and Practice.

Google Maps/Google Earth (2016). Densu Delta protected area. Map data (c) 2016 Google. maps.google.com. Accessed on January 05, 2016 from https://www.google.de/maps/place/Densu+Delta+Protected+Area/@5.5301685,-0.3602193,12z/

Kuma, J., & Ashley, D. (2008). Runoff estimates into the Weija reservoir and its implications for water supply to the Accra area, Ghana. *Journal of Urban and Environmental Engineering, 2* (2), 33–40. doi:10.4090/juee.2008.v2n2.033040

Kundzewicz, Z. W., Kanae, S., Seneviratne, S. I., Handmer, J., Nicholls, N., Peduzzi, P., et al. (2013). Flood risk and climate change: Global and regional perspectives. *Hydrological Sciences Journal, 59*(1), 1–28. doi:10.1080/02626667.2013.857411

Kusimi, J. M. (2008a). Analysis of sedimentation rates in the Densu River channel: The result of erosion and anthropogenic activities in the Densu basin. *West African Journal of Applied Ecology, 14.*

Kusimi, J. M. (2008b). Stream processes and dynamics in the morphology of the Densu River channel in Ghana. *The International Archives of the Photogrammetry, Remote Sensing and Spatial Sciences, XXXVII*(Part B8 Beijing), 1177–1182.

McGranahan, G., Balk, D., & Anderson, B. (2007). The rising tide: Assessing the risks of climate change and human settlements in low elevation coastal zones. *Environment and Urbanization, 19*(1), 17–37. doi:10.1177/0956247807076960

Mensah, S. O. (2003). The institutional capacity for land utilization for development: Constraints and Suggested solutions. In Ghana Academy of Arts and Sciences (Ed.), *Land as a resource for development*. Accra, Ghana: Ghana University Press.

Nicholls, R. J., Hoozemans, F. M. J., & Marchand, M. (1999). Increasing flood risk and wetland losses due to global sea-level rise: Regional and global analyses. *Global Environmental Change, 9*, 569–587.

Nicholls, R. J., Wong, P. P., Burkett, V., Woodroffe, C. D., & Hay, J. (2008). Climate change and coastal vulnerability assessment: Scenarios for integrated assessment. *Sustainability Science, 3*, 89–102.

Newton, A., Carruthers, T. J. B., & Icely, J. (2012). The coastal syndromes and hotspots on the coast. *Estuarine, Coastal and Shelf Science, 96*, 39–47.

Nyarko, B. K. (2000). Flood risk zoning of Ghana: Accra experience. *International Archives of Photogrammetry and Remote Sensing, 33*(B7/3; PART 7), 1039–1050.

Osei, J., Nyame, F., Armah, A. K., Osae, S., Dampare, S. B., Fianko, J. R., Adomako, D., Bentil, N. (2010). Application of multivariate analysis for identification of pollution sources in the Densu Delta wetland in the vicinity of a landfill site in Ghana. *JWARP, 02*(12), 1020–1029. doi:10.4236/jwarp.2010.212122

Osei, J., Osae, S., Adamako, D., Laar, C., Anim, A. K., Ganyaglo, S. Y., et al. (2011). The impact of Oblogo landfill site in Accra-Ghana on the surrounding environment. *Research Journal of Environmental and Earth Sciences, 3*(6), 633–636.

Oteng-Ababio, M. (2013). Unscripted (in)justice: Exposure to ecological hazards in metropolitan Accra. *Environment and Planning A, 45*(5), 1199–1218. doi:10.1068/a45256

Oxford University Press. (2015). Oxford dictionaries. Accessed June 8, 2015 from www.oxforddictionaries.com

Wisner, B., Blaikie, P., Cannon, T., & Davis, I. (2004). *At risk: Natural hazards people's vulnerability and disasters* (2nd ed.). Oxon, New York, NY: Routledge.

Chapter 21
Competing Narratives of Water Resources Management in Ethiopia

Catherine Fallon Grasham

Abstract In Ethiopia, urban water supply and irrigation are competing for water resources. The Millennium Development Goals have spurred large donor investment in water supply resulting in a rapid increase in coverage for health and human development. At the same time, most of Ethiopia's population is engaged in low-productivity rainfed agriculture and the government has made smallholder irrigation an investment priority for food security and poverty alleviation. In areas where water is physically scarce, there is fierce competition between water supply and irrigation resulting in unsustainable abstraction from common pool water resources. In the Haramaya watershed in Eastern Ethiopia, this has resulted in the severe depletion of Haramaya Lake, once an important water source for urban water supply for the historical town of Harar. Unregulated smallholder irrigation has expanded significantly and has contributed to the displacement of the urban water supply to over 72 km away. Water developments have been influenced by land-use change, international, national and local institutions and biophysical changes in the watershed. This chapter employs the nascent concept of the waterscape in order to explore how competition for water resources plays a role in the mediation of land-use change and vice versa.

Keywords Food security · Poverty · Land-use change · Scale · Waterscape

21.1 Introduction

This chapter is concerned with competition between irrigation and urban water supply for limited water resources in a semi-arid waterscape in Eastern Ethiopia. One of the greatest challenges of water resources management is to balance water demands from different sectors, namely irrigation, industry and domestic use (World Bank 1993). In addition to human demands on water, the environment needs water to

C.F. Grasham (✉)
School of International Development, University of East Anglia, Norwich, UK
e-mail: C.Grasham@uea.ac.uk

J. Niewöhner et al. (eds.), *Land Use Competition*,
Human-Environment Interactions, DOI 10.1007/978-3-319-33628-2_21

sustain vital ecosystems. However, in a developing economy such as Ethiopia, it is institutionally challenging to balance demands from human uses and the environment in water resources management planning. The case study in this chapter explores what can arise as a result of unintegrated management of water resources; in this case, it resulted in the complete disappearance of a tropical lake, Lake Haramaya, in Eastern Ethiopia in 2004. Previous to its disappearance, the lake provided urban water supply for the town of Harar, three smaller towns and a number of small villages, surrounding rural communities for domestic use and livestock and smallholder irrigation. Currently, the lake has partially returned and is providing water for irrigation and the proximal rural communities. Due to the limited amount of water available in the lake, many irrigators and urban water supply have converted to groundwater utilisation. This chapter will explore how competing narratives of water resources management in Ethiopia, namely water for urban water supply and water for irrigation, have contributed to the decline of available water.

In Ethiopia, there are two important narratives that are competing. One exists at the national level and one at the international level and together they are providing a backdrop to a number of political-economic processes that have directly translated into significant investment for urban water supply and smallholder irrigation in Ethiopia. First, at an international scale, the target of the Millennium Development Goals to halve the number of people without sustainable access to clean drinking water has spurred donor investment in Ethiopia for increased water supply. There is currently a nationally managed One WASH National Programme (OWNP) with a budget of $431 million financed by the World Bank, DfID, African Development Bank and UNICEF. Launched in 2013, the primary goal of the programme is to increase coverage of drinking water supply access and sanitation in both rural and urban areas of the country (FDRE 2013). Second, at a national level, the Ethiopian government has recently established irrigation development institutions, increasing their annual budgetary investment in the irrigation sector with a particular focus on smallholder irrigation to increase food security (MoFED 2006). This is because 85 % of the country is engaged in low-productivity rainfed agriculture that is vulnerable to the spatial and temporal variability of rainfall and endemic drought in many parts of the country. The government's overarching Agricultural Development Led Industrialization (ADLI) strategy (MoPED 1993) and more recent policies that promote the expansion of small-scale irrigation are providing a favourable environment for the recent increase of smallholder irrigation in Ethiopia. Water supply and irrigation are both expanding at a high rate and hence, scarce water resources are vulnerable to over-exploitation.

Water resources are being continually shaped by human intervention and are increasingly understood as socio-natural hybrids that are coproduced by humans and nature (Swyngedouw 1999; Linton and Budds 2014). A paradigm shift in the academic literature has seen academics move away from a purely material understanding of water resources to one that draws on political ecology. Water is governed and utilised by humans, meaning that it is shaped by actors, institutions, infrastructure and power relations. In this chapter, the concept of the waterscape is employed to enable an integrated discussion of a physical watershed and the social arrangements that govern it.

This chapter draws on empirical evidence collected during fieldwork conducted by the author in Ethiopia in 2014–15. The data collection methods include an urban household and business survey, gathering of quantitative data sets and semi-structured interviews with key actors involved in water governance from the grassroots to the federal level. The first section of the chapter establishes the theoretical framework for the case study, namely the governance of water resources as a common pool resource (CPR) and the governance context of Ethiopia. In the final two sections, the case study is introduced with a historical analysis of the waterscape and the waterscape today and how it has evolved over time. The chapter attempts to navigate the complexities of water governance in Ethiopia by assessing how competing narratives have spurred political processes and translated into a backdrop of policies that are enabling the rapid development of urban water supply projects and irrigation in Ethiopia. Moreover, it aims to show that employing the concept of the waterscape within a study of the governance of CPRs can enable a holistic analysis of competition for water resources, water–land coupling and how they change over time.

21.2 Water as a Common Pool Resource

This chapter draws on CPR theory in order to explore what factors have led to the over-abstraction of water resources in a semi-arid waterscape in Ethiopia. CPRs are natural resources that are utilised by more than one user of which water resources are an archetypal example. If water resources are conceptualised as CPRs, when water is removed from the common pool, by biophysical processes or through human use, it becomes fully or partially consumed rendering it unavailable for utilisation by another user. In a situation where there are too many demands on the resource, the resource is susceptible to degradation and over-abstraction. Therefore, restriction of some users' access may be required and not all demands on the common pool may be met. The governance of CPRs has been studied extensively, in particular with the influential framework of Elinor Ostrom that illuminated how the governance of CPRs can succeed and fail in different contexts due to many different factors (Ostrom 1990). In the case of Ethiopia, the constitution states that water resources are the property of the Ethiopian people and hence, water is a publicly owned and an open-access resource. The challenges of CPR governance are particularly acute where there is competition between water users for open-access resources and there is a need for a strong, contextually sensitive institutional framework in order to make decisions about who gets access to CPRs and to how many. CPR theory is particularly relevant to employ in the analysis of a waterscape since, "all CPR systems are inextricably bound to the social systems in which they are managed and exploited" (Kimbrough and Vostroknutov 2015: 39).

21.3 Water Governance in Ethiopia

This chapter is concerned with competition between irrigation and urban water supply for CPRs and this section will begin by defining urban water supply in the context of water governance in Ethiopia. The processes and actors involved in water governance in Ethiopia exist within a policy framework, legal framework and institutional framework which include NGOs, the private sector and government actors; the latter of which will be discussed at more depth in the case study. The overarching policy framework for governing and managing water resources in Ethiopia is made up of two main documents: first, the Ethiopian Water Resources Management Policy (FDRE 1999) and second, the Ethiopian National Water Sector Strategy, developed to translate the policy into action (FDRE 2001). In the policy, water supply is characterised as "water supply for human as well as animal consumption, industrial and other uses outside irrigation and hydropower" (FDRE 1999: 19). Water supply is understood as an overarching term but, in isolation, water for drinking is given the highest policy priority. Due to economies of scale, providing an urban water supply can be a highly cost-effective way of rapidly increasing drinking water coverage. In the government's first economic Growth and Transformation Plan (GTP-1), the country's overarching economic growth strategy, there is a target to achieve 100 % urban potable water coverage (access within 0.5 km) by 2015 from a baseline of 91.5 % in 2010 (FDRE 2007). In the GTP-I strategy document, urban water supply is primarily understood as potable water coverage. However, urban water supply consists of water for drinking, domestic use, sanitation, animals, services, industry and irrigation (for urban agriculture and gardens) and hence, it is important for the economic and social viability of an urban centre in many different ways, not only for drinking water.

Agricultural development is at the heart of Ethiopia's economic growth strategy and this is reflected in the water policy, GTP-I, and will continue to be in the GTP-II, the government's second economic growth strategy that will come into effect in 2016. In the water policy, the second priority for water resources management is small-scale irrigation for "food self-sufficiency at the household level" (FDRE 1999: 30). The GTP-1 highlights the importance of smallholder irrigation for poverty reduction and food security and also that "irrigation development, in particular small-scale irrigation will be accelerated" (FDRE 2010: 41). In 2010, the total irrigated area in the country was estimated at 640,000 and the national irrigation potential calculated at 5.3 million ha (Awulachew 2010). Irrigation is of high importance nationally to meet the agricultural GDP targets of the GTP-I. This strategy document influences how the contribution of water resources to economic growth is understood in Ethiopia and hence, how water resources should be allocated between sectors. The Ethiopian Ministry of Water, Irrigation and Electricity (MoWIE) plans to "increase the developed irrigable [land] aggressively up to 5 million ha" (MoWIE 2013: 3) throughout the period of the GTP-II. Therefore, within policy, Ethiopia is set for continued irrigation expansion up to and including 2020.

The policy framework supports a two-pronged approach to water governance: water for water supply and water for irrigation, providing a backdrop that facilitates direct competition between water for drinking and water for small-scale irrigation. Overall, drinking water and small-scale irrigation have the highest two policy priorities in Ethiopia. These priorities exist not only in policy, but pervade at all levels of water governance in Ethiopia, from the grassroots woreda (district) level to the federal level. Due to the dichotomous approach to water governance, water supply and irrigation are being steered independently at all levels of government. Actors and institutions are competing at different scales and this is resulting in the non-integrated development of water resources in Ethiopia.

In addition to the policy framework, there is also a legal framework made up of government proclamations enacted to regulate the governance of water resources. The Ethiopian Water Resources Management Proclamation, ratified in 2000, outlines how and by whom water resources are to be managed. Water resources are primarily managed by the administrations of Ethiopia's nine regional states, but when the resource crosses a regional boundary, it is managed by the federal state or a nominated body (FDRE 2000). The government enacted the River Basin Proclamation of Ethiopia, which mandates the development of River Basin Authorities (RBAs) in a move towards Integrated Water Resources Management (IWRM) (FDRE 2007). The river basin proclamation states that IWRM, "requires arrangements for reconciling the different uses of water resources within a river basin to achieve balance and sustainable development of water resources as economic as well as environmental resources" (ibid: 3655). It highlights the importance of stakeholders, with different interests, working together to manage the river basin. Therefore, balancing demands from competing sectors is an important role for RBAs. Moreover, the proclamation argues that water resources should be understood as economic as well as environmental resources and seeks to introduce a common property regime by issuing permits for use and collecting fees based on the amount of water abstracted from the common pool. This has the potential to significantly change the nature of common pool water resources in Ethiopia from open access. However, currently only three RBAs exist for the 12 river basins in Ethiopia and they are newly established and lack the capacity to carry out their mandate to integrate the stakeholders and sectors involved in water resources management (Mosello et al. 2015).

21.4 Competition in the Waterscape

The concept of the waterscape helps to better understand human competition for water resources and to delineate the geospatial scale of the physical material water resource under examination and the social arrangements governing it. The waterscape (Swyngedouw 1999, 2004) is a nascent concept that is increasingly adopted in literature that explores biophysical processes and social arrangements around water in conjunction (see Ahlers et al. 2014). This chapter explores two competing narratives that, via different processes, have spurred direct water competition between different users at a local level. The waterscape defines water as a resource that is continually coproduced by society and nature and enables the scales of water governance in terms of ownership of the resource, institutional governance arrangements and power dynamics that influence water use to be discussed concurrently (Budds and Hinojosa 2012). Examining human competition over CPRs demands an analysis that includes, but also goes beyond, physical water and land resources to assess the social arrangements that govern them. These arrangements, narratives and policies exist within the waterscape and, moreover, within its history. Its history is important in understanding the current waterscape; for example, it is necessary to know how and why new actors have entered and left the waterscape in order to understand the social processes behind the changing nature of competition for CPRs.

Within a waterscape, competition for CPRs exists at different spatial scales and in semi-arid environments, where water is scarce, land-use and water-use competitions are closely coupled. Competition can exist, inter alia, between narratives contending to enter policy, policies themselves, actors and institutions. These competitors exist at different scales such as international, national and local. Different competitors compete across and between scales, and competition is exacerbated when resources are scarce. A political economy understanding of scarcity (see Mehta 2010) is important when discussing scarcity in a waterscape because, similarly to the resources themselves, scarcity is continually coproduced by nature and society through biophysical processes, actors, institutions and infrastructure and is embedded within power relations (Budds 2008). Many land-based activities such as agriculture, urban centres and industry require water as an input and hence are shaped by the available water resources and vice versa. In the next two sections, this chapter will explore the history and current state of the waterscape in semi-arid Ethiopia exploring competition for CPRs and how water scarcity exacerbates land–water coupling within a waterscape.

21.5 The Changing Waterscape

This section introduces the Haramaya waterscape and explores changing land use and water use with biophysical changes, and as new actors, water users and infrastructure have entered the waterscape over time. The physical resources of the waterscape under examination are the land and water within Haramaya watershed (previously known as Alemaya watershed), which is located in the semi-arid highlands of Eastern Ethiopia. Geographically, the watershed lies at 9023–9026 North latitude and 41059–42002 East longitude. The watershed is in the Wabi Shebele River basin and the average annual rainfall is 800 mm with an altitude range of 2006–2436 m above sea level. Administratively, Haramaya watershed lies mostly in the Haramaya woreda and partially in the Kombolcha woreda in the East Hararghe Zone of the Oromia region in Ethiopia. The majority of people in the watershed are engaged in agricultural activities in mixed crop and livestock systems including rainfed and irrigated agriculture (Abebe et al. 2014) meaning that their livelihoods are inherently linked to land and water. Within the watershed,

Haramaya Lake flourished for many years and was a highly important resource for urban water supply for the towns of Harar, Awoday, Haramaya and Hamaressa, smallholder irrigation, local industries and for domestic use and livestock in the surrounding rural communities. In 2004, the lake completely dried due to the unsustainable abstraction of water resources and biophysical changes in the watershed. Today, the lake is partially returned, but the surface water resources in the watershed are far fewer than they once were.

Figure 21.1 shows the fluctuation of the lake level of Haramaya Lake from 1975 to 2009. The Ethiopian Ministry of Water, Irrigation and Electricity (MoWIE), the federal body responsible for collecting hydrological data in Ethiopia, abandoned their hydrological monitoring station in 2009. Therefore, there are no data available to determine the current lake level. Up until 2009, the station was collecting daily water level data. There are some significant gaps in the daily readings but the general trend of a highly fluctuating lake with eventual disappearance is apparent in the data and has been captured in Fig. 21.1. In addition to water level diminution, the size of the lake over time also decreased from 3.9 km^2 in 1965 to 2.3 km^2 in 2002 (Setegn et al. 2009).

The decline of Haramaya Lake has been attributed to both biophysical changes and human intervention and existing evidence suggests a number of explanations for the diminished lake. First, biophysical processes: increased temperatures in the watershed led to increased evaporation from the lake over time and annual rainfall has been reducing in the watershed (National Meteorology Agency, Addis Ababa, direct communication.). Second, human influences: the over-abstraction of water by

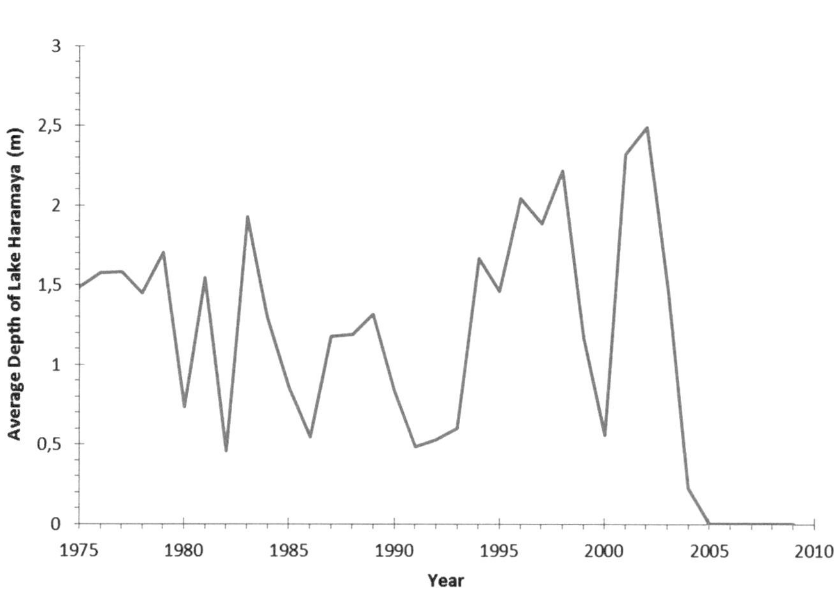

Fig. 21.1 Average depth in metres of Lake Haramaya from 1975 to 2009. Author's own. *Data Source* Ministry of Water, Irrigation and Electricity (MoWIE), Addis Ababa

irrigators, industry and a growing population resulted in a water deficit in the water balance of the lake (Setegn et al. 2011). There was increased siltation of the lake due to deforestation and subsequent soil erosion (Muleta et al. 2006) and social inaction by governing bodies to introduce measures to restrict abstraction (Kassa 2014). Therefore, from the existing evidence and in the context of the waterscape, it can be inferred that the decline of the lake has been coproduced by human and biophysical processes.

New water users entering the waterscape over time has led to significant land-use land-cover change which has exacerbated water scarcity. Land use and water use are coupled in Haramaya watershed by both biophysical processes and human activities. Figure 21.2 shows the land-use land-cover change of Haramaya watershed between 1973 and 2010. In 1973, the majority of the land was under rainfed cultivation, there was a large amount of shrub land, and the lake occupied a significant area of the watershed. There were also small amounts of bare land, plantation forest and grassland. Over time, the lake began to shrink as did the area of plantation forest. By 2010, the majority of land had become intensively irrigated cultivation and the lake had completely disappeared. Many smallholder rainfed farmers have converted to intensive irrigation for the cultivation of khat, a leafy narcotic plant that is a high-value export crop and of strong cultural significance in the area. Farmers can improve their income by growing khat since the nearby town of Awoday is an important khat market for local consumption and for export. The

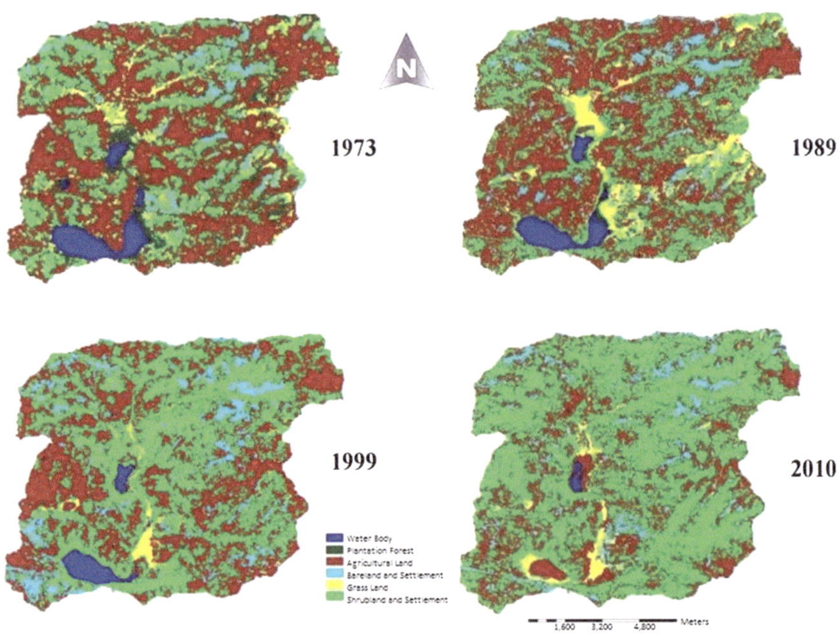

Fig. 21.2 Land-use land-cover change of Haramaya watershed, 1973–2010. *Source* Haramaya University, 2015

land once occupied by the lake is now predominantly occupied by agricultural land. Farmers engage in furrow irrigation, abstracting a large amount of water of which a significant amount evaporates from the land. This highly inefficient irrigation technique is compounding the contribution of increased irrigation activity to water scarcity in the waterscape.

21.6 Social Arrangements in the Waterscape

In addition to the physical resources in the Haramaya watershed, the waterscape is comprised of narratives, political-economic processes, policies/policy priorities, governance actors, institutions, power structures, infrastructure and water users. The social arrangements around the Haramaya water resources will be illuminated in this section. First, in Ethiopia in general, there has been recent increased investment in water resource development for drinking water. This has primarily been spurred by the international Millennium Development Goal to halve the number of people without safe access to drinking water. The highest priority in the Ethiopia water policy is water for drinking water, which, in the case of urban centres, is in the form of urban water supply. In the case of Haramaya watershed, in 1966 a water supply system for Harar and three smaller towns was constructed on Haramaya Lake. The total population using the supply at that time was around 20,000 and the system was designed to abstract 60 l of water per second. In 2004, when the lake dried, the system failed and five boreholes were drilled in a state of emergency in order to meet the urban water demand. The urban water supply was and continues to be coordinated by one actor, the Harari regional water supply office. The office was responsible for the construction of the original water supply from Haramaya Lake and also for drilling the five boreholes in the watershed. It has facilitated the increase in water supply coverage over time from 20,000 to a population today of 150,000 in Harar alone.

Similarly, the Ethiopian government's drive to increase smallholder irrigation for food security and poverty reduction (see FDRE 2011) has resulted in the nearly doubling of land under irrigation in the waterscape from 107 km^2 in 2013 to 204 km^2 in 2015 (Haramaya Woreda Irrigation Office, Haramaya, direct communication.). Smallholder irrigation is the Ethiopian government's second policy priority for water resources development in the country. The only criterion required by a farmer to access the water resource is the purchasing power to buy or rent a motor pump. Smallholder irrigation activities in the waterscape have not been regulated or restricted over time, due to the open-access nature of water resources in Ethiopia and a weak institutional framework. Irrigation development in the waterscape is uneven and dependent upon farmers' wealth meaning that irrigation activities develop on a first come, first serve basis. Therefore, there are important decisions around equitable access to water resources that need to be regulated, for water allocation not only between sectors, but also within sectors.

Haramaya watershed has been providing water to different sectors for decades and as more users have entered the waterscape, competition has become more fierce. In addition to smallholder irrigation and urban water supply, other significant water users in the waterscape are the Harar beer factory and Haramaya University. Haramaya University, situated adjacent to the watershed, was established in 1952 and is growing year on year. The Harar beer factory was established in 1984 and is highly important for the local economy providing on average 218 jobs in Harar town (Harar Beer Factory, Harar, direct communication.). The factory is a significant water user abstracting, on average, around 139 l per second (ibid.). Both the university and the factory are geographically located outside of the watershed but they are actors that exist within the waterscape. The water users and the social arrangements around water have changed the nature of competition for water and land resources in the waterscape over time.

21.7 The Waterscape Today

The waterscape today is characterised by the partial reformation of Haramaya Lake, a new water source and water infrastructure to meet urban water demand, newly established institutions to facilitate irrigation expansion and continued water abstraction by competing users. Land–water coupling is more observable in the waterscape as urbanisation, population growth and irrigation expansion have exacerbated water scarcity. The return of the lake is difficult to explain considering that the intensive irrigation of khat and vegetables has been expanding up to and including 2015. However, the graph in Fig. 21.1 shows severe fluctuation of the lake level over time suggesting that its reformation could be due to biophysical processes such as an increase in rainfall and reduction in temperature. Moreover, the land in the watershed is now saturated with khat cultivation and further irrigation expansion is now limited by the amount of land available for further conversion and hence, irrigation abstraction could have stabilised. In terms of urban water supply, less water is now being abstracted from the Haramaya watershed, since the water source for Harar and surrounding towns is now coming from a well field around 72 km away. However, the sustainability of the return of the lake is compromised by irrigators returning to utilise the lake waters, population growth and rapid expansion of Haramaya University.

The disappearance of Haramaya Lake severely compromised the Harar and surrounding towns' water supply; hence, there are some significant competition outcomes that have resulted from the changing waterscape. The urban water supply for Harar and surrounding towns is now being pumped from a well field along a 72-km pipeline to Harar over an elevation of 1000 m. The well field has been developed at huge expense, with a loan from the African Development Bank, in order to meet the urban water demand. The electricity costs of pumping the water from the well field to the towns are not being covered with the water user fees and hence, the situation is economically unsustainable. Moreover, the project was designed to abstract 300 l of water per second from the well field, the calculated urban demand, but frequent, crippling power cuts mean that an average of only 90 l per second is travelling along the pipeline to the towns. Not only is the project economically unsustainable, it is also not functioning in the way it was designed. Urban dwellers are receiving insufficient water to meet their domestic and livelihood demands. Therefore, despite drinking water being the policy priority for Ethiopia, in this case study, irrigation has monopolised the CPR and urban water supply has been compromised.

With the introduction of the new water source from the well field 72 km away, the waterscape has now expanded to include this new water source and pipeline infrastructure. This is because the same institution, the Harari regional water supply office, is managing the urban water supply both from the well field and from the Haramaya watershed. In addition, the shifting of the urban water supply has resulted in changes in the nature of water-use competition as irrigating farmers and urban water supply are directly competing via the new pipeline. In an interview with the Harari regional water supply office, this story was exposed:

"There was one very clever farmer. He built a house on his farm and we asked him, 'Why are you building a house here?' He said that he needed to sleep with his [khat] farm. What he did was, he dug a hole to the pipe and paid someone to fix in a tap on the pipeline. Then, you know what he did? He filled the hole and even, he laid a carpet on the floor of the house. He moved the pipe to make it look like [the water was coming from] a spring and all year his farm was green. The man who fixed the tap said to the farmer that if he didn't pay him some money he would tell the government what he was doing and if it wasn't for that, we would never know." (Harari Regional Water Supply Office, Harar, July 2015).

As a waterscape changes overtime, so does the nature of competition for water. This example illustrates that water resource developments can have a significant impact on the nature of competition for water resources within a waterscape. In this case, the introduction of new infrastructure, a new pipeline, has enabled a new form of competition to emerge. The behaviour of the farmers is illegal and condemned by institutions managing both irrigation and water supply, but it is still taking place. The scarcity of water resources has contributed to the severity of competition in the waterscape.

The current social arrangements around water are significantly changed. The most significant water users in the waterscape are still: Harar beer factory; Haramaya University; smallholder irrigators and Harar and surrounding towns' water supply, but irrigation in the waterscape is now governed by a number of new institutions. Figure 21.3 shows the current institutional arrangement of governing water resources in the waterscape for irrigation and urban water supply. The Ethiopian government implements a top-down approach to water governance starting from the federal level through to local woreda level. There is strong vertical communication between the different levels of government. However, there were none or very weak lateral linkages observed between the institutions. At the federal level, large- and medium-scale irrigation is managed by the Irrigation and Drainage Directorate at the Ministry of Water, Irrigation and Electricity (MoWIE) and

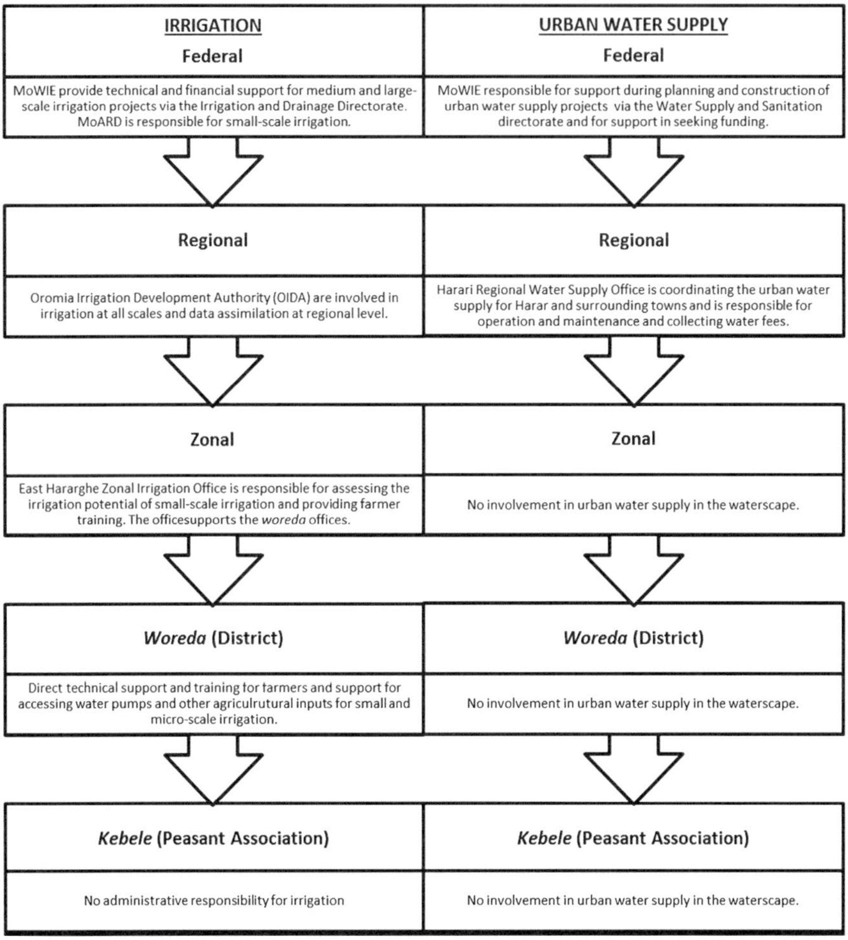

Fig. 21.3 Governance structure for irrigation and urban water supply in the waterscape. Author's own

small-scale irrigation by the Ministry of Agriculture and Rural Development (MoARD). Urban water supply is managed by the Water Supply and Sanitation Directorate at MoWIE. There is limited collaboration between the three actors resulting in a lack of integration between the irrigation and water supply sectors.

A similarly fractured situation is observed at the lower levels of government. Established in 2013 are the Haramaya Woreda Irrigation Office, East Hararghe Zonal Irrigation Office and the Oromia Irrigation Development Authority (OIDA). These new institutions, operating at three different levels of government, have been established in order to better coordinate irrigation development activities. Their establishment is in line with the targets of the government's Growth and Transformation Plan (GTP-I) and reflects their drive to expand small-scale irrigation. The Harari regional water supply office is managing urban water supply at a regional level and there are no actors at zonal or woreda level involved in the governance of urban water supply. Similarly to the federal level, the institutions are not communicating in order to integrate their water development activities in the waterscape. Moreover, the expansion of small-scale irrigation activities is sporadic and unregulated and there is not an operational governance arrangement that gives any actor(s) the power to regulate over-abstraction of water resources if it becomes unsustainable or if it is having an impact on other users or uses of water. Stronger institutional integration and an enforced regulatory framework to ensure the sustainability of water resources are crucial in semi-arid waterscapes where water use is mediating land-use competition and vice versa.

21.8 Conclusion

Material water resources are continually shaped by the social arrangements that govern them and the concept of the waterscape enables a holistic analysis of the socio-hydrological dynamics of common pool water resources over time. Changes in a waterscape have many drivers but in the case of the Haramaya watershed the most important ones are: (1) the target to achieve food security, and economic growth by rapidly increasing irrigation and (2) the need to increase water supply coverage for health and human development. These drivers have resulted in competition at different scales between irrigation and water supply, in policy, institutional arrangements and at a local level, contributing to the over-exploitation of Haramaya watershed and the drying of Haramaya Lake. The water policy of Ethiopia demands an integrated approach to the management of water resources, but this study has found that, from the grass roots to the federal level, water is predominantly managed in two siloes, namely water supply and irrigation, contributing to the unsustainable abstraction of water resources to the detriment of urban water supply. The water supply for Harar and surrounding towns has been resourced to an economically unsustainable and dysfunctional situation and irrigation has flourished.

In this case, the competition for water resources is exacerbated because: water resources are physically scarce; actors have unregulated access to water resources and the governing institutional framework is weak meaning that it is not currently possible to curtail unsustainable abstraction. For the sustainability of waterscapes, there must be an integrated institutional framework where actors managing different sectors are working together to balance competition. Competition for water resources is inherently linked to land use, which is acutely observed with irrigation since it is a large land and water user. As irrigation increased in the waterscape, less water became available for other users and uses and hence, the availability of water resources is now dictating how land can be utilised in the waterscape. Also, as land use changed to irrigation and the lake shrunk, there was more land available for irrigation. Therefore, irrigation expansion plays a significant role in the mediation of land-use competition and vice versa.

Acknowledgments The author would like to thank the International Water Management Institute, The Hilda Martindale Trust and Funds for Women Graduates who funded this research. Additionally, the author would like to express sincere thanks for comments received from the proof readers who helped to improve the chapter.

References

Abebe, S., Haji, J., & Mengistu, K. (2014). Impact of disappearance of Lake Haramaya on the livelihood of the surrounding community: The case of Haramaya District in Oromia national regional state. *Ethiopia Journal of Economics and Sustainable Development, 5*(18), 141–147.

Awulachew, B. (2010). *Irrigation potential in Ethiopia: Constraints and opportunities for enhancing the system*. Unpublished Report to the Bill and Melinda Gates Foundation.

Ahlers, R., Cleaver, F., & Schwartz, K. (Eds.). (2014). Informal space in the urban waterscape. *Water Alternatives, 7*(1) 1–159 (Special Issue).

Budds, J. (2008). Whose scarcity? The hydrosocial cycle and the changing waterscape of La Ligua Basin, Chile. In M. K. Goodman, M. T. Boykoff & K. T. Evered (Eds.), *Contentious geographies: Environmental knowledge, meaning, scale* (pp. 59–78). Aldershot: Ashgate.

Budds, J., & Hinojosa, L. (2012). Restructuring and rescaling water governance in mining contexts: The co-production of waterscapes in Peru. *Water Alternatives, 5*(1), 119–137.

FDRE (The Federal Democratic Republic of Ethiopia). (1999). *Ethiopian water resources management policy*. Addis Ababa: Ministry of Water Resources.

FDRE (The Federal Democratic Republic of Ethiopia). (2000). Proclamation No. 197/2000. Water resources management proclamation. Addis Ababa.

FDRE (The Federal Democratic Republic of Ethiopia). (2001). *Ethiopian water sector strategy*. Addis Ababa: Ministry of Water Resources.

FDRE (The Federal Democratic Republic of Ethiopia). (2007). Proclamation No. 534/2007. River basin councils and authorities proclamation. Addis Ababa.

FDRE (The Federal Democratic Republic of Ethiopia). (2010). *Growth and transformation plan 2010/11–2014/15*. Ministry of Finance and Economic Development (MoFED): Addis Ababa.

FDRE (The Federal Democratic Republic of Ethiopia). (2011). *Small-scale irrigation capacity building strategy for Ethiopia*. Addis Ababa: Natural Resources Management Directorate, Ministry of Agriculture and Deutsche Gesellschaft für Internationale Zusammenarbeit (GIZ).

FDRE (The Federal Democratic Republic of Ethiopia). (2013). *One WASH national program: A multi-sectoral SWAp program document*. Addis Ababa: One WASH National Program.

Kassa, K. (2014). Action, inaction and environmental destruction: Socionatural determinants of the disappearance of Lake Alemaya (Haromaya), eastern Ethiopia. *Research Journal of Agriculture and Environmental Management, 3*(8), 361–369.

Kimbrough, E. O., & Vostroknutov, A. (2015). The social and ecological determinants of common pool resources. *Sustainability Journal of Environmental Economics and Management, 72*, 38–53.

Linton, J., & Budds, J. (2014). The hydrosocial cycle: Defining and mobilizing a relational-dialectical approach to water. *Geoforum, 57*, 170–180. doi:10.1016/j.geoforum.2013.10.008

Mehta, L. (Ed.). (2010). *The limits to scarcity: Contesting the politics of allocation*. London: Earthscan.

MoFED (Ministry of Finance and Economic Development). (2006, September). *Ethiopia: building on progress. A plan for accelerated and sustained development to end poverty 2005/6–2009/10.*

MoPED (Ministry of Planning and Economic Development). (1993). *An economic development strategy for Ethiopia*. Addis Ababa: Ministry of Planning and Economic Development.

MoWIE (Ministry of Water, Irrigation and Electricity). (2013). *Irrigation potential and agri-business investment opportunities in Ethiopia*. Irrigation and Drainage Directorate: Addis Ababa.

Mosello, B., Calow, R., Tucker, J., Parker, H., Alamirew, T., Kebede, S., et al. (2015). *Building adaptive water resources management in Ethiopia*. London: Overseas Development Institute (ODI).

Muleta, S., Yohannes, F., & Rashid, A. M. (2006). Soil erosion assessment of lake Alemaya watershed, Ethiopia. *Land Degradation and Development, 17*, 333–341.

Ostrom, E. (1990). Governing the commons: The evolution of institutions for collective action. In *Political economy of institutions and decisions*. Cambridge: Cambridge University Press.

Setegn, S. G., Chowdary, V. M., Mal, B. C., Yohannes, F., & Kono Y. (2011). Water balance study and irrigation strategies for sustainable management of a tropical Ethiopian lake: A case study of lake Alemaya. *Water Resources Management, 25*, 2081–2107. doi:10.1007/s11269-011-9797-y

Setegn, S. G., Yahannes, F., Quraishi, S., Chowdary, V. M., & Mal, B. C. (2009). Impact of land use/land cover transformations on Alemaya Lake, Ethiopia. *Journal of Indian Water Resources Society, 29*(3), 40–45.

Swyngedouw, E. (1999). Modernity and hybridity: Nature, regeneracionismo, and the production of the Spanish waterscape 1890–1930. *Annals of the Association of American Geographers, 89* (3), 443–465.

Swyngedouw, E. (2004). *Social power and the urbanization of water: Flows of power*. Oxford: Oxford University Press.

World Bank. (1993). *Water resources management*. A World Bank policy paper. Washington D.C.: World Bank.

Chapter 22
Competition in Transition: An Exploration of Water and Land Use in the Wien River Valley Through the Eyes of Mid-Nineteenth-Century Engineers

Christina Spitzbart-Glasl and Gudrun Pollack

Abstract Between 1847 and 1857 engineering students under the civil engineer and architect, Josef Stummer (1808–1891) created a rich cartographic and written description of Wien River, the largest tributary of the Viennese Danube. The highly dynamic flow regime, the riverbed and shores are described as intensively used in manifold ways by different actors. Townspeople used the water's energy to drive mills and process water for dyeing works, intanneries and laundries. Others extracted gravel from the riverbed or dried laundry on the shores, people grew vegetables and animals grazed. While the city grew along the shores of the river, water and land uses were still influenced by old manorial regulations. The river itself, with its natural dynamic change between periods of water scarcity and torrential floods, had an important bearing on the possible uses. In the 1850s, Wien River valley was a waterscape on the verge of change from an agrarian to an industrial socio-metabolic regime. Only a few years after the engineers had conducted their survey, the mill creeks were abandoned. 40 years later city authorities intervened with extensive regulation and constructed a railway line in the riverbed. We use Stummer's material to tell an environmental history of competition over water and land in the Wien River waterscape during this transitional period.

Keywords Environmental history · Land–water nexus · Governance · Waterscape

22.1 Introduction

In a river valley water and land use are always inseparably connected. Still the modes of competition over the limited natural resources of water and land and the values attributed to them by (potential) users change over time. Big transformation processes such as the industrialization and urbanization of Vienna encompassed the

C. Spitzbart-Glasl (✉) · G. Pollack
Institute of Social Ecology, Alpen-Adria University, Vienna, Austria
e-mail: Christina.Spitzbart@aau.at

J. Niewöhner et al. (eds.), *Land Use Competition*,
Human-Environment Interactions, DOI 10.1007/978-3-319-33628-2_22

remaking of the small Wien River valley in the middle of the nineteenth century, its use, perception, and hence competition over water and land. The society-wide transition from an agrarian to an industrial socio-metabolic regime characterized by a shift from solar energy to fossil fuels, as described, e.g., by Sieferle et al. (2006), Fischer-Kowalski and Haberl (2007), Krausmann and Fischer-Kowalski (2010), is the framework for our case study. This transition is widely considered as a major driver for land-use competition, e.g. by Haberl et al. (2014: 60) effective in the past and until today in developing regions and cities (ibid: 58).

We conceptualize Wien River valley as a socio-natural hybrid following the concept of waterscapes by Swyngedouw (2009: 56) and the environmental history concept of socio-natural sites (Winiwarter and Schmid 2008: 160). The materialist approach of socio-natural sites sees them as formed by the co-evolution of practices and material arrangements. Swyngedouw emphasizes the importance of power relations between different user interests and agency to understand socio-natural transformations. Both concepts stress the significance of historical studies for understanding material and immaterial conditions, consequences and legacies of past activities (Winiwarter et al. 2013: 120; Swyngedouw 1999: 460): Bygone competitions on water and land—be they mediated by ecological, economic, political or social logics—have long-lasting effects, often discernible until the present.

Vienna is situated at the upper Danube in a large alluvial basin (Wiener Becken). The historical city is located on the right bank of the Donaukanal, the former Wiener Arm, nowadays a channelled side arm of the Danube. About 70 streams originate in the Wienerwald, a hilly region in the west and south-west of the city, and discharge into the Viennese Danube. Most of them were integrated into the city-wide waterborne sewage system in the nineteenth and twentieth century (Gantner 1991: 21).

Wien River (Wienfluss), delineated in Fig. 22.1, is the largest of these tributaries. It flows through the city from the west and mounts into the Donaukanal. At the beginning of the twenty-first century, about one-third of its 230-km^2 basin lies in more densely populated urban area and is drained by sewers. Wien River has alpine character with highly fluctuating discharge between 0.3 m^3/s at low water level and 300 m^3/s or more in case of floods (BMLFUW 2011). Due to the low infiltration capacity of the soils in the upper catchment, water levels rise immediately, when it rains. At present, mean annual discharge lies at about 1 m^3/s, but we assume about 2–3 m^3/s prior to the extensive regulation and the implementation of a sewage system (unpublished reconstruction by Severin Hohensinner, based on Pollack 2013). On its (today) 35-km route, the river drops more than 400 m with a gradient of about 4.4 ‰ in the urban area (Stummer 1847–1857).

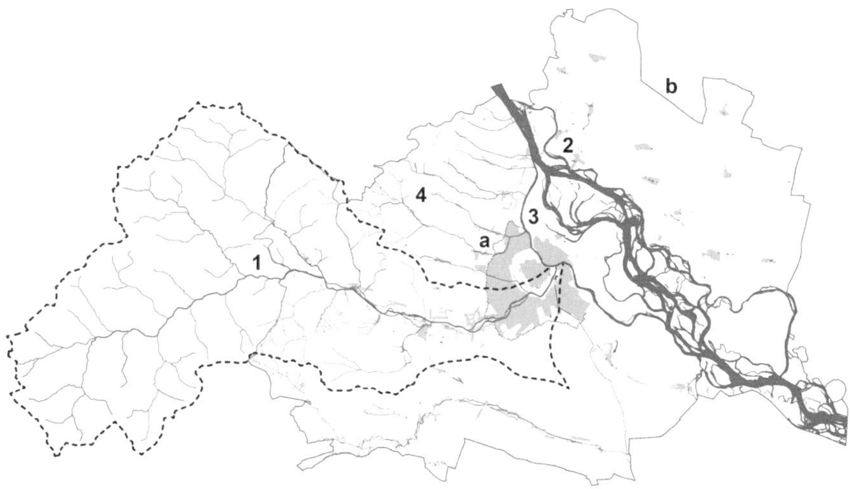

Fig. 22.1 Wien River and its location within the urban aquatic network around 1825. *1* Wien River, *2* Danube, *3* Donaukanal, *4* Area of the Wienerwaldbäche, **a** Linienwall, **b** present-day city limit. Wien River catchment highlighted. Graphic design by Friedrich Hauer based on the Franciscean Cadastre, 1817–1829, digitized by Stadtarchaeologie Wien

22.2 Sources and Methods

The starting point and major source for this case study is an inventory called "Topographical and hydro-technical survey of Wien River" (Die topographische und hydrotechnische Aufnahme des Wienflusses unter Prof. Josef Stummer, 1847–1857—short "Wienflussaufnahme"). It is preserved in the Municipal and Provincial Archives of Vienna (WStLA, Kartographische Sammlung, P1.253). Professor Josef Stummer (1808–1891), an engineer and architect who taught hydraulic construction at k. k. Polytechnisches Institut (today Vienna University of Technology), is named as the lead author of the work (Kurdiovsky and Offenthaler 2012: 2). The survey was the final assignment of his students who worked in groups guided by experienced engineers to prove their practical qualification. Stummer drew up detailed instructions for his students. They were to imagine the planning of a new railway line in the Wien River valley as reason for their records (Stummer 1847a). The Wienflussaufnahme consists of elevation tables, one detailed longitudinal and 139 transversal sections of Wien River between Purkersdorf, at that time the first post station west of Vienna, and the confluence with Donaukanal as depicted in Fig. 22.2. Besides these tables and maps, it includes detailed descriptions structured along the transversal sections.

We do not know whether the Wienflussaufnahme had any direct implication on river regulation or water management practices carried out in the second half of the nineteenth century. As far as we know it is the first source that comes up with the idea of a railway line in the river valley—an idea that has later been widely discussed (Atzinger and Grave 1874) and implemented at the turn of the century.

Fig. 22.2 Wien River valley from Mariabrunn to the Donaukanal around 1850. Mill creeks are highlighted in *dark grey*, former mill creeks are marked with *dotted lines*, weirs with a *bar* and mills with a *dot*. The map shows the stretch of the valley described in the Wienflussaufnahme. The village of Purkersdorf is located to the west of Mariabrunn, just outside of this map. *1* Mariabrunn and Mariabrunn weir, *2* Mariabrunn mill creek, *3* Hacking, *4* Penzing, *5* Meidling weir, *6* Gumpendorf mill creek, *7* Gumpendorf weir, *8* Wieden mill creek, *9* Karlsplatz, *10* Glacis. Graphic design by Friedrich Hauer based on the Franciscean Cadastre, 1817–1829, digitized by Stadtarchaeologie Wien

The survey allows us to experience Wien River through the eyes of young engineers—a group of actors who largely determined its transformation at the turn of the century. It describes an urban waterscape at the verge of change from an agrarian to an industrial socio-ecological regime in an assemblage of soon to be abandoned and just emerging practices and arrangements. By analysing these practices and arrangements together with a set of drivers contributing to the profound transformation of the river at the turn of the century we ask whether such a transition can be understood as competition over water and land.

22.2.1 The Wien River Waterscape in Stummer's Survey

The Wienflussaufnahme focuses on river morphology, fluvial dynamics and the built environment, such as embankments, bridges, settlement areas, mill and transport infrastructure. Besides technical details such as gradient, water depth, height of banks, distances and the like, it touches on a wide variety of social and economic information. The waterscape described is clearly natural and social at the same time: People intervene and change the river and its floodplain while at the same time the river works according to its own dynamics (cp. White 1995: 109). Wien River's dynamic flow regime forced people to adapt practices of water- and land-use to flooding and water scarcity, which materializes in the design of arrangements. Natural and social processes are inextricably entangled, as causes and effects turn into one another (Winiwarter et al. 2013: 107).

22.2.2 River Dynamics and Regulation

Near-constant water flow secured in one narrow channel (Normalprofil), flood control and land reclamation were the desired outcomes of most engineering efforts at Wien River during the nineteenth century (Atzinger and Grave 1874: 61). Prof. Stummers young engineers were also guided by this idea and depicted already existing regulation measures and the deviation of the river bed from the Normalprofil. They described the dynamic flow regime and its influence on the landscape in detail: the river splits into several arms, relocates its main arm from one bank to the other, erodes its shores, undermines bridges and bank reinforcements and deposits huge gravel bars. Regulation measures existing at the time were pictured as patchwork, as each of the adjoining land owners was responsible for their own protection (Stummer 1852–1855b: 38). While the lower stretch of Wien River within the outer fortification line (Linienwall) was already highly altered and the banks "quite conserved" (Stummer 1847b: 5) the patchwork of regulation measures was more scattered upstream. A variety of interventions from spur dikes, hurdle works and ripraps to the plantation of willows are described. The engineers recorded new reinforcements erected after the flood of 1851 next to flood-proof as well as degraded older infrastructure. For the village of Hacking the engineers noted how the remainders of a bridge, destroyed by flooding, had been used in turn to protect the shores around a new bridge (Stummer 1852–1855a: 43).

Still, these scattered regulation measures had an impact on the river as they shifted the eroding powers and flood risks to other spots upstream, downstream or on the opposite shore. That way new regulation measures could easily interfere with existing ones, the engineers noted for example for the village of Penzing: The river had eroded a 20-m-wide strip of the communal grazing grounds because of a regulation measure on the opposite shore (Stummer 1855c: 15). The engineers approvingly recorded successful reclamation of floodplains and bemoaned the loss of land due to the dynamics of the river. Apart from the aim to stabilize the relations between land and water, flood protection for extensive and vulnerable infrastructure such as mill creeks, weirs and canals was an important issue. Numerous small channels diverted overabundant water from the mill creeks back into the river (Stummer 1852–1855a: 54). An example for the protection of a mill creek with dykes is given in Fig. 22.3 (Stummer 1855c: 23).

Building practices were respectful of the river, especially outside of the Linienwall. Only few houses and businesses along the river were described as prone to flooding. Still the vulnerability in this dense socio-natural setting is clearly visible from the survey. The engineers exemplified the situation of a bone factory in one of the Vororte (villages outside of Linienwall) close to Meidling weir. After the weir had been destroyed by flooding in 1847 and not been rebuilt, gravel deposited

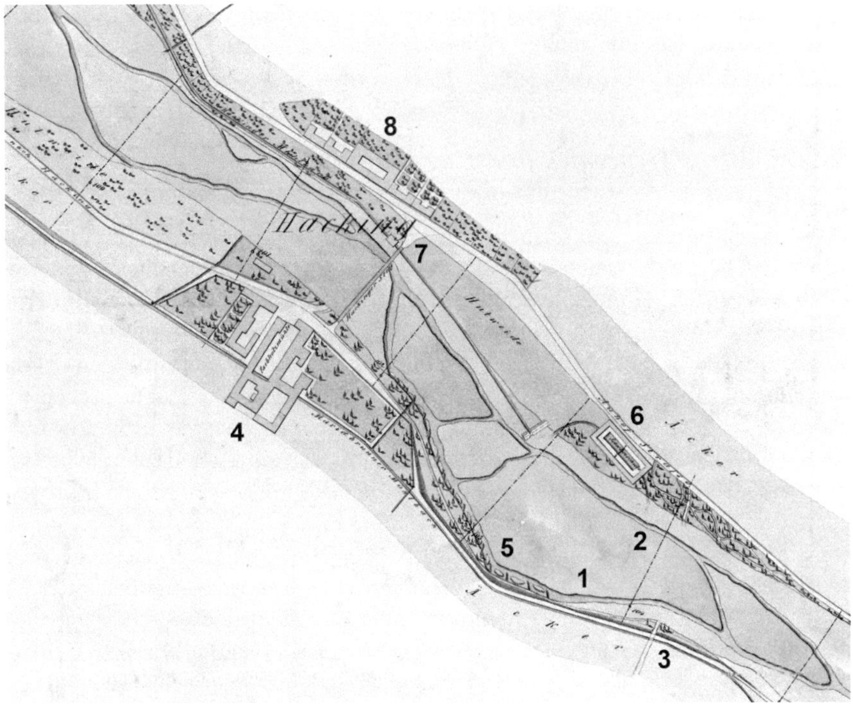

Fig. 22.3 Detail of Stummer's Wien River survey maps, showing the village of Hacking, one of Vienna's surrounding villages. Wien River is shown with two arms (*1*, *2*) in a wide bed that comes up close to the Mariabrunn mill creek (*3*), supplying Hacking mill (*4*). In order to prevent Wien River to erode the bank of the mill creek, the miller built spur dykes (*5*). The map also shows a swimming pool (*6*), a washing house (*7*) and a printing factory (*8*) all using water from Wien River (WStLA, Kartographische Sammlung, P1.253.8, Stummer, 1847–1857)

by the river behind the weir was gradually redistributed and raised the bed downstream. In order to protect this factory from the subsequently increased flood risk the owner invested in a solid stonewall (Stummer 1852–1855b: 9–10).

22.2.3 Water Uses and Users

The Wienflussaufnahme provides a vivid picture of the importance and diversity of water uses along the river, even though only those regulated by chartered water rights and user contracts and bound to permanent infrastructure are depicted. Informal, transient uses were rarely included but have been documented by other authors (Atzinger and Grave 1874: 50). Amongst the water uses the engineers described, mills using the water's energy received most attention. At the time of the survey, an extensive network of mill creeks paralleled the river. It powered nine grain mills, one brewery, one polishing mill and three tanning factories (see Fig. 22.2). Most of these mills and mill creeks dated back to the Middle Ages (Lohrmann 1980: 14). Although it is likely that the mill creeks originated from natural side arms of the river (Hofmann et al. 2003), it becomes clear from the survey that the amount of building material and labour needed to secure them from fluvial dynamics had been significant (Stummer 1852–1855d: 3). Weirs were prone to destruction by flooding but also increased flood risks for the surrounding area. This was especially an issue for Gumpendorf weir (Fig. 22.4), the only one located within the densely populated urban area. Strategies to deal with flooding involved constructional measures such as the introduction of large sluice gates and management practices such as the employment of a guard to observe water levels and open the sluices in case of rising water levels (Stummer 1847b: 12). By order of the city of Vienna millers had to set up common financial reserves for occasions when flooding could not be prevented in order to pay for damage related to their weir (Stummer 1847b: 24).

Millers sharing one mill creek had to cooperate to maintain their infrastructure. They split labour input and maintenance costs, although they competed over water flow and available head. Overabundance and water scarcity were equally problematic.

Fig. 22.4 A detailed picture of Gumpendorf weir is featured amongst other important infrastructure such as bridges in the Wienflussaufnahme. Gumpendorf weir was frequently destroyed by floods and rebuilt according to the latest technical knowledge. The pictured design is from 1818 and was considered especially flood prove due to its bottom outlets. A sluice chamber is located on the left side of the weir, adjacent Wieden mill creek is diverted from the river in the far left of the image (WStLA, Kartographische Sammlung, P1.253.7, Stummer 1847–1857)

During dry periods, almost the entire flow of Wien River was diverted to the mill creeks, leaving the riverbed almost waterless, and still, the operation of mills was severely hampered. Millers had to shut down one or more gears. Some of them turned to alternative power sources such as horse capstans (Stummer 1847b: 21).

Tanneries, dyeing works, laundries and textile printers also used Wien River and especially the mill creeks as they often contained more water than the river itself for washing or rinsing of raw materials and products (Pollack et al., in preparation). Water use was strictly regulated and bound to the payment of rents, participation in maintenance work or cooperation agreements such as to keep a space on both sides of the mill creek clear to make room for cleaning operations (Stummer 1852–1855b: 18–19). Extensive drying and bleaching areas were located in the floodplains. Factory owners discharged their waste water directly into the mill creeks or the river (Stummer 1852–1855a: 53) and deposited waste on the shores (Stummer 1852–1855b: 21). These practices were not regulated, resulted in a large amount of pollution with bacteria, oxygen-depleting substances and toxins and were seen as a big problem during the nineteenth century (Pollack 2013: 52).

Swimming schools and baths also diverted water from the river, while vegetable gardeners used it for irrigating their parcels in the floodplain. Gravel was extracted as building material, for example to maintain streets in the river valley, while river water was used to spray them. Figure 22.3 gives an example of the crowded users and uses on one small stretch of the river.

This dense network of users and uses poses questions on conflicts related to water supply and quality. However, the engineers did not mention any. Other sources, e.g. the yearly reports of the urban health authorities (Stadtphysikat) or the records of Wassergrafenamt (OeStA, FHKA AHK VDA Mühlen), an authority responsible for water rights, make clear that such conflicts were ever-present.

22.3 Transformation of the Wien River Waterscape 1850–1910

The waterscape described in so much detail in the Wienflussaufnahme was subject of major transformations associated with the transition from an agrarian to an industrial metabolic regime. Industrialization, urbanization and hygiene have been recognized as main drivers affecting urban rivers in Europe in the nineteenth century (e.g. Deligne and others in Castonguay and Evenden 2012). Hence, we also touch on the role of available energy sources and technology in relation to river transformation.

The biggest and most long-lasting intervention into Wien River was its extensive regulation between 1894 and 1904, which massively changed appearance, hydrology, ecology and possible uses. It stands at the end of a century of intensive use and pollution while the river's role in the city was heavily debated.

Over the course of the nineteenth century, Vienna expanded and an agglomeration formed (Eigner and Schneider 2005). Between 1830 and 1869, the population in the city and its Vorstädte (suburbs inside Linienwall) doubled, from about 300,000 to 600,000. With the incorporation of the Vororte outside of the Linienwall in 1890, Vienna counted more than 1 million people (1.75 million in 1900) (VID 2014).

Between Karlsplatz and its mouth into Donaukanal, Wien River flowed through the open Glacis in front of the fortification walls. To protect the city, an order posed after the first Ottoman siege (1526–1555) demanded to keep this area free of buildings. The fortification was demolished in 1850 and the building ban was lifted shortly after in 1859. In the following decades, public authorities erected a series of representative buildings along a ring boulevard (Ringstraße) and laid out a city park next to the river (Czeike 1992–2004). The area further upstream within the Vorstädte was already densely settled when it was incorporated in the territory of the city in 1850. The area outside of the Linienwall was the primary site of urbanization and industrialization (Pollack et al., in preparation).

From the late eighteenth century onwards, Wien River valley became a major axis for proto-industrial and industrial development. Small enterprises such as textile dyers, which depended on water for their production processes, settled in the Vororte in proximity of the river. Later in the nineteenth century, this development continued further upstream with the establishment of larger industries, which needed more space and took advantage of looser emission regulations (ibid).

Urbanization and industrialization in the Wien River valley led to river pollution by domestic and industrial waste and wastewater. Miasmatic theory, the main interpretative framework for infectious disease since antiquity, led to the perception of faeces and other badly smelling matter as serious health risk. The heavily polluted Wien River was therefore seen as a dangerous source of diseases such as typhoid fever or cholera (Knelz 1843). During the first cholera epidemic in Vienna from 1831 to 1832, more than 7000 people contracted the infection and more than half of them died (Birkner 2002: 48). Cholera epidemics in 1854 and 1873 followed. The first epidemic led to a turn in city planning with hygiene becoming a very prominent issue. The newly installed urban health authority (Stadtphysikat) started to publish yearly reports about the sanitary situation in Vienna. Its work has to be seen in the context of a Europe-wide hygienic movement, which believed in space, clean fresh air and the banning of bad smell (Büschfeld 1997). Between 1831 and 1839 intercepting sewers were constructed along Wien River between Linienwall and Donaukanal, taking the wastewater from most buildings in the Vorstädte and transporting it to the Donaukanal (Birkner 2002: 42). Still river pollution remained a problem throughout the century, because of continued population growth, new industries in the former villages upstream and the direct use of the river by tanners, textile dyers and printers despite bans to do so (Pollack 2013: 54). Although city authorities improved water supply and discharge infrastructure, many houses still relied on water from house wells, which were often situated close to potentially leaking cesspools and sewers (Pollack et al., in preparation). Only after 1873, when the first alpine water pipeline (Hochquellenleitung) started to bring fresh water via an aqueduct from springs 100 km south of Vienna to the city, water ceased to be scarce and bad water quality slowly became obsolete. In the following three decades, the city-wide supply and discharge infrastructure was further improved. A waterborne sewage system was installed, and intercepting sewers along Wien River were extended to the surrounding villages (Meißl 2001). Upon insistence of industrialists, a Wien River valley pipeline was built using water from a reservoir in the upper catchment of the Wien River. It opened in 1898 and supplied the western parts of the city with process water until 1964 (Czeike 1992–2004).

Mills utilizing Wien River by diversion of mill creeks (see Fig. 22.2) had been important parts of the area's economy and contributed to its commercial character (Hauer et al., in preparation). Their main product, flour, was a core necessity of the Viennese. Nevertheless, in 1847, when Meidling weir, located upstream of the Linienwall and supplying four mills, was damaged by a flood, it was not reconstructed. Instead, city authorities ordered the filling of the mill creek. In 1855, they demanded the abandonment of Wieden mill creek for sanitary concerns in connection with the second cholera epidemic. Gumpendorf weir was dismounted, the mill creek filled and three mills closed in 1856 (Protokollbücher 1847–1857).

Wien River had lost its significance as energy source for production processes while mills were increasingly perceived as sanitary and flood risks. A bundle of technical innovations, socio-economic changes, new sanitary paradigms and the availability of coal as a new energy source explain this transformation (Reynolds 1983: 321; Melosi 2006: 6). Improvements in transport infrastructure (steam ships, roads and railroads) lifted transportation constraints and enabled Vienna to become part of an emerging global market for flour (Persson 1999). Large-scale flour industries driven by steam engines, new processing and storage technologies demanded high investments only feasible for big investors, who focused their activities on Budapest rather than Vienna (Österreichisch-Ungarische Müller-Zeitung 1890; Beer 1901).

In the first half of the nineteenth century, the bed of Wien River was already straightened, deepened and stabilized within the fortification lines by order of Vienna city authorities. Upstream small-scale local regulation measures still allowed for a dynamic riverbed, as described in the Wienflussaufnahme. One reason for the patchwork of regulative interventions was the ownership structure in the areas outside of the Linienwall. In contrast to the Vorstädte, which were already under administrative control by the municipality of Vienna, the riverbanks upstream were administered by several aristocratic or clerical landlords who took different measures to secure their property (Atzinger and Grave 1874: 19).

Comprehensive plans for the regulation of the whole river were discussed by city authorities (Gemeinderat, Stadtbauamt) starting in the 1870s. Proposals ranged from the transformation into a shipping canal (suggested by Atzinger and Grave in 1874) to the diversion of the river to another, more southerly, river basin (proposed by Theodor Geiger in 1873). Regulation projects aimed at improving the sanitary situation, flood protection, making room for the construction of a railway in or next to the riverbed and creating space for buildings. The river as such had lost its value in the city and some planners and engineers held that it should better disappear. It took until the 1890s for city authorities to agree on the extensive regulation of Wien River, which was carried out between 1894 and 1904 (Pollack et al., in preparation). These major changes resulted in a functional segregation and dislocation of the river. Functions the river had fulfilled were separated by means of technical infrastructure, as apparent in the transversal sections in Fig. 22.6: The new socio-technical system consisted of retention basins for flood control; pipes for water supply; sewers for waste water disposal; a straightened and partly vaulted concrete bed for the discharge of the remaining water; a railway in the riverbed for transport; and vaulted areas for streets, markets and squares (Pollack 2013). Wien River's hydrology and ecology were significantly changed as the catchment area was reduced. Habitat was lost, access to the river and most pre-industrial uses such as washing and drying of laundry, extraction of gravel and watering of animals were made impossible as long-lasting material arrangements were created. The transformation of Wien River valley becomes visible when comparing the ground figure plans from 1825 and today (Fig. 22.5). It is bound to the new, fossil-based energy regime in two ways. The emergence of coal decreased its significance as an

Fig. 22.5 Ground figure plan depicting the transformation of Wien River valley from 1825 to 2010. The map of 1825 (*top*) is based on the Franciscean cadastre, 1817–1829, digitized by Stadtarcheologie Wien. The map of 2010 (*bottom*) is based on the online maps of the city of Vienna and open street maps. Graphic design by Friedrich Hauer

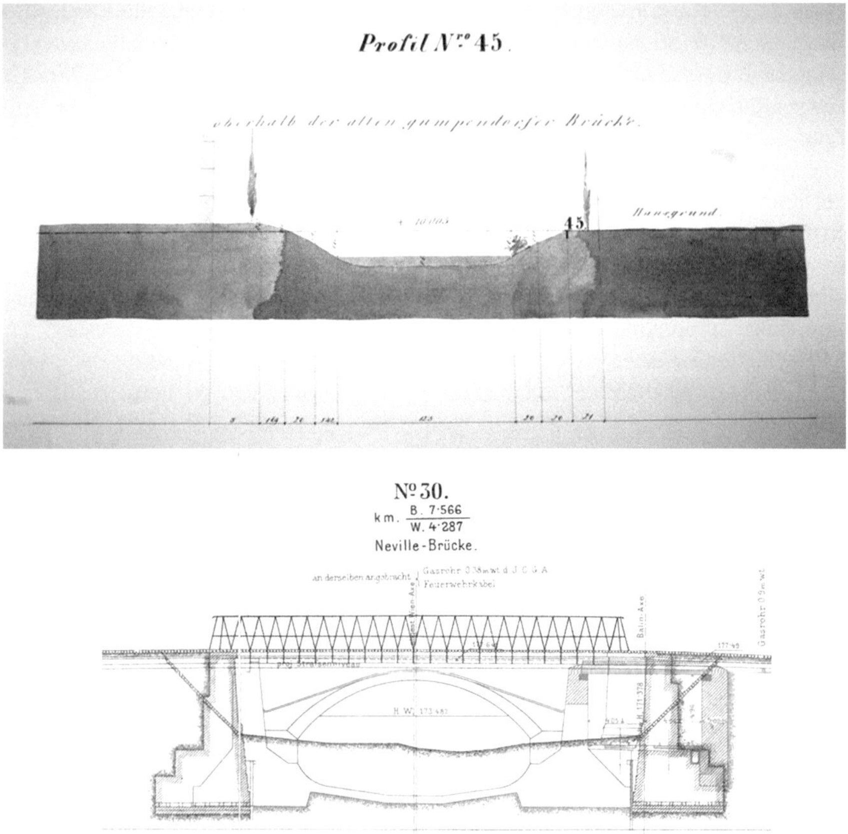

Fig. 22.6 Transversal section of Wien River from the 1850s (top, WStLA, Kartographische Sammlung, P1.253.5, Stummer 1847–1857) and 1897 (*bottom*, Archiv der Wiener Linien, B61 Bau- und Anlagendokumentation). Since this section is located within the Linienwall, the picture at the *top* already shows a narrowed and deepened riverbed in the 1850s. The picture at the *bottom* shows how the former river bed has been divided into a concrete river bed *(left side)* and railway lines *(right side)*.

energy source, and a comprehensive regulation needed coal-powered machines for the construction.

22.4 Competition and the Land–Water Nexus in the Wien River Waterscape

In the unique source of the Wienflussaufnahme, a transitional period in the utilization and perception of an urban river becomes visible in remarkable detail. Practices and material arrangements typical for an agrarian regime continued to exist while interventions characteristic of an industrial regime had already started. In the agrarian regime, Wien River's waters were a desirable resource for material and energetic use. Landowners attempted to protect built-up area and particularly pastures and fertile land for cultivation against natural dynamics (Stummer 1847–1857). Competition played out mainly over water use, access to and protection from water. Practices and arrangements of river utilization were confronted with the constraints of a torrential river with periods of flooding and water scarcity. Continuous maintenance, localized measures of flood protection and evasion technologies such as animal capstans as additional drivers of machinery were common strategies for dealing with natural dynamics. Floods and dry periods were equally challenging, which is why early regulation projects focused on stabilization of water flow, for example by constructing reservoirs on the tributaries (e.g. Bayer and Brequin in the 1780s, see Atzinger and Grave 1874: 14).

Competition patterns shifted with the new energy regime. While the river's waters lost significance due to depletion and alternatives provided by the networked city (Tarr and Dupuy 1988)—such as fresh water pipelines and sewage systems—the river bed as spatial resource became increasingly important for city authorities and engineers. To create room for a railway line, a market, streets and squares in the expanding city a river bed small enough to maintain a constant flow and big enough to absorb floods, the Normalprofil was engineered. Space reclaimed from the rivers former floodplains had to be compensated for in form of flood retention basins at the periphery.

The "complexity of interrelationships" (Edgeworth 2011: 53) between material arrangements, users and their practices with its high potential for competition, conflicts and need for negotiations and compromises is typical for the use of a pre-industrial urban river. The transformation and segregation of river functions resulting in homogenization of once heterogeneous water- and landscapes (Haberl et al. 2014: 41) in course of industrialization has been described for many rivers (Castonguay and Evenden 2012). It is associated with a tendency to reduce river functions (Winiwarter et al., in preparation) and the establishment of new power relations (Swyngedouw 2009: 57). In the agrarian system, practices of Wien River use were administered mainly through the manorial system backed by state institutions such as the Wassergrafenamt (Lohrmann 1980: 5). City authorities gained influence after the incorporation of the Vorstädte and Vororte and were able to enforce their sanitary concerns. They could establish new infrastructure and were able to spatially regulate urban expansion. Institutional change, therefore, was a major driver in the transformation of Wien River valley (Hauer et al., in preparation). As engineers carried out surveys and proposed solutions to the city administration, they became decisive for the transformation of urban spaces (Oliver 2000).

The Wienflussaufnahme, drawn up in the spirit of technological control over nature, a paradigm characteristic for the nineteenth century (ibid: 228), represents valuable evidence on how the engineers' point of view contributed to the transformation of hydro-social conditions by providing knowledge and planning imperatives. By collecting information not only on biophysical characteristics of the river but also on its different users and their social and economic situation, Stummer and his students created a powerful knowledge base on interests tied to the river. They were aware that a profound river regulation would change morphology and hydrology of the river as well as availability of and access to resources alike. A river restricted to a Normalprofil, flowing between secured and deeply incised banks could not be used in the same way as the wide floodplain with the patchwork of regulation measures. By restricting the survey to chartered water rights and large-scale infrastructure, the engineers made a deliberate selection on whose interests were to be considered in future planning and building processes. Thereby, they contributed to the decision on who would have access to and who would be excluded from access to water, the river and its shores. Being the work of students the Wienflussaufnahme is also an evidence for the dissemination of certain perceptions of the river that were most likely influencing the future work of the young hydraulic engineers. The meticulous survey, although never explicitly mentioning any conflicts, is therefore a valuable source on what has been defined as land-use competition: "a contest between different purposes, or functions" for which a river —and its fluid and solid resources—could be used (Haberl et al. 2014: 37).

Environmental history investigates the conditions and consequences of changes in socio-natural sites. The profound transformation of the Wien River waterscape is a consequence of industrialization and urbanization and the preference of certain water- and land-use functions over others by city administration and engineers. The consequences of this competition are preserved in material arrangements still in place today. These arrangements became conditions for present-day competition as they enable certain uses and disable others. They influence how present actors perceive and interact with a river, which underwent a profound remake at the turn of the nineteenth century and constrict their (future) scope of action. People and institutions are bound to the maintenance of such arrangements (Winiwarter et al. 2013: 107) and find it difficult to reverse them (Haberl et al. 2014: 42) as they are embedded in larger infrastructures and functional interactions. The Wienflussaufnahme sheds light on the conditions of this major transformation and points to the immaterial consequences, the preclusion of a variety of pre-industrial and early-industrial river functions of water and land uses. Today, many urban dwellers may favour a less restrictively regulated riverbed with floodplains, which act as biodiversity habitats and are accessible for recreation—they favour a different set of river functions. Restoration projects motivated by ideas of conservation and inclusion can bring back some of the characteristics of the historic river's water- and landscape. Still, all future negotiations about water and land use in the Wien River valley will need to take the socio-natural legacies of river regulation into account. Sources like the Wienflussaufnahme help us to consider material and immaterial conditions and consequences of competition over water and land and base management decisions on the firm ground of historical knowledge (Winiwarter et al. 2013: 120).

Acknowledgments We especially thank Christoph Sonnlechner of the City Archives of Vienna (Wiener Stadt- und Landesarchiv), for support of our research on the Wienflussaufnahme; Verena

Winiwarter and Robert Groß, for their valuable comments; and Friedrich Hauer, for providing us with his maps of the area.

References

Atzinger, F., & Grave, H. (1874). *Geschichte und Verhältnisse des Wien-Flußes sowie Anträge für dessen Regulierung und Nutzbarmachung mit Rücksichtnahme auf die jetzigen allgemeinen und lokalen Anforderungen*. Wien: Alfred Hölder, Beck'sche Universitäts-Buchhandlung.

Beer, A. (1901). *Müllerei und Mehlhandel Oesterreich-Ungarns von 1850 bis Ende 1900*. Wien: Verlag von Albert Beer.

Birkner, O. (2002). *Die bedrohte Stadt. Cholera in Wien*. Forschungen und Beiträge zur Wiener Stadtgeschichte 35. Franz Deuticke, Wien.

Büschfeld, J. (1997). *Flüsse und Kloaken: Umweltfragen im Zeitalter der Industrialisierung (1870–1918)*. Stuttgart: Klett-Cotta.

Castonguay, S., & Evenden, M. (Eds.). (2012). *Urban rivers. Remaking rivers, cities, and space in Europe and North America*. Pittsburgh: University of Pittsburgh Press.

Czeike, F. (1992–2004). *Historisches Lexikon Wien*. Wien: Verlag Kremayer & Scheriau.

Deligne, C. (2012). Brussels and its rivers, 1770–1880: Reshaping an urban landscape. In S. Castonguay & M. Evenden (Eds.), *Urban rivers. Remaking rivers, cities, and space in Europe and North America* (pp. 17–33). Pittsburgh: University of Pittsburgh Press.

Edgeworth, M. (2011). *Fluid pasts. Archaeology of flow*. London: Bristol Classical Press.

Eigner, P., & Schneider, P. (2005). Verdichtung und Expansion. Das Wachstum von Wien. In K. Brunner & P. Schneider (Eds.), *Umwelt Stadt: Geschichte des Natur- und Lebensraumes Wien* (pp. 11–53). Wien: Böhlau.

Fischer-Kowalski, M., & Haberl, H. (2007). Conceptualizing, observing and comparing socioecological transitions. In M. Fischer-Kowalski & H. Haberl (Eds.), *Socioecological transitions and global change. Trajectories of social metabolism and land use* (pp. 1–30). Cheltenham: Edward Elgar Publishing Limited.

Gantner, C. (1991). *Die Als - vom Bach zum Bachkanal: Die Geschichte eines Wasserlaufes*. Wiener Geschichtsblätter, Beiheft 4, Verein für Geschichte der Stadt Wien, Wien.

Haberl, H., Mbow, C., Deng, X., Irwin, E. G., Kerr, S., Kuemmerle, T., et al. (2014). Finite Land Resources and Competition. In K. T. Seto & A. Reenberg (Eds.), *Rethinking global land use in an urban era* (pp. 35–69). Cambridge: MIT Press.

Hauer, F., Hohensinner, S., & Spitzbart-Glasl, C. (in preparation). *How water and its use shaped the spatial development of Vienna*.

Hofmann, T., Pfleiderer, S., & Stürmer, F. (2003). *Digitaler angewandter Geo-Atlas der Stadt Wien*. GEO-Modul Endbericht Teil I, Geologische Bundesanstalt, Wien

Knelz, J. (1843). *Darstellung der Brechruhr-Epidemie in der k. k. Haupt- und Residenzstadt Wien*. Wien.

Krausmann, F., & Fischer-Kowalski, M. (2010). Gesellschaftliche Naturverhältnisse. Globale Transformationen der Energie- und Materialflüsse. In R. Sieder & E. Langthaler (Eds.), *Globalgeschichte 1800–2010* (pp. 39–66). Wien: Böhlauer.

Kurdiovski, R., & Offenthaler, E. (2012). *Stummer von Traunfels, Josef Mauritius Ritter (1808–1891)*. In OeBL—Österreichisches Biographisches Lexikon 1815–1950, Bd. 14 (p. 2). \http://www.biographien.ac.at/. Accessed 03 Feb 2015.

Lohrmann, K. (1980). *Die alten Mühlen an der Wien*. Wiener Bezirkskulturführer 26, Verlag Jugend und Volk, Wien.

Meißl, G. (2001). Hochquellleitung und Unratsschiffe. Zur Geschichte der Wiener Wasserver- und Entsorgung. In S. Hahn & R. Reith (Eds.), *Umwelt-Geschichte: Arbeitsfelder, Forschungsansätze, Perspektiven*. Wien: Verlag für Geschichte und Politik.

Melosi, M. (2006). Energy transitions in historical perspective. In B. Dooley (Ed.), *Energy and culture. Perspectives on the Power to Work* (pp. 3–18). Aldershot: Ashgate.
OeStA (Österreichisches Staatsarchiv) FHKA AHK VDA Mühlen. (1554–1780). A 12, B 2, G 8, H 9, K 4, M 3, R 11, S 4, V 2, W 7, W 9, W 10.
Oliver, S. (2000). The Thames embankment and the disciplining of nature in modernity. *The Geographical Journal, 166*(3), 227–238.
Österreichisch-Ungarische Müller-Zeitung. (1890). Fünfzig Jahre! 16. März 1890, Jahrgang 24, Wien, Berlin.
Persson, K. G. (1999). *Grain markets in Europe, 1500–1900. Integration and deregulation.* Cambridge: Cambridge University Press.
Pollack, G. (2013). Verschmutzt – verbaut – vergessen. Eine Umweltgeschichte des Wienflusses von 1780 bis 1910. *Social Ecology Working Paper* 138, Institute of Social Ecology, Vienna.
Pollack, G., Gierlinger, S., & Haidvogl, G. (in preparation). *Using and abusing a small urban river. The Wien River before and during industrialization.*
Protokollbücher. (1847–1857). WStLA (Wiener Stadt- und Landesarchiv), Hauptregistratur, series 1.2.2. B1—Politica, Department Q (Bau).
Reynolds, T. S. (1983). *Stronger than a hundred men. A history of the vertical water wheel.* Baltimore, London: The John Hopkins University Press.
Sieferle, R., Krausmann, F., Schandl, H., & Winiwarter, V. (2006). *Das Ende der Fläche: Zum gesellschaftlichen Stoffwechsel der Industrialisierung*. Köln: Böhlau.
Stummer, J. (1847–1857). Die topographische und hydrotechnische Aufnahme des Wienflusses unter Prof. Josef Stummer (Wienflussaufnahme). WStLA (Wiener Stadt- und Landesarchiv), Kartographische Sammlung, P1.253.
Stummer, J. (1847a). Instruktionen, betreffend die hydrotechnische Aufnahme 1847–1852 und 1857. WStLA (Wiener Stadt- und Landesarchiv), Kartographische Sammlung, P1.253.2.
Stummer, J. (1847b). Protokoll über die Erhebung des Wienflusses. WStLA (Wiener Stadt- und Landesarchiv), Kartographische Sammlung, P1.253.3.
Stummer, J. (1852–1855a) Situations-Beschreibung. WStLA (Wiener Stadt- und Landesarchiv), Kartographische Sammlung, P1.253.4.
Stummer, J. (1852–1855b) Protokoll über die Erhebung des Flusses. WStLA (Wiener Stadt- und Landesarchiv), Kartographische Sammlung, P1.253.4.
Stummer, J. (1852–1855c) Protokoll über die topographischen Erhebungen. WStLA (Wiener Stadt- und Landesarchiv), Kartographische Sammlung, P1.253.4.
Stummer, J. (1852–1855d) Protokoll über die Erhebung des Mariabrunner Mühlbaches. WStLA (Wiener Stadt- und Landesarchiv), Kartographische Sammlung, P1.253.4.
Swyngedouw, E. (1999). Modernity and hybridity: Nature, regeneracionismo, and the production of the Spanish waterscape, 1890–1930. *Annals of the Association of American Geographers, 89*(3), 443–465.
Swyngedouw, E. (2009). The political economy and political ecology of the hydro-social cycle. *Journal of Contemporary Water Research and Education, 142*, 56–60.
Tarr, J. A., & Dupuy, G. (Eds.). (1988). *Technology and the rise of the networked city in Europe and America.* Philadelphia: Temple University Press.
VID Vienna Institute of Demography. (2014). *Historisches Ortslexikon. Statistische Dokumentation zur Bevölkerungs- und Siedlungsgeschichte Wien.* Accessed February 03, 2015 from http://www.oeaw.ac.at/vid/download/histortslexikon/Ortslexikon_Wien.pdf
White, R. (1995). *The organic machine. The remaking of the Columbia River.* New York: Hill and Wang.
Winiwarter, V., Haidvogl, G., Hohensinner, S., Hauer, F., & Bürkner, M. (in preparation). *The long-term evolution of urban waters and their 19th century transformation in European cities. A comparative environmental history.*

Winiwarter, V., & Schmid, M. (2008). Umweltgeschichte als Untersuchung sozionaturaler Schauplätze? Ein Versuch, Johannes Colers "Oeconomia" umwelthistorisch zu interpretieren. In T. Knopf (Ed.), *Umweltverhalten in Geschichte und Gegenwart* (pp. 158–173). Vergleichende Ansätze: Attempto Verlag, Tübingen.

Winiwarter, V., Schmid, M., Hohensinner, S., & Haidvogl, G. (2013). The environmental history of the Danube River Basin as an issue of long-term socio-ecological research. In S. J. Singh, H. Haberl, M. Chertow, M. Mirtl & M. Schmid (Eds.), *Long term socio-ecological research. studies in society: Nature interactions across spatial and temporal scales* (pp. 103–122). Dordrecht: Springer.

MIX
Papier aus verantwortungsvollen Quellen
Paper from responsible sources
FSC® C105338

If you have any concerns about our products,
you can contact us on
ProductSafety@springernature.com

In case Publisher is established outside the EU,
the EU authorized representative is:
Springer Nature Customer Service Center GmbH
Europaplatz 3, 69115 Heidelberg, Germany

Printed by Libri Plureos GmbH
in Hamburg, Germany